Integrated Green Energy Solutions Volume 1

Scrivener Publishing
100 Cummings Center, Suite 541J
Beverly, MA 01915-6106

Publishers at Scrivener
Martin Scrivener (martin@scrivenerpublishing.com)
Phillip Carmical (pcarmical@scrivenerpublishing.com)

Integrated Green Energy Solutions Volume 1

Edited by
Milind Shrinivas Dangate
W.S. Sampath
O.V. Gnana Swathika
and
P. Sanjeevikumar

Scrivener Publishing

WILEY

This edition first published 2023 by John Wiley & Sons, Inc., 111 River Street, Hoboken, NJ 07030, USA and Scrivener Publishing LLC, 100 Cummings Center, Suite 541J, Beverly, MA 01915, USA
© 2023 Scrivener Publishing LLC

For more information about Scrivener publications please visit www.scrivenerpublishing.com.

All rights reserved. No part of this publication may be reproduced, stored in a retrieval system, or transmitted, in any form or by any means, electronic, mechanical, photocopying, recording, or otherwise, except as permitted by law. Advice on how to obtain permission to reuse material from this title is available at http://www.wiley.com/go/permissions.

Wiley Global Headquarters
111 River Street, Hoboken, NJ 07030, USA

For details of our global editorial offices, customer services, and more information about Wiley products visit us at www.wiley.com.

Limit of Liability/Disclaimer of Warranty
While the publisher and authors have used their best efforts in preparing this work, they make no representations or warranties with respect to the accuracy or completeness of the contents of this work and specifically disclaim all warranties, including without limitation any implied warranties of merchantability or fitness for a particular purpose. No warranty may be created or extended by sales representatives, written sales materials, or promotional statements for this work. The fact that an organization, website, or product is referred to in this work as a citation and/or potential source of further information does not mean that the publisher and authors endorse the information or services the organization, website, or product may provide or recommendations it may make. This work is sold with the understanding that the publisher is not engaged in rendering professional services. The advice and strategies contained herein may not be suitable for your situation. You should consult with a specialist where appropriate. Neither the publisher nor authors shall be liable for any loss of profit or any other commercial damages, including but not limited to special, incidental, consequential, or other damages. Further, readers should be aware that websites listed in this work may have changed or disappeared between when this work was written and when it is read.

Library of Congress Cataloging-in-Publication Data

ISBN 9781119847434

Front cover images supplied by Pixabay.com
Cover design by Russell Richardson

Set in size of 11pt and Minion Pro by Manila Typesetting Company, Makati, Philippines

Printed in the USA

10 9 8 7 6 5 4 3 2 1

Contents

Preface		xvii
1	**Green Economy and the Future in a Post-Pandemic World**	**1**
	Luke Gerard Christie and Deepa Cherian	
	1.1 Intergovernmental Panel on Climate Change	2
	1.2 The Need to Question How we Do Business and the Evolution of Green Policies	3
	1.3 The Shift from Fossil Fuels to Nuclear Energy for a Cleaner, Sustainable Environment	4
	1.4 Significance of Emergent Technologies in the Reduction of Global Warming and Climate Change	6
	Conclusion	8
	Bibliography	9
2	**Home Automation System Using Internet of Things for Real-Time Power Analysis and Control of Devices**	**11**
	Richik Ray, Rishita Shanker, V. Anantha Krishnan, O.V. Gnana Swathika and C. Vaithilingam	
	2.1 Introduction	12
	2.2 Methodology	14
	2.3 Design Specifications	15
	2.3.1 Components Required	15
	2.3.2 Circuit Diagram and Working	18
	2.3.3 Blynk GUI (Graphical User Interface) for Smartphone	19
	2.3.4 PCB (Printed Circuit Board) Design	20
	2.4 Results and Discussion	20
	2.4.1 Prototype Design Completion	20
	2.4.2 Testing and Observations	22
	2.4.3 Future Prospects	23
	2.5 Conclusion	24
	References	25

3 Energy Generation from Secondary Li-Ion Batteries to Economical Na-Ion Batteries 27
R. Rajapriya and Milind Shrinivas Dangate
    3.1 Introduction 28
    3.2 Li-Ion Battery 29
    3.3 Sodium-Ion Batteries 33
    3.4 Conclusion 40
    References 41

4 Hydrogen as a Fuel Cell 45
R. Rajapriya and Milind Shrinivas Dangate
    4.1 Introduction 45
    4.2 Operating Principle 48
        4.2.1 Types of Fuel Cells 49
    4.3 Why Hydrogen as a Fuel Cell? 50
        4.3.1 Electrolyte 52
        4.3.2 Catalyst Layer (At the Cathode & Anode) 52
        4.3.3 Bipolar Plate (Cathode & Anode) 52
    4.4 Hydrogen as an Energy-Vector in a Long-Term Fuel Cell 53
    4.5 Application 55
    4.6 Conclusion 56
    References 57

5 IoT and Machine Learning–Based Energy-Efficient Smart Buildings 61
Aaron Biju, Gautum Subhash V.P., Menon Adarsh Sivadas, Thejus R. Krishnan, Abhijith R. Nair, Anantha Krishnan V. and O.V. Gnana Swathika
    5.1 Introduction 61
    5.2 Methodology 63
    5.3 Design Specifications 65
        5.3.1 NodeMCU 65
        5.3.2 Relay 65
        5.3.3 Firebase 66
        5.3.4 Raspberry Pi 66
        5.3.5 Camera 66
    5.4 Results 66
    5.5 Conclusion 69
    References 69

6	**IOT-Based Smart Metering**		**71**
	Parth Bhargav, Umar Ansari, Fahad Nishat		
	and O.V. Gnana Swathika		
		Abbreviations and Nomenclature	72
	6.1	Introduction	72
		6.1.1 Motivation	72
		6.1.2 Objectives	73
	6.2	Methodology	73
		6.2.1 Advent of Smart Meter	73
		6.2.2 Modules	77
		6.2.3 Energy Meter	77
		6.2.4 Wi-Fi Module	78
		6.2.5 Arduino UNO	78
		6.2.6 Back End	78
	6.3	Design of IOT-Based Smart Meter	81
		6.3.1 Energy Meter	81
		6.3.2 Arduino UNO	82
		6.3.3 Wi-Fi Module	83
		6.3.4 Calculations	84
		6.3.5 Units	84
	6.4	Results and Discussion	84
		6.4.1 Working	84
		6.4.2 Readings Captured in the Excel Sheet	85
		6.4.3 Predication Using Statistical Analytics	86
		6.4.4 Quantitative Analytics	86
		6.4.5 Predication of Missing Data	87
		6.4.6 Hardware Output	87
	6.5	Conclusion	88
		References	89
7	**IoT-Based Home Automation and Power Consumption**		
	Analysis		**93**
	K. Trinath Raja, Challa Ravi Teja, K. Madhu Priya		
	and Berlin Hency V.		
	7.1	Introduction	94
	7.2	Literature Review	94
	7.3	IoT (Internet of Things)	96
	7.4	Architecture	96
	7.5	Software	97

	7.5.1	IFTTT	97
	7.5.2	ThingSpeak	97
	7.5.3	Google Assistant	98
7.6	Hardware	98
	7.6.1	DHT Sensor	98
	7.6.2	Motor	98
	7.6.3	NodeMCU	99
	7.6.4	Gas Sensor	99
7.7	Implementation, Testing and Results	99
7.8	Conclusion	102
	References	103

8 Advanced Technologies in Integrated Energy Systems — 105
Maheedhar and Deepa T.

8.1	Introduction	106
8.2	Combined Heat and Power	107
	8.2.1	Stirling Engines	107
	8.2.2	Turbines	108
	8.2.3	Fuel Cell	110
	8.2.4	Chillers	112
	8.2.5	PV/T System	113
8.3	Economic Aspects	114
8.4	Conclusion	115
	References	116

9 A Study to Enhance the Alkaline Surfactant Polymer (ASP) Process Using Organic Base — 119
M.J.A. Prince and Adhithiya Venkatachalapati Thulasiraman

9.1	Introduction	119
9.2	Materials and Methods	121
9.3	Similarity Study of NA in the Saline Water Containing Cations Having a Valency of 2	122
9.4	Results and Discussion	123
	9.4.1	Alkalinity Contributed by NA for Intensifying the IFT Characteristics	123
	9.4.2	Interfacial Tension Properties	124
	9.4.3	The Similarity of NA + Polymer	124
	9.4.4	Traits of Adsorption	125
	9.4.5	Economics	125
	9.4.6	Regular NA Injection Recommendation	125
9.5	Conclusions	126
	References	126

10 Flexible Metamaterials for Energy Harvesting Applications — 129
K.A. Karthigeyan, E. Manikandan, E. Papanasam and S. Radha

- 10.1 Introduction — 130
- 10.2 Metamaterials — 131
  - 10.2.1 Energy Harvesting Using Metamaterials — 132
  - 10.2.2 Solar Energy Harvesting — 132
    - 10.2.2.1 Numerical Setup — 133
  - 10.2.3 Acoustic Energy Harvesting — 135
  - 10.2.4 RF Energy Harvesting — 137
- 10.3 Summary and Challenges — 138
- References — 138

11 Smart Robotic Arm — 141
Rangit Ray, Koustav Das, Akash Adhikary, Akash Pandey, Ananthakrishnan V. and O.V. Gnana Swathika

- Abbreviations and Nomenclature — 141
- 11.1 Introduction — 142
  - 11.1.1 Motivation — 142
  - 11.1.2 Objectives — 143
  - 11.1.3 Scope of the Work — 143
  - 11.1.4 Organization — 143
- 11.2 Design of Robotic Arm with a Bot — 144
  - 11.2.1 Design Approach — 144
    - 11.2.1.1 Codes and Standards — 144
    - 11.2.1.2 Realistic Constraints — 144
  - 11.2.2 Design Specifications — 149
- 11.3 Project Demonstration — 152
  - 11.3.1 Introduction — 152
  - 11.3.2 Analytical Results — 153
  - 11.3.3 Simulation Results — 153
  - 11.3.4 Hardware Results — 154
- 11.4 Conclusion — 155
  - 11.4.1 Cost Analysis — 155
  - 11.4.2 Scope of Work — 155
  - 11.4.3 Summary — 155
- References — 156

12 Energy Technologies and Pricing Policies: Case Study — 157
Shanmugha S. and Milind Shrinivas Dangate

- 12.1 Introduction — 157
- 12.2 Literature Review — 159

12.3	Non-Linear Pricing	161
12.4	Agricultural Water Demand	162
12.5	Priced Inputs and Unpriced Resources	163
12.6	Proposed Set Up on Paper	164
12.7	Empirical Model	167
12.8	Identification Strategy	168
12.9	Data	170
12.10	Empirical Results	171
12.11	Counterfactual Simulation A	173
12.12	Counterfactual Simulation B	174
12.13	Counterfactual Simulation: Costs of Reduced Groundwater Demand	176
12.14	Conclusion	180
	References	181

13 Energy Availability and Resource Management: Case Study 185
Shanmugha S. and Milind Shrinivas Dangate

13.1	Introduction	185
13.2	Literature Review	187
13.3	Study Area	189
	13.3.1 Producer Survey	192
13.4	Empirical Model of Adoption	193
13.5	Material and Methods	196
13.6	Results	198
13.7	Conclusion	203
	References	204

14 Energy-Efficient Dough Rolling Machine 207
Nerella Venkata Sai Charan, Abhishek Antony Mathew, Adnan Ahamad Syed, Nallavelli Preetham Reddy, Anantha Krishnan V. and O.V. Gnana Swathika

14.1	Introduction	208
14.2	Methodology	208
14.3	Specifications	210
	14.3.1 Motor	210
	14.3.2 Switch Mode Power Supply (SMPS)	210
	14.3.3 Speed Reduction	211
	14.3.4 Coupler	212
	14.3.5 Main Base Structure	212
	14.3.6 Rotating Platform and Rollers	212

	14.3.7	Rotating Platform	213
	14.3.8	Rollers	213
14.4		Result and Discussion	215
14.5		Conclusion	215
		References	215

15 Peak Load Management System Using Node-Red Software Considering Peak Load Analysis — 217
Mohit Sharan, Prantika Das, Harsh Gupta, S. Angalaeswari, T. Deepa, P. Balamurugan and D. Subbulekshmi

15.1		Introduction	218
15.2		Methodology	219
	15.2.1	Peak Demand and Load Profile	219
	15.2.2	Need of Peak Load Management (PLM)	220
	15.2.3	Data Analysis	220
	15.2.4	Need to Flatten the Load Curve	221
	15.2.5	Current Observations	221
	15.2.6	Equations	221
15.3		Model Specifications	221
15.4		Features of UI Interface	225
	15.4.1	App Prototype	225
15.5		Conclusions	227
		Bibliography	227

16 An Overview on the Energy Economics Associated with the Energy Industry — 229
Adhithiya Venkatachalapati Thulasiraman and M.J.A. Prince

16.1		Time Value of Money	230
	16.1.1	Present Value of an Asset	230
	16.1.2	Future Value of an Investment	230
	16.1.3	Rule of 72	231
16.2		Classification of Cost	232
	16.2.1	Fixed Cost of an Asset (FCA)	232
	16.2.2	Variable Cost of a Plant (VCP)	232
	16.2.3	Total Cost of a Plant (TCP)	232
	16.2.4	Break-Even Location (BEL)	232
16.3		Economic Specification	233
	16.3.1	Return on Cost (ROC)	233
	16.3.2	Payback Span	233
	16.3.3	Net Present Worth	233

		16.3.4	Discounted Money Flow (DMF)	234
		16.3.5	Internal Charge of Returns (ICR)	234
	16.4	Analysis		234
		16.4.1	Incremental Analysis (IA)	234
			16.4.1.1 Pertinent Cost (PC)	234
			16.4.1.2 Non-Pertinent Cost (NPC)	235
		16.4.2	Sensitivity Analysis (SA)	235
		16.4.3	Replacement Analysis (RA)	237
	16.5	Conclusion		239
		Bibliography		240

17 IoT-Based Unified Child Monitoring and Security System   241
A.R. Mirunalini, Shwetha. S., R. Priyanka and Berlin Hency V.

17.1	Introduction		242
17.2	Literature Review		243
17.3	Proposed System		247
	17.3.1	Block Diagram	247
	17.3.2	Design Approach	249
	17.3.3	Software Analysis	249
	17.3.4	Hardware Analysis	252
		17.3.4.1 Experimental Setup	253
17.4	Result and Analysis		256
17.5	Conclusion and Future Enhancement		259
	17.5.1	Conclusion and Inference	259
	17.5.2	Future Enhancement	260
	References		260

18 IoT-Based Plant Health Monitoring System Using CNN and Image Processing   263
Anindita Banerjee, Ekta Lal and Berlin Hency V.

18.1	Introduction		264
18.2	Literature Survey		265
18.3	Data Analysis		268
	18.3.1	Convolutional Neural Network	268
	18.3.2	Phases of the Model	269
	18.3.3	Proposed Architecture	269
18.4	Proposed Methodology		271
	18.4.1	System Module and Structure	271
	18.4.2	System Design and Methods	272
	18.4.3	Plant Disease Detection and Classification	272

		18.4.3.1	Dataset Used	272
		18.4.3.2	Preprocessing and Labelling Methods	273
		18.4.3.3	Procedure of Augmentation	273
		18.4.3.4	Training Using CNN	273
		18.4.3.5	Analysis	275
		18.4.3.6	Final Polishing of Results	275
	18.4.4	Hardware and Software Instruments		275
18.5	Results and Discussion			275
18.6	Conclusion			286
	References			286

19 IoT-Based Self-Checkout Stores Using Face Mask Detection — 291
Shreya M., R. Nandita, Seshan Rajaraman and Berlin Hency V.

19.1	Introduction		292
19.2	Literature Review		292
	19.2.1	Self-Checkout Stores	292
	19.2.2	Face Mask Detection	293
19.3	Convolution Neural Network		295
19.4	Architecture		298
19.5	Hardware Requirements		299
	19.5.1	PIR Sensor	299
	19.5.2	LCD	299
	19.5.3	Arduino UNO	299
	19.5.4	Piezo Sensor	299
	19.5.5	Potentiometer	300
	19.5.6	LED	300
	19.5.7	Raspberry Pi	300
19.6	Software		300
	19.6.1	Jupyter Notebook	300
	19.6.2	TinkerCAD	300
19.7	Implementation		300
	19.7.1	Building and Training the Model	301
	19.7.2	Testing The Model	302
19.8	Results and Discussions		303
19.9	Conclusion		306
	References		306

20 IoT-Based Color Fault Detection Using TCS3200 in Textile Industry — 309
T. Kalavathidevi, S. Umadevi, S. Ramesh, D. Renukadevi and S. Revathi

20.1	Introduction	310
20.2	Literature Survey	311
20.3	Methodology	313
	20.3.1 Sensor	314
	20.3.2 Microcontroller	315
	20.3.3 NodeMCU and Wi-Fi Module	317
	20.3.4 Servomotor	317
	20.3.5 IoT-Based Data Monitoring	318
	20.3.6 IR Sensor	318
	20.3.7 Proximity Sensor	319
	20.3.8 Blynk	319
20.4	Experimental Setup	321
20.5	Results and Discussion	322
20.6	Conclusion	324
	References	324
21	**Energy Management System for Smart Buildings**	**327**
	Shivangi Shukla, V. Jayashree Nivedhitha, Akshitha Shankar, P. Tejaswi and O.V. Gnana Swathika	
21.1	Introduction	328
21.2	Literature Survey	328
21.3	Modules of the Project	331
	21.3.1 Data Collection for Accurate Energy Prediction	331
	21.3.2 ML Prediction	332
	21.3.3 Web Server	332
	21.3.4 Hardware Description and Implementation	332
21.4	Design of Smart Energy Management System	334
	21.4.1 Design Approach	334
	21.4.1.1 ML Algorithm	334
	21.4.1.2 EMS Algorithm	334
	21.4.2 Design Specifications	336
21.5	Result & Analysis	337
	21.5.1 Introduction	337
	21.5.2 ML Model Results	337
	21.5.3 Web Page Results	337
	21.5.4 Hardware Results	339
21.6	Conclusion	346
	References	346

22	**Mobile EV Charging Stations for Scalability of EV in the Indian Automobile Sector**		**349**
	Mohit Sharan, Ameesh K. Singh, Harsh Gupta, Apurv Malhotra, Muskan Karira, O.V. Gnana Swathika and Anantha Krishnan V.		
	22.1	Introduction	350
	22.2	Methodology	350
		22.2.1 Design Specifications	351
		22.2.2 Block Diagrams	356
	22.3	Result	357
	22.4	Conclusions	358
		Bibliography	358
About the Editors			**361**
Index			**363**

Preface

Renewable energy supplies are of ever-increasing environmental and economic importance all over the world. A wide range of renewable energy technologies has been established commercially and recognized as growth industries. World agencies, such as the United Nations, have extensive programs to encourage renewable energy technology.

This two-volume set, *Integrated Green Energy Solutions*, will bridge the gap between descriptive reviews and specialized engineering treatises on particular aspects. It centers on demonstrating how fundamental physical processes govern renewable energy resources and their application. Although the applications are being updated continually, the fundamental principles remain the same, and we are confident that this book will provide a useful platform for those advancing the subject and its industries. We have been encouraged in this approach by the ever-increasing commercial importance of renewable energy technologies.

Integrated Green Energy Solutions is a numerate and quantitative text covering subjects of proven technical and economic importance worldwide. Energy supply from renewables is an essential component of every nation's strategy, especially when there is responsibility for the environment and sustainability. These books will consider the timeless renewable energy technologies' timeless principles yet seeks to demonstrate modern applications and case studies. This volumes will stress the scientific understanding and analysis of renewable energy since we believe these are distinctive and require specialist attention.

The five most important topics covered in these two books are:

1. Education in Energy Conversion and Management
2. Integrated Energy Systems
3. Energy Management Strategies for Control and Planning
4. Energy economics and environment
5. World Energy demand

1

Green Economy and the Future in a Post-Pandemic World

Luke Gerard Christie[1]* and Deepa Cherian[2]

[1]VIT University, Chennai Campus Chennai, Tamil-Nadu, India
[2]Indian School of Business, Hyderabad, India

Abstract

The future to geo-political and geo-economic conundrums is by transforming current economies into inclusive and sustainable societies. In this race for global dominance and hegemony, policy makers must be wary of not forgetting institutional practices of conserving and preserving ecosystems and biospheres with pro-active and proper thinking. Governments that are in power must be sensible to realize that economies will eventually grow when more people join the formal and informal sectors, but the challenge is to have a planet that sustains our needs rather than addressing our greed. Legal systems must work harder in the 21st century to embed proper and critical thinking driven by an ecological conscience to preserve, conserve and protect the environment that sustains us. The technology that is being built and fashioned to drive businesses must submit to stringent ecological standards. With the rapid spread of Covid19, scientists are aware that humanity will be afflicted with more such zoonotic diseases primarily brought on by the global warming and climate change. Third world governments in their search for competing and contributing with the global economy forget the impending dangers of a cataclysmic warmer, hotter and unsustainable planet that will deprive burgeoning populations of food and clean water furthering a health scare. Across the globe, we have witnessed government's response to Covid19 especially in the third world and the loss of lives that could have been prevented. This affliction is bound to endure owing to the inadequate policies that fail to create low–carbon economies or submit to Sustainable Development Goals that could mitigate the debilitating effects of a globally warmer planet. In all of this, the future will be fought not over oil but wars are bound to be fought over water and food and lack of immediate or urgent healthcare support. It is

---

Corresponding author: lukegerard.christie@vit.ac.in

Milind Shrinivas Dangate, W.S. Sampath, O.V. Gnana Swathika and P. Sanjeevikumar (eds.) Integrated Green Energy Solutions Volume 1, (1–10) © 2023 Scrivener Publishing LLC

observed painfully, that the people most affected or afflicted with the mostly the marginalized, the poor, the disadvantaged. In this paper, I propose how governments of the day must transform their economies to be sustainable and inclusive, ameliorate global warming, promote healthy agricultural practices, constantly set higher moral standards for a low-carbon economy and build on a healthcare system that is robust and flexible to everyone's needs. The globe after observing many discussions at Copenhagen is now becoming familiar with the reality of a resource-efficient economy and natural capital as an invaluable economic asset.

Keywords: Inclusive environment, green economy, energy economics, Fourth Industrial Revolution, low-carbon, agriculture, freshwater resources, governmental policies

1.1 Intergovernmental Panel on Climate Change

A report released in August 2021 stated that the ecological damage to the planet is irreversible and there is an ambitious requirement to reduce carbon-dioxide and other greenhouse gas emissions from accelerating climate change. The bigger problem is that it would take at least 80 to 100 years for climate to improve and stabilize conditions and temperatures. The 6th Assessment Report, which will be released in 2022, reflects that the climatic scenario is not as it looks and is more adverse in nature, but it can be mitigated with global engagement of governments and citizens with proper and effective value building measures in negotiations and decision-making with all global stake-holders from governance to business. All regions face stupendous challenges in the upward trajectory of increasing temperatures and climate changes that will be problematic, as that would bring changes to cyclonic patterns, affect agricultural practices causing droughts, reducing harvest yields and causing flooding and changes to monsoon and precipitation that will vary from regions. There will be swifter melting of polar ice caps in the Arctic and Antarctic regions, causing ocean temperature to increase, threatening marine life and normal or natural functioning of biodiversity in ecosystems. There will be continued sea level rise throughout the 21st century in coastal areas with frequent and severe coastal flooding in low-lying areas and coastal erosion. The prediction is that sea level events that occurred once in 100 years could happen every year by the end of this century. It is most observed that cities will be amplified with heatwaves and cold waves with heavy precipitation. The thawing of the permafrost is bound to threaten 4 million people and there is a desperate need for drastic measures to reduce emissions.

1.2 The Need to Question How we Do Business and the Evolution of Green Policies

The world today requires an enterprising and proactive vision towards mitigating global warming and climate change that will build a sustainable planet. Governments and policy makers are not the only stake holders responsible for implementation of a robust, green economy; it is through a collective consciousness of engendering an environment and ecological mindset in parallel with technology and newer advances in technology. Countries have to harness emergent technologies to prove their technological resilience in global economics, trade and commerce and in agriculture and for fresh water resources. The most developed countries have been able to accomplish a sustainable greener initiative with constant amendments to their policies and with systemic engagement with their citizenry. Most countries are acutely aware of greening the economy and creating employment opportunities in renewable energy and technologies. It is also to be found that the future to global trade and supply chains will be driven with more economic opportunities in the green sector.

Policy makers have woken up to this reality of ameliorating emissions and the many positives of decelerating global warming and climate change. Yet, implementation of green policies seems not to have become a social and economic reality owing to the existing and engendered policies and older technologies that have harnessed growth of organizations and economies alike. Concurrently, countries are forced to swiftly change their predilections to avoid pitfalls of global warming and climate change. The Bretton Woods institutions such as the World Bank, International Monetary Fund, and the International Trade Organization have recalibrated economic strategies to prevent inflation but also to improve global economic outlook. The oil crisis fostered the path for the creation of the International Energy Agency to manage oil disruptions, creating policy awareness for global security supplies and initiatives. The major organizations from focusing on preventing economic shocks and setbacks are not coerced to consider the biggest elephant in the room, global warming and climate change brought on by the ignorance of previous policies. Past decisions have benefitted economies and expanded supply and the demand footprint but they inflated a bigger problem on everyone's hands and in global political discourse of taking effective measures to ameliorate or reduce global warming and climate change and greenhouse emissions that propel climate change.

In the post-globalized world, from automobile manufacturers, air-conditioning makers to all diverse products that reach markets have to submit to reducing the amount of emissions to building a cleaner and safer environment. The well-advanced countries have shifted to guaranteeing a low-carbon economy with cleaner and sustainable ways of doing business but enforcing those global greening laws on other countries whom they do business with to bring about a reinforced vision of inclusive and sustainable societies. With the advent of newer policies with global citizens waking up to a necessity of cleaner environments, organizations have lent credence to ensuring that all products that reach shores across the globe are in line with the provisions of the requirements of global policy on emission reduction. Citizens with directives from their governments across the globe have been campaigning to buy greener products, avoid lethal products, practice waste reduction and optimize on energy saving. With the constant serious intervention of policy makers for a low carbon economy, most major businesses have taken the plunge, keeping in mind economic downturns, and sensitized their business practices toward environmental benefits and have positioned themselves to promote environmental goals or to being precursors due to strong legislations for industry with government commitments. In several global firms and global supply chains, primary importance is being given to submit to environmental practices in the manufacturing process of products and commodities, which results in managing costs and furthering environmentally friendly practices. In short, most policies from the government and organizationally have corporate environment responsibility projects for better and enhanced environmental sustainability. Economic incentives like taxes or tax exemptions, and subsidized permits, are often encouraged by governments for oganizations to comply with environmental standards. Trilateral or bi-lateral agreements between businesses across the globe and between governments has shifted the focus onto reducing emissions and making commitments to eco-friendly practices.

1.3 The Shift from Fossil Fuels to Nuclear Energy for a Cleaner, Sustainable Environment

Fossil fuels are creating even more pollution with the growing number of cars on the roads. When India uses nuclear power, pollution can be confronted and electricity can be generated at triple the amount that is being generated with burning of coal. The future for countries like India is to utilize nuclear power. The developed societies have moved onto renewable energy, but in a country like India, and with climate change and global warming, unpredictable weather patterns makes it an impossibility to rely on solar or wind technology alone to power large-scale industries. We have to utilize nuclear power as we are a growing economy where we have reached the 2 trillion dollar mark and by 2025 we will cross the 4 trillion dollar mark. This is mainly due to the exports being driven by the demand of the global economy and businesses operating at full flow keeping up to stiff competition with the developed world. There have been just four accidents involving nuclear power in 60 years and the main causes for the accidents were due to improper safety mechanisms, gross human error or utilization of poor technology. The technology used had not been upgraded and countries that embrace nuclear energy must be well educated before installation of the plants to avoid accidents.

The idea of setting up nuclear power plants in areas that threaten livelihoods and displace people should be taken into consideration and plants must be set up where agriculture land and people's livelihoods are not threatened. It is seen and has been proved that the developed world has used or is using nuclear power in tandem with renewable sources of energy making their economies low-carbon. If India can use the same principles—as the weather in India is unpredictable—the economy can reduce the carbon footprint and in turn make societies more adaptable and efficient. The greatest advantage for India to shift to nuclear power is that the waste material can be contained in an area if she shifts to thorium. The waste it seems can be contained in an area the size of a football field which can be destroyed after a few decades. The soil can then be revitalized, unlike the waste that comes from a uranium enrichment nuclear plant. India has large reserves of thorium, which is the nuclear fuel of the future. However, the technologies needed to extract thorium for the plants need to be designed, which may take some time. What must be noted is that fossil-based fuels or fossil fuels are not sustainable and even more important is that with the scarcity of fossil fuels, there will definitely be geo-political instability. We have seen that in the case of Exxon, which is a leading producer of petroleum and is on a new trajectory in trying to find shale gas from complex rock formations in the Arctic to counter the problems of not being able to find petroleum. They had exploited the earth, all parts of the globe, from Indonesia to Africa,

Latin America, Russia, Iraq and now to the Arctic region trying to find shale gas under tough weather conditions. Exxon Mobil had been tough in criticizing the global community's concern about changing weather patterns and global warming, until in 2009 it did agree that the planet is warming due to manmade disasters. When we keep in mind a warmer planet, the size of the carbon footprint growing larger, agricultural practices being disrupted due to global warming and climate change, we must realize that the people who will have to struggle will be those who live on the fringes of society or who live on state benefits.

It is a far superior idea to shift to nuclear energy, harness thorium; tackle global warming and climate change building sustainable societies solving agricultural problems without soil degradation or the crops having the potential of a qualitative yield when harvested. The only challenge is even though we have thorium, we must be able to manufacture the fissile material which is necessary in sustaining a chain reaction when bombarded by neutrons.

The fact remains that countries will have to shift to nuclear energy and ensure that nuclear plants are well established with adequate and proper safety measures. Nuclear power and energy is the best alternative to fossil fuel, and the technology that countries use in these plants, through improved and advanced with shifts in nano-technology, have benefitted developed economies and that influence has caught up with the developing societies. The United Nations Millennium Development Goals and World Bank's goals in recognition of inclusive and sustainability as essential global practices for the environment are the force driving the policy agenda forward. A synergy is created in energy, business and transport where environmental policies are integrated in their framework of manufacturing procedures rather than solely pursuing the agenda individually. For example: automobiles will have to follow a certain standard of emissions they are allowed to release into the atmosphere, abiding by government norms and policies on a global scale as all countries have agreed to reduce emissions and pollutants into the atmosphere by 2030.

1.4 Significance of Emergent Technologies in the Reduction of Global Warming and Climate Change

AI (Artificial Intelligence), Big Data, and 3D printing are used extensively to develop solutions to mitigate or offer solutions to reduce global warming and climate change. Yet, despite the challenges and investments, there is no stopping of temperature rise in the Arctic of 3-5 degree Celsius by 2050. Technologies like Big Data and Artificial Intelligence can be used to collect or curate vast amounts of data and be used for insightful information in an intelligent manner.

Cloud computing is another disruptive technology that has been extensively used and continues to identify its possibilities to boost a green economy. For instance, the recent pandemic times have witnessed an exponential increase in data usage, as companies resorted to work from home. Enormous rise in data led to an increase in cloud providers for the storage and management of data. Cloud Computing Environment (CCE) has been a widely recognized technology during work from home. Organizations such as the Word Economic Forum and the Organization for Economic Cooperation and Development (OECD) have called for a "green reset" following COVID-19. In that regard, cloud and green computing can help progress towards a more sustainable green future by reducing carbon emissions to the ecosystem through various energy-saving digital modes.

Further, Remote Working has reduced the carbon footprint in the environment during the pandemic and this recently evolved work model has equipped organizations to valiantly face contingencies by not trading off on a green economy. Remote working that utilizes a green cloud technology offers the flexibility to work anytime and anywhere, and it has improved productivity and abridged the daily commute of employees to the office. This decrease in commuting has reduced fuel consumption and carbon emission to the atmosphere and has furthered organizations to cut down on various marginal expenses such as rent and land costs while reducing energy consumption at the office premises. Many organizations have decided to take forward this work model post-pandemic for a greener economy and to save costs.

In businesses today, Artificial and Big Data is used by resource personnel to take challenging decisions and to achieve functional excellence. These new or emergent technologies can inform governments or policy makers to develop impactful plans in reducing emissions and climate change. Google, the world's largest search engine, has already invested in newer technologies and uses big data to estimate greenhouse gas emissions to inform citizenry and governments of the increase that enables them to educate society about the pitfalls. Conversely, Artificial Intelligence can sense their ecosystem, think, and adapt to their programmed initiatives and be used intensively in energy-saving initiatives by incorporating data from smart sensors, smart meters and Internet of Things to forecast energy demand and its surplus. AI can help electricity grid providers to optimize energy production and reduce any loss or to mitigate impact on the climate. Technology providers across the globe seeing and realizing the importance of mitigating emissions are working on simulation-based technologies to aid in planning future cities that are smart, sustainable with a low-carbon footprint. IBM as of now has developed technology that informs cities about heat waves and how to prevent future heat waves. The technology developed informs citizens on the best locales to plant trees that are susceptible to heat waves and reduce the intensity of the heat waves. Governments have been developing a serious awareness of the dangers of global warming and climate change, and are now working on technologies and Artificial Simulation programmes for accurate weather forecasting and city planning to ameliorate climate change and its impact. 3D printing in dynamic mode across the developed economy actually reduces manufacturing costs and significantly reduces carbon emissions. This major innovative accomplishment is achieved by reduction of raw material or using recycled material for manufacturing building constructing purposes. 3D printing can guarantee a truly innovative way to dispose of trash in a constructive manner that can be used for the efforts in city planning. 3D technology aims to be more ascendable and disrupt the construction industry as this seems more a viable option for reducing waste than having waste recycled for infrastructure and city building projects. Another such technology in the modern era that can be used efficiently is Augmented Reality or Virtual Reality, where cities and countries can identify the impacts of climate change and can inform policy makers to prioritize efforts in the drive against global warming by economic development and climate adaption initiatives. Augmented reality helps visualize disaster prone areas by embedding Artificial Intelligence and 3D technology, by which knowledge of the repercussions is gained. The visualization of ecological degradation areas with viable solutions will help all governments to make better decisions on climate change. Covid-19 has largely witnessed enormous increase in data and the use of cloud-enabled services which resulted in low carbon emissions.

Conclusion

The primary goal to reduce global warming and climate change rests today with assertive and ambitious policy makers working and collaborating with new-age technology companies that can effectively create an impact on the mitigating of emissions by involvement and participative engagement with all stakeholders in order to create a less polluted, livable planet. There is a serious demand for strong cooperation and with efficient public-private partnership, innovative solutions can be formed to solve a crisis of our own making. This will reduce the anxieties for the coming generations and allow them to live healthier lives in a cleaner, greener sustainable planet, fulfilling the promise of the 196 countries who have committed to fight global warming and climate change. It is a vision that can be achieved with everyone's unbiased participation for clean fresh water resources, better agricultural yields and healthier breathing ecosystems that sustain life. The only way forward is to act now without delay.

Bibliography

Anderson, Dennis, and Catherine D. Bird. 1992. "Carbon Accumulations and Technical Progress: A Simulation Study of Costs." *Oxford Bulletin of Economics and Statistics* 54(1).

Annan, R. H. 1992. "Photovoltaic Energy, Economics and the Environment." In B. Abeles, A. J. Jacobson, and P. Sheng, eds., *Energy and the Environment.* Singapore: World Scientific.

Ahmed, K., & Anderson, D. (1994). Renewable energy technologies: a review of the status and costs of selected technologies. World Bank technical paper (ISSN 0253-7494, (240).

Apak, S., Atay, E., & Tuncer, G. (2011). Financial risk management in renewable energy sector: Comparative analysis between the European Union and Turkey. *Procedia-Social and Behavioral Sciences.*

Bhandari, A. (2015, October). India's Power Utilities Owe Banks $90 billion, Cripple Future.

Bhatia, M., & Banerjee, S. G. (2011). Unleashing the potential of renewable energy in India. World Bank Publications.

Bhattacharya, S. C., & Jana, C. (2009). Renewable energy in India: historical developments and prospects. *Energy*, 34(8), 981-991.

Bulfotuh, F. (2007). Energy efficiency and renewable technologies: the way to sustainable energy future.

Cavallo, Alfred J., Susan M. Hock, and Don R. Smith. 1993. "Wind Energy: Technology and Economics." In Thomas B. Johansson and others, eds., *Renewable Energy: Sources for Fuel and Electricity*. Washington, D.C.: Island Press.

Costello, D., and P. Rappaport. 1980. "The Technological and Economic Development of Photovoltaics." *Annual Review of Energy.*

Development Sciences, Inc. 1981. "The Economic Costs of Renewable Energy." Report prepared for the Agency for International Development.

Deloitte, IEC (2013). Securing tomorrow's energy today: Policy and Regulations Meeting the Financing Challenge in the Energy Sector in India.

Devadas R.P. (1988). Management of development Programmes for Women and Children, Vol. II. Coimbatore.

Drennen, T. E., Erickson, J. D., & Chapman, D. (1996). Solar power and climate change policy in developing countries. *Energy Policy*.

Nelson, D., Shrimali, G., Goel, S., Konda, C., & Kumar, R. (2012). Meeting India's renewable energy targets: The financing challenge. CPI-ISB Report, Climate Policy Initiative.

Richardson, R; and Wilkins, M. (2010). Can Capital Markets Bridge the Climate Change Financing Gap?

Schwabe, P., Mendelsohn, M., Mormann, F., & Arent, D. (2012). Mobilizing Public Markets to Finance Renewable Energy Projects.

William, C.A. and Heins, M.R. (1976). *Management and Insurance*. New York: McGraw-Hill Books Co.

2

Home Automation System Using Internet of Things for Real-Time Power Analysis and Control of Devices

Richik Ray, Rishita Shanker[1], V. Anantha Krishnan[1]*, O.V. Gnana Swathika[2]† and C. Vaithilingam[1]

[1]School of Electrical Engineering, Vellore Institute of Technology, Chennai Campus, Tamil Nadu, India
[2]Centre for Smart Grid Technologies, School of Electrical Engineering, Vellore Institute of Technology, Chennai, Tamil Nadu, India

Abstract

Conventional power systems have been in use since the beginning of centralized power generation and distribution arrangements. These systems supply energy to human settlements and their workplaces and are responsible for a reliable delivery of power and in cases of faults, also to fix them. With time and considerable advancements in technology, the world has seen a necessary shift from the conventional power systems to Smart Grids that are based on principles of Distributed Generation and similar crucial methods like Customer Interaction. The need for smarter grids arose due to change in the everyday lifestyle of people that in turn led to an increase in power consumption, which had to be coupled with Renewable Resources of Energy since conventional resources such as fossil fuels have been depleting drastically. Now alongside smarter grids, the Automation Industry has paved its own path into our lifestyles, with the integration of Internet of Things (IoT) and microcontrollers/microprocessors. From Smart Cars to Smart Refrigerators, this industry is booming and certainly depicts that it is the future of technology for a world depending on clean fuels and more reliable and user-friendly technology. A Smart Home is similar to the previous examples, as it depends on multiple smart devices like Smart Energy Meter, Alarm System, Garage Parking, etc., that all combine and form the Home Automation System, powered by a Smart Grid supply.

---

*Corresponding author: ananthakrishnan.v@vit.ac.in
†Corresponding author: gnanaswathika.ov@vit.ac.in

Milind Shrinivas Dangate, W.S. Sampath, O.V. Gnana Swathika and P. Sanjeevikumar (eds.) Integrated Green Energy Solutions Volume 1, (11–26) © 2023 Scrivener Publishing LLC

In this paper, a functional prototype for the same is designed, but this system carries out the forementioned functions on a single platform by integrating the control of these separate devices on a user-friendly application directly from our smartphones based on IoT, rather than having separate protocols for each of the applications, and further, the future prospects for this model are presented.

Keywords: Internet of Things (IoT), smart grid, distributed generation, power BI, Blynk, renewable energy, automation

2.1 Introduction

A Home Automation System is the basic building block of a fully functional Smart Home that is connected to a local power station, which is a part of a larger smart grid. It is the distributed generation principles that are applied across the whole network that make the change from a conventional grid to one that is smarter. Based on IoT protocols and Machine Learning, this leads to the development of an Artificial Intelligence (AI) dependent system that monitors the whole grid from generation to final distribution in homes and workstations. The Home Automation System is responsible for integrating all the essential functions that a smart home consists of as discussed, and with its direct connection to the grid, it results in higher customer interaction, allowing the users to gain access to data like consumption levels, consumption pattern graphs from the supply end for peak load analysis, tariff rates, etc.

Multiple devices can be utilised to build a fully functional smart home automation system. This includes devices like a Smart Power-Strip. It works as a wall outlet adapter and can be controlled using a microcontroller and allows the user to upgrade to an automated home without any additions. The Wi-Fi connection or sensors collect the data and then the microcontroller passes instructions using the same (e.g., programmed parameters) [1]. Internet of things is used in every industry and machinery for sharing intel and completing tasks within short time frames, via Internet, in comparison to the more time-consuming manual operation [2]. Under the IoT fog computing paradigm, exchange of messages is possible with the integration of IoT-based nodes and a common data cloud. Wi-Fi and ZigBee amongst these are the most ideal communication technologies for smart homes. Wi-Fi although popular, has a restricted application due to high energy consumption and also it lacks standard mesh networking competences for low-power devices. Hence, ZigBee was selected as a more preferable option for wireless automation devices in the industries [3].

As of now, smart home systems are based on primitive technology, and users will be unable to control home appliances since their functionality will be limited. Moreover, there are concerns about the security of these systems since they are prone to hacking, potentially leading to major problems [4]. Multiple layers are included in an IoT-based system, and each of these layers has an additional Security layer for protection purposes and privacy protocols. A smart home will include [5] energy and water consumption that is to be monitored to recommend appropriate use of resources. Remote appliance control will also prevent accidents and energy wastage. Intruders can be detected, and storerooms be observed. In the market, there are various types of Single Board Computers (SBCs) other than Raspberry Pi such as Galileo and Arduino. Thus, the need for flexible data across heterogeneous systems is critical [6].

An IoT-based smart home is an integrated connection of electronics, sensors and software, building a network of physical devices inside a house that provides functionality [7]. Automated homes have built-in detection and control devices, such as air conditioning, lighting and security systems. Modern systems interconnect switches and sensors, and are called "gateways", which basically control the system through a device, i.e., phone or computer via a user interface [8]. Systems that integrate IoT have a key role in the incipient smart home environment. By 2023, the smart home market is forecasted to rise to US$137.91 billion, growing at 13 percent annually with respect to the compound annual growth rate (CAGR) between 2017 and 2023 [9].

The proposed model in this paper discusses the reduction of high installation and maintenance costs of a smart home automation system by the elimination of separate smart devices and integrating them on a single platform. For example, in the existing billing system, companies cannot keep track of the dynamic changes in customer demand. Consumers often receive bills for services they already paid before due dates. This is accompanied by poor reliability and quality often too. Keeping track of the consumer side with respect to load will ensure a more accurate billing system, allow tracking the maximum demand, and also help to calculate and note the threshold values. These are crucial features that must be incorporated in the design of an efficient energy billing system [10]. The smart energy meter is hence nullified by real-time power consumption being displayed directly on a smartphone.

With the influx of the smart grid era and the dawn of advanced communication and information infrastructures, bidirectional communication, energy storage systems and home area networks will alter the outlines of electricity usage and energy preservation at the consumption sites. Coupled with the rise of vehicle-to-grid machineries and distributed generation, there is a profound transition from the conventional centralized infrastructure towards the new and adaptive, autonomous and cyber-physical energy systems with renewable sources [11]. Installing various sensors in the house, environmental and regular devices are operated in automated remote and GSM instructions. Green renewable source of energy is utilised in power generation for smart appliance functioning in order to sustain the automated home [12].

In addition to these, smart homes connected to a smart grid depend on renewable energy as their primary source, making the setup more ecologically and commercially beneficial, and thus the smart energy meter does the needful, as mentioned earlier, and helps monitor power usage and live data.

The prototype discussed in this paper is a model designed to show the proposed idea at a smaller scale. Based on IoT and microcontroller configuration, the circuit can be connected with a smartphone using a Wi-Fi module and the application on the phone can be used to control two devices, get live readings for temperature and humidity, and also give power consumption by the devices in real time.

The next segment, i.e., segment 2, covers the methodology for designing the prototype.

2.2 Methodology

The prototype that is discussed in this paper is based on the principles of IoT and functions on the integration of multiple electronic components and a cellular device, i.e., a smartphone, over a shared wireless connection, allowing the user to monitor and control the prototype from the application designed on the cell phone. The fundamentals of IoT allows the establishment of the private network which is embedded with the sensor, microcontroller, and other similar components and hence allows connecting and exchanging data and instructions over the wireless connection for the prototype.

It can be observed from Figure 2.1 that the Wi-Fi module or the microcontroller for the prototype that is responsible for sending and receiving instructions for the circuit based on IoT, is connected to a pre-registered smartphone via Hotspot tethering over the same network. Once a command has been given by the user on the application created on the smartphone, the data is processed by NodeMCU Wi-Fi module, and the necessary actions are taken with respect to device control, which includes a sequence of LED lights and a fan. The power consumption of the devices is calculated using the pre-set code in the module. In a larger scale for a

Home Automation System for Power Analysis 15

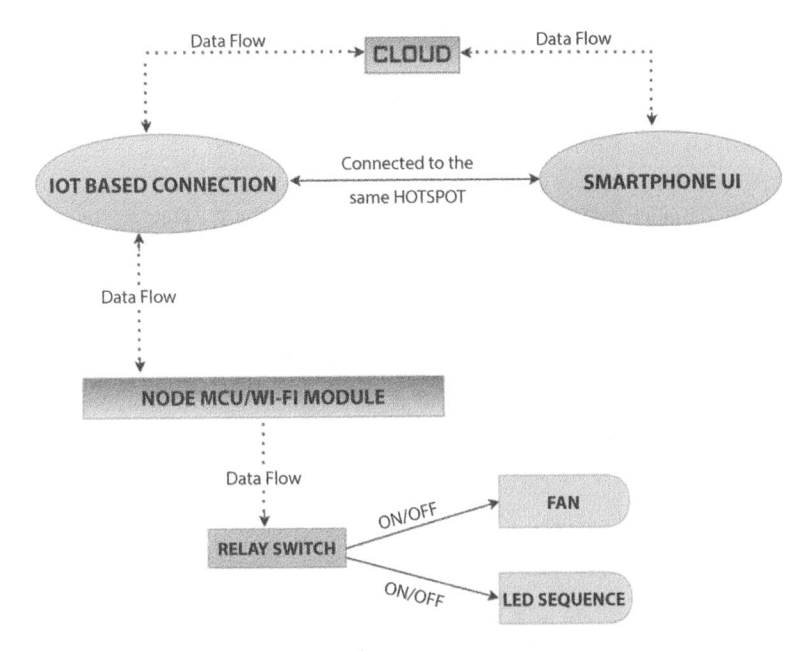

Figure 2.1 Block diagram for the prototype.

bigger project, this data is sent to the cloud after which it is displayed on the smartphone. All the exchange of data in the case of this prototype is through the hotspot connection.

The next segment, i.e., segment 3, has the design specifications.

2.3 Design Specifications

In this segment, the various topics regarding the building of the prototype are covered, including the components required, circuit assembly and working, application for the user interface on the smartphone, and the PCB design.

2.3.1 Components Required

The Wi-Fi module is the key component in making the prototype as it is what works on the fundamentals of IoT and connects the smartphone and the circuit leading to the sharing of data via a common connection. In specific cases, the NodeMCU microcontroller running on ESP8266 Wi-Fi SoC, is what connects the system to the cloud and is responsible for the sharing of intel between users via devices connected to the same (Figure 2.2).

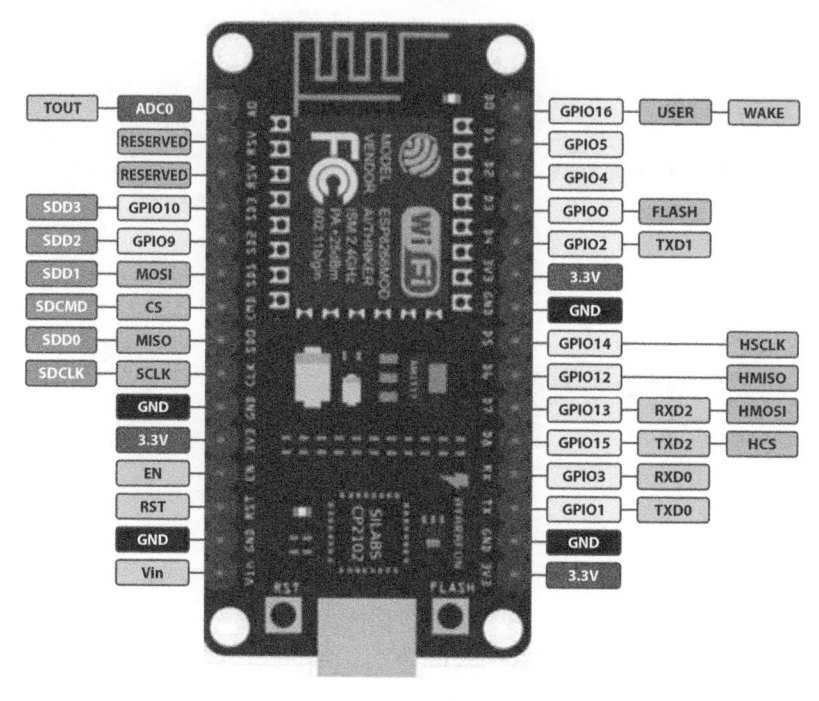

Figure 2.2 NodeMCU controller pin description.

Table 2.1 NodeMCU controller pin description.

PIN	Description
Micro USB, 3.3V, GND, Vin	Voltage is supplied to Micro-USB, 3.3V & Vin pins to power the board. GND is the ground pin.
EN, RST	These are the control pins that enable and reset the microcontroller.
A0	Analog pin for voltage measurement in the range of 0-3.3V.
GPIO Pins	16 pins.
SPI Pins	4 pins to configure the SPI communication.
UART Pins	Uploading the firmware/program.
12C Pins	Supporting functions.

Table 2.1 includes the most frequently used pins in the Wi-Fi module used for the prototype along with their description. Furthermore, all sorts of coding required for the control of the circuit too is encoded in this microcontroller using Arduino. The software is readily available for all versions of Windows and Mac (Figure 2.3).

As mentioned in Table 2.1, all the pins and their functional descriptions are unique and have been presented.

The rest of the components required for the prototype are as given in Table 2.2.

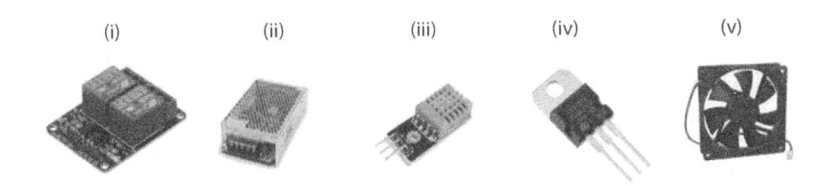

Figure 2.3 Components from Table 2.2.

Table 2.2 Components required besides Wi-Fi module.

Component	Quantity	Description
Relay Module (i)	1	A 5V Relay Module is used as an automatic switch. Low current signal utilized for high current control.
SMPS (ii)	1	It is an AC-DC converter which gives an output voltage of 12V DC.
Temperature Sensor (DHT11) (iii)	1	DHT11 is a sensor module that senses the temperature and humidity.
Regulator IC (iv)	2	It is used to regulate the voltage for smooth functioning of electronic devices.
Fan (v)	1	A very low current consumption fan with a voltage requirement of 12V is used.
LED	8	LEDs with power consumption of 0.48 W are used.

2.3.2 Circuit Diagram and Working

Working

The power supply of 230V AC is stepped down to 12 V DC for the relay switch to function by using the SMPS as a step-down voltage component or a transformer. Voltage regulators step down the voltage to 9 V and then 5 V from 12 V for Node MCU to function. The DHT11 sensor is directly integrated with the Wi-Fi module for constant data transfer to display real-time temperature and humidity values on the smartphone application. The module has to be connected with the pre-registered smartphone via hotspot tethering over the same network. The registration of the smart phone beforehand is essential for security purposes, and maintains a private control over the network and its system. Using the app, we can control the devices by sending a signal through the same network, that operates the relay and the devices (Figure 2.4). The devices are connected parallel to

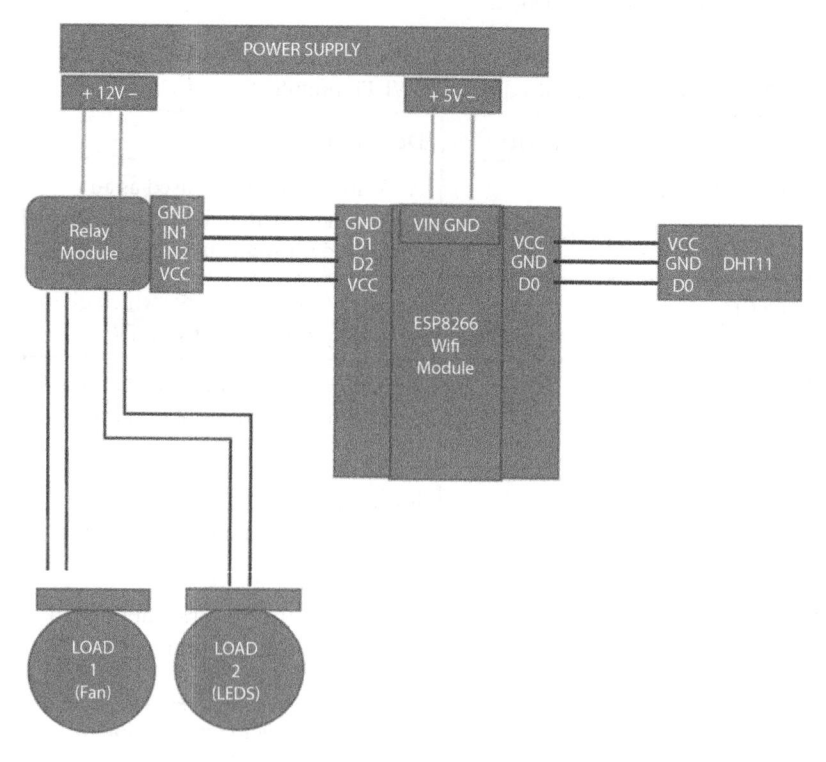

Figure 2.4 Circuit diagram.

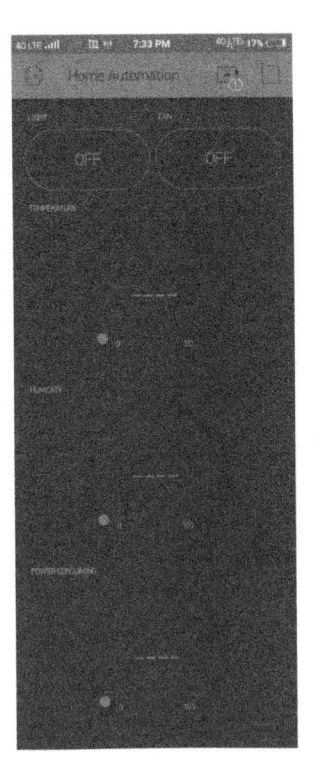

Figure 2.5 Blynk GUI.

each other with the relay working as a switch that operates after receiving instructions from the module. The devices and the relay work on 12 V itself. Further, NodeMCU is coded accordingly to calculate the power consumed by the devices that can be accessed by the same app on our smartphone (Figure 2.5).

2.3.3 Blynk GUI (Graphical User Interface) for Smartphone

Blynk is a hardware controlling application which is an IoT platform that allows the user to create interfaces for their personal or public projects through the combination and control of various widgets. Readily available on Google Play Store, this application can be configured easily to work with any Wi-Fi modules and is flexible for IoT-based applications. This is the user interface for the application that has been developed for the prototype. This has 7 slots (V1-V7) including switches for the devices and gauges for temperature, humidity and power consumption (Figure 2.6).

2.3.4 PCB (Printed Circuit Board) Design

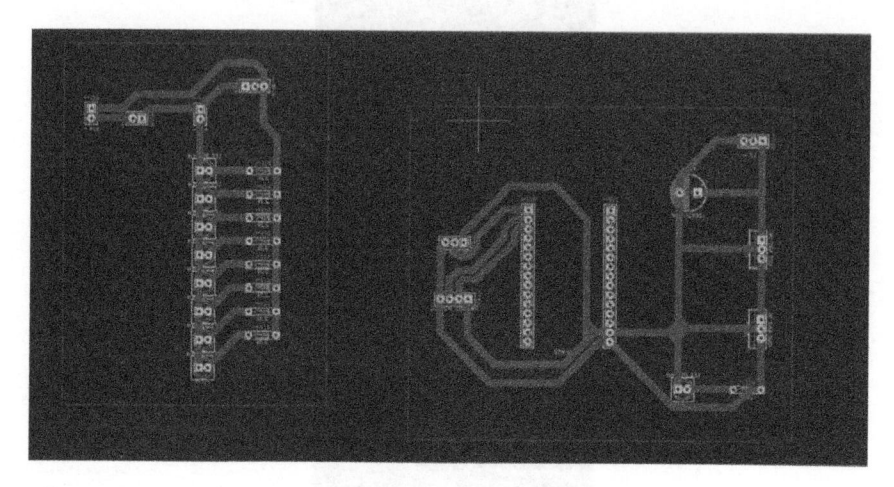

Figure 2.6 PCB.

The PCB is designed using EAGLE software, which is a product by CadSoft Computer GmbH.

The next segment, i.e., segment 4, covers the results and discussion.

2.4 Results and Discussion

As presented in the previous segment, the hardware model for the prototype is designed and tested in this segment, and further the results are displayed. This includes setting up the components and connecting them in order to form the prototype circuit, and also creating the NodeMCU Wi-Fi module code on Arduino application which will be the instructional unit for the system. Further, the GUI is designed on Blynk application on the smartphone, and the hotspot connection is pre-registered along with the device details for security purposes.

2.4.1 Prototype Design Completion

All the components as mentioned previously are assembled according to the circuit diagram. Powered by home supply voltage, i.e., 230 V, the circuit becomes operational automatically when connected to the registered smart phone. Henceforth, the user can control the LED sequence and the fan by using virtual switches on their smartphone GUI, and at the same time receive real-time values for power being consumed and the temperature and humidity levels in the surroundings (Figure 2.7).

Further, we have the Blynk GUI showing us results for a particular test case where the two appliances are switched off (Figure 2.8).

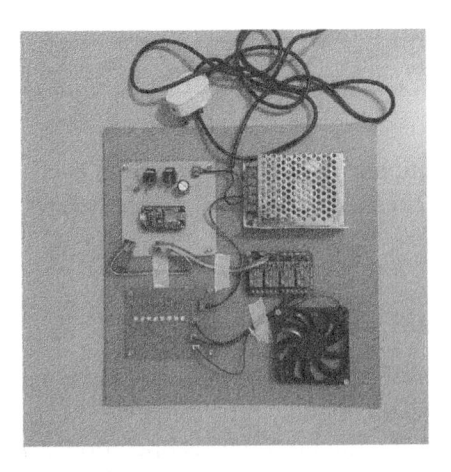

Figure 2.7 Live hardware model of prototype.

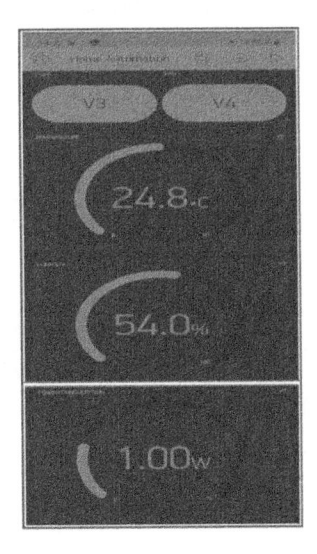

Figure 2.8 Blynk GUI (operational view).

Table 2.3 Test cases for prototype functionality check.

Serial no.	Light	Fan	Temperature (Celsius)	Humidity (%)	Power consumption (Watts)
1	Off	Off	30.2	84.3	1.00
2	On	Off	32.5	81	1.48
3	Off	On	32.3	84	2.92
4	On	On	31.9	84.6	3.40

There can be multiple combinations from the user's end for the prototype's functionality check. All test cases for these have been visualised in Table 2.3.

2.4.2 Testing and Observations

As observed from Table 2.3, the prototype gives varying results for the differing test cases and hence is fully functional.

As seen in Figure 2.9, the prototype is operational as per the 4th test in Table 2.3, and has both the LED sequence and fan working at the same time, thereby giving real-time power consumption, temperature and humidity readings on the user's smartphone.

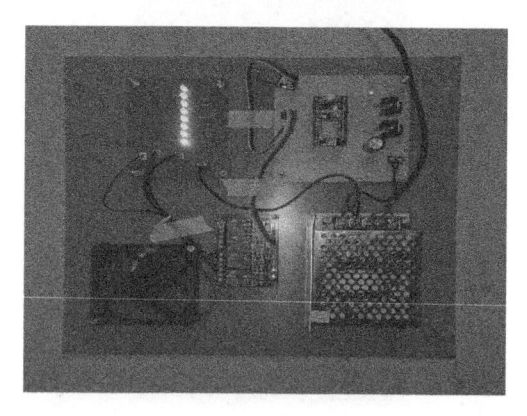

Figure 2.9 Live hardware model of prototype (operational).

2.4.3 Future Prospects

The discussed prototype can be implemented on a large scale for people to install and use in their smart homes. Further integration with a direct link to the smart grid that ensures the supply of power to the homes can increase the functionality and benefits of the same.

Power BI is an application available on Play Store for Android users, alongside its web version, that allows live streaming for data and updates dashboards made based on IoT in real time, usually presented in a graphical format. This application can be utilised in this prototype if applied on a commercial scale with the aim to increase customer interaction/engagement with the smart grid. The home automation system can be integrated with the Power BI app that can be made available for all smart home users directly by the electricity supply board. This will allow all the consumers to easily understand the pattern of different variants like temperature, previous dataset values, on and off-peak hours based on power demand, etc. This further can help them get a hold of the forecasting of power consumption done by the board daily to supply ample power to the users. The data can be received on the same application that was designed for the smart home in partnership to increase the ease of access by adding the live stream feature that allows users to make informed decisions to utilize power efficiently. Tariff rates for different hours based on on-peak and off-peak hours can be added as another feature, and this overall proposal will make it easy for a common consumer to become a Prosumer in the smart grid, that is, a user that consumes as well as sells power to the grid during high-demand situations by storing energy in off-peak hours.

As mentioned amongst the advantages of implementing the proposed model, there is a large scope to extend the features of the system and the smart home itself by adding more slots for different functions on the fully-flexible application, that is Blynk. A smart alarm system can detect smoke in case of a fire when the house is empty, and via the cloud, the data received by the smoke detecting sensors can be sent to the user on their smartphone in terms of an alarm, and an added option to call the emergency services can be visualized correspondingly.

As available readily, an integration with voice control can also be done using Google Alexa to make it easier for all age groups to interact with the system. Hence, with the addition of a new application in collaboration with the previous, and with a few more additions in features, the prototype can be made into a successful commercial product for smart home users to install.

Humidity
BY TIME

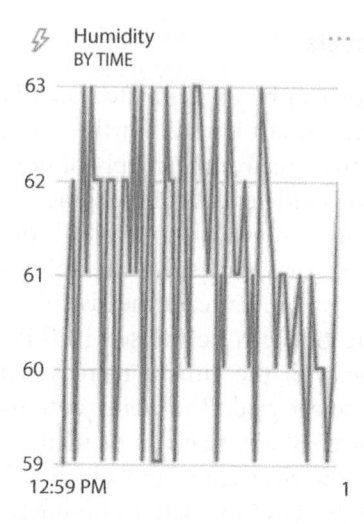

[Graph showing humidity values ranging from 59 to 63 over time starting at 12:59 PM]

Figure 2.10 Power BI graph for humidity in 24 hours of a day as recorded on the prototype.

The Power BI Graph for humidity in 24 hours of a day as recorded on the prototype is shown in Figure 2.10.

2.5 Conclusion

The prototype proposed in this paper can be successfully upgraded and installed as a fully functional and efficient commercial model for smart homes. The flexibility in its design and the overall user-friendly interface makes it a reliable product for the automation market that is growing exponentially every year with new advancements in technology; for example, a single click on an option for the home security system can give direct live feed access to all the cameras in the house, on the smartphone. If implemented on a larger scale, it will eliminate the requirement for multiple smart devices like the smart energy meter, or a smart thermostat, as these functions are successfully carried out on a single platform. Further, installation and maintenance costs can be dropped significantly, and due to pre-registration of the network, security and privacy can be increased and sustained. With the future prospects in consideration, the system can work parallel with the smart grid as its source and via machine learning and IoT applications, a stronger and more resilient network can be built with a smart decision making A.I. arrangement handling the tasks from the source to the user end effectively.

References

1. Y. Lee, J. Jiang, G. Underwood, A. Sanders and M. Osborne, "Smart power-strip: Home automation by bringing outlets into the IoT," *2017 IEEE 8th Annual Ubiquitous Computing, Electronics and Mobile Communication Conference (UEMCON)*, 2017, pp. 127-130, doi: 10.1109/UEMCON.2017.8249007.
2. Vanmathi U., Jadhav H., Nandhini A., Rajesh Kumar M. (2020) Accelerometer Based Home Automation System Using IoT. In: Das K., Bansal J., Deep K., Nagar A., Pathipooranam P., Naidu R. (eds.) *Soft Computing for Problem Solving. Advances in Intelligent Systems and Computing*, vol. 1057. Springer, Singapore.
3. Froiz-Míguez, I.; Fernández-Caramés, T.M.; Fraga-Lamas, P.; Castedo, L. Design, Implementation and Practical Evaluation of an IoT Home Automation System for Fog Computing Applications Based on MQTT and ZigBee-WiFi Sensor Nodes. *Sensors* 2018, 18, 2660.
4. Nagaraja G.S., Srinath S. (2020) Security Architecture for IoT-Based Home Automation. In: Satapathy S., Bhateja V., Mohanty J., Udgata S. (eds.) *Smart Intelligent Computing and Applications. Smart Innovation, Systems and Technologies*, vol. 159. Springer, Singapore.
5. Wan Nor Naema Wan Aziz, Muhammad Shukri Ahmad, Muhammad Mahadi Abdul Jamil, "Development of Novel Home Automation System via Raspberry Pi", *IEEE International Conference on Control System, Computing and Engineering*, 28–30 November 2014, Penang, Malaysia.
6. Patil, S. A., & Pinki, V. (2017). Home Automation Using Single Board Computing as an Internet of Things Application. *Proceedings of International Conference on Communication and Networks*, 245–253.
7. Mussab Alaa, A.A. Zaidan, B.B. Zaidan, Mohammed Talal, M.L.M. Kiah, A review of smart home applications based on Internet of Things, *Journal of Network and Computer Applications*, Volume 97, 2017, pp. 48-65, ISSN 1084-8045.
8. Galinina, O., *et al.*, Smart home gateway system over Bluetooth low energy with wireless energy transfer capability. *EURASIP Journal on Wireless Communications and Networking*, 2015. 2015(1): p. 1-18.
9. Y. Meng, W. Zhang, H. Zhu and X. S. Shen, "Securing Consumer IoT in the Smart Home: Architecture, Challenges, and Countermeasures," in *IEEE Wireless Communications*, vol. 25, no. 6, pp. 53-59, December 2018.
10. Birendrakumar Sahani, Tejashree Ravi, Akibjaved Tamboli, Ranjeet Pisal, "IoT Based Smart Energy Meter", in *International Research Journal of Engineering and Technology* (IRJET), vol. 04, issue 04, pp. 96-102, April 2017.
11. Bin Zhou, Wentao Li, Ka Wing Chan, Yijia Cao, Yonghong Kuang, Xi Liu, Xiong Wang, Smart home energy management systems: Concept, configurations, and scheduling strategies, *Renewable and Sustainable Energy Reviews*, Vol. 61, 2016, pp. 30-40, ISSN 1364-0321.

12. D. Nag et al., "Green energy powered smart healthy home," *2017 8th Annual Industrial Automation and Electromechanical Engineering Conference (IEMECON)*, 2017, pp. 47-51.

3

Energy Generation from Secondary Li-Ion Batteries to Economical Na-Ion Batteries

R. Rajapriya and Milind Shrinivas Dangate*

Chemistry Division, School of Advanced Sciences, Vellore Institute of Technology, Chennai, India

Abstract

In the future, batteries are predicted to be important enablers of a low-carbon economy. With the expanding diversity of electrical energy sources, particularly in the renewable energy pool, and the increasingly complex of electrical energy usage, a need for storage systems to manage demand and supply is unavoidable. Batteries comprise electrochemical cells that use ongoing exergonic electrochemical reactions to transfer chemical energy stored inactive materials into electrical energy. The Na-ion rechargeable batteries are an important technological development of energy conversion and its electrochemical stable batteries. The Li-ion battery dramatically raising the demand is likely to put serious pressure on high resource and supply while Na-ion secondary (or) rechargeable batteries have been considered to provide a cheaper, less resource and supply by the use of various cheaper and more economic materials than Li-ion batteries. Because lithium-ion batteries have a greater energy density compared to sodium-ion batteries, a smaller lithium-ion battery will last longer between charges. This chapter is focused on why Lithium-ion batteries are better than Sodium-ion batteries.

Keywords: Sodium-ion batteries (SIB), lithium-ion batteries (LIB), energy storage system (ESS), high power density

---

*Corresponding author: milind.shrinivas@vit.ac.in

3.1 Introduction

Batteries are expected to be critical enablers of a low-carbon economy in the future-decay. The current worldwide economic and environmental situations have increased demand for an energy storage system, lowering the cost of Li-ion batteries and allowing their use to gradually expand beyond mobility to a grid. The primary source of energy is fossil fuels, which are the world's most commonly used source of energy. The difficulties of access to resources, contamination of the atmosphere, and political turmoil in the development of fossil fuel have resulted in the rapid growth of various unreliable sources for renewable energy, and cleaner sources like wind, solar and waves. A large-scale energy storage system (ESS) is crucial for high-speed change operations to incorporate these renewable energy sources within the grid. Because of this stability, high power conversion and easy protection, the use of an electrical secondary battery be a capable system designed for large-scale electricity stocking [1]. The primary components of stationary electric storage batteries vary greatly from the power batteries used in electric vehicles. The main criteria are long cycle, cost-effectiveness and high protection. A large, non-toxic, stable, as well as low sprain electrode substance, therefore is required to be produced to ensure large and long-term applications and to reduce battery costs for management. Sodium-Ion batteries and Lithium-ion batteries to date have been considered as possible energy storage equipment for ESS [2]. The technology has been a leading energy storage solution in many portable devices since Sony commercialized rechargeable Li-ion batteries in 1991 because of the high energy density [3]. With rising lithium costs, attention has been focused on making alternative carriers such as Na-ion more easily available, cheaper and environmentally friendly [4]. Because of their great energy density as well as environmental friendliness, Li-ion batteries be measured as the best option in the last couple of decades for many energy storage applications such as electric cars and energy storage systems. Recent concerns about insufficient Li and Co supplies, together with rising demand for $LiCoO_2$-cathode materials used for LIBs, have increased costs. New types of storage systems have been proposed as an alternative to LIBs to solve these problems [5]. Because of the low cost and availability of Na resources in each crust, as well as the similar energy storage mechanism to LIBs, sodium-ion has gotten a lot of attention as LIB alternatives (Figure 3.1).

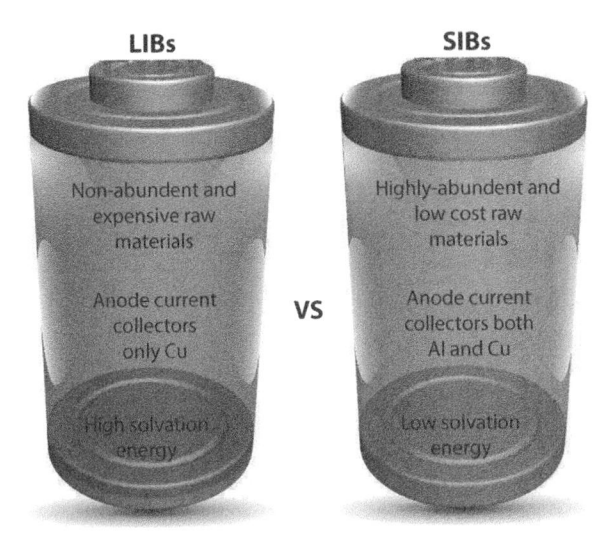

Figure 3.1 Advantages of lithium-ion batteries vs. sodium-ion batteries [6].

3.2 Li-Ion Battery

A Li-ion battery is a kind of rechargeable battery wherein when discharge, lithium ions go to the positive electrode from the negative electrode and then return to the negative electrode throughout charging. The chemistry of lithium-ion batteries varies in performance, cost, and safety. In consumer electronics, LIBs are widely used. They have such a high open-circuit voltage, a low self-discharge rate, no memory effect, and a slow loss of charge while not in use, making them one of the finest energy-to-weight ratios available, and among the most common rechargeable battery for convenient gadgets. As a result of their great energy density, Lithium-ion batteries are becoming increasingly used within military, electric vehicles, as well as aerospace applications. By converting electrical energy into electrochemical energy, a lithium ion battery can serve as an energy storage device. The cathode, anode and electrolyte are the three main components of a LIB device. During charging and discharging Li-ions migrate from the cathode to the anode via the electrolyte/separator, whereas electrons flow out of the external circuit to provide electrical power (Figure 3.2).

While several factors determine the efficiency of energy conversion in lithium-ion batteries, their actual performance is significantly in connection with the properties of the materials as well as the structure used.

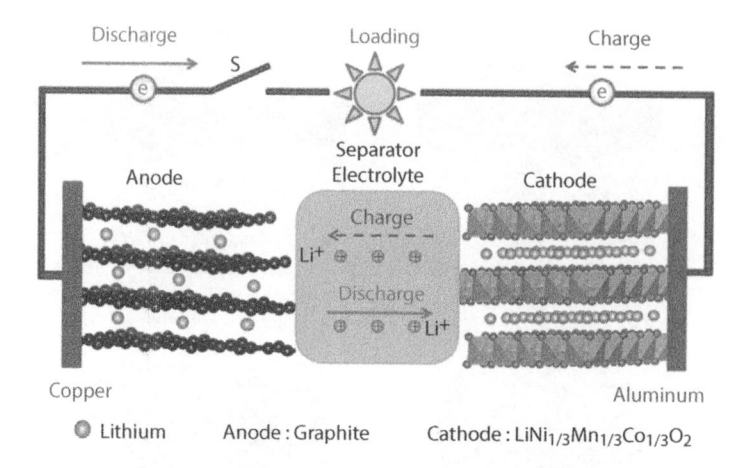

Figure 3.2 Diagrammatic illustration of part of lithium-ion battery [7].

The separator, a microporous membrane to enable the electrolyte to pass through while preventing shorting between the two electrodes separates the cathodes and anode regions. In the theory, the electrolyte will be both ionically and electrically conductive; however, the electrolyte's real properties are more difficult [4]. The work of Noble Laureates has changed electrical energy storage, a field that deals with storing electrical energy. In the 1970s, Professor M. Stanley Whittingham invented titanium disulphide; like a lithium-ion batteries cathode material it can efficiently Lithium ions are hosted as well as intercalated. Furthermore, in the 1980s, Professor John B. Goodenough's metal oxide research, namely lithium cobalt oxide, gave an alternative of lithium-ion batteries; the cathode material is converted to metal oxide, allowing them to run at higher voltages (3.5-4V) and hence provide increased density of energy/power. The anode was made of metallic lithium in these studies, and it is a very flammable and reactive substance that is not commonly employed in battery applications. Professor Akira Yoshino later developed a carbon anode on a lithium-ion battery (petroleum coke) instead of lithium as well as the cathode is lithium cobalt oxide material about 1985 [8]. In 1980, $LiCoO_2$ was first shown to serve as a potential cathode material for lithium-ion rechargeable batteries [9]. The $LiCoO_2$ layered structure was first discovered in 1980, followed by $LiMn_2O_4$ spinels in 1986, and the $LiMPO_4$ (M=Fe, Mn, etc.) [10, 11] olivine family in 1997 [12]. Traditional cathode materials are classified by structure and contain layered compounds $LiMO_2$ (M=Co, Ni, Mn, etc.). All of these compounds have been extensively explored as well as successfully utilized to the improvement of commercial Li-ion batteries in a relatively brief length of time [13]. For high-energy systems, layered materials are employed as cathodes, whereas Spinel oxides and olivines are evaluated such as high-intensity Li-ion batteries because of their inexpensive price as well as long lifespan, respectively [14]. The crystal structures of layered, Spinel and olivine compounds are shown in Figure 3.3.

Lithium-ion batteries' great energy density, as well as a huge proportion of the number of discharge cycles, have been the most crucial features that make them essential within mobile phones as well as electric vehicles. Furthermore, they outperform typical battery chemistries in several other ways. In addition to having a greater energy density, those cells that can discharge at high power can then be recharged considerably faster. In applications where charging powers (or) time are not constrained, such as solar PV system (or) stop-start autos, operating lithium-ion cells at a partial state of charge is not harmful. This allows them to operate more freely than lead acids. Operating voltage and current, energy density, cyclability, and stability, above all are factors that influence the performance of a lithium battery. The operating voltage of a lithium-ion battery be influenced via means of band gap-related electronic state, for example, are electrical properties of cathode materials locations. The most commonly used in cathode materials is $LiCoO_2$ [16]. The first and most important economically victorious kind of coated transition metal oxide cathodes; $LiCoO_2$ (LCO) was introduced via Goodenough [9]. Sony was the first to market it and it is silent utilized Li-ion batteries are utilized in most commercialized Lithium-ion batteries today. Co as well as Li, which are found in octahedral locations, occupy opposing strata as well as have hexagonal symmetry [17].

Despite the fact the $LiCoO_2$ is a viable cathode material, other materials are being explored to reduce costs and increase stability. Cobalt is less common and therefore more expensive than other transition metals like

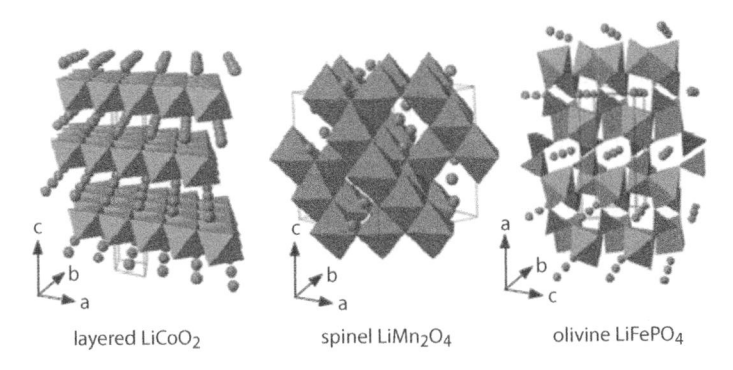

layered $LiCoO_2$ spinel $LiMn_2O_4$ olivine $LiFePO_4$

Figure 3.3 Crystal structures of layered, spinel and olivine compounds [15].

manganese, nickel, and iron. Combining $LiNiO_2$ and $LiMnO_2$ in a 1:1 ratio, generating layered $LiNi_{0.5}Mn_{0.5}O_2$, and generating Li-Co-Ni-Mn-O layered compounds are some of the methods used [18]. Ohzuku et al. published the first good-quality electrochemical data on $LiNi_{0.5}Mn_{0.5}O_2$ in 2001 [13]. $LiNiO_2$ (LNO) does have a crystal structure identical to $LiCoO_2$ as well as a theoretical specific capacity of 275mAhg^{-1}. The key research driving forces are its comparatively high energy density, also reduced cost as compared to Co-based materials. Pure $LiNiO_2$ cathodes, on the other hand, are unsuitable because Ni^{2+} ions prefer to substitute Li^+ sites through synthesis also delithiation, obstructing Li diffusion routes [19]. Since Ni^{3+} is more easily reduced than Co^{3+}, LNO is much more thermally unstable than LCO. It was discovered that partially replacing Ni with Co was an efficient strategy to minimize cationic disorder [20]. $LiCoO_2$ and $LiFePO_4$ are currently the majority extensively utilized in commercialized Lithium-ion batteries. $LiCoO_2$ is simple to make in large quantities which are stable in air. On a full charge, it has a practical ability of 140mAh/g and a theoretical capability of 274mAH/g. Aside from its poor practical capacity, $LiCoO_2$ has other notable drawbacks, including its expensive material cost and cobalt toxicity.

$LiFePO_4$ based cathode materials, on the other hand, have gotten a lot of attention in the last decade because of their low cost and low environmental effect. $LiFeO_4$ has a variety of advantages over $LiCoO_2$, including sturdiness, long cycle life as well as thermal stability (-20 to 70°C). Furthermore, $LiFePO_4$ have weak electrical and ionic conductivity, with values of 10^{-10} s/cm and 108 cm^2/sec, etc., and also a limited capacity [21]. Another problem is one-dimensional Li-ion diffusion channels that are frequently obstructed by defects and contaminants. The lower lithium-ion relocation force is found mostly with channels running parallel to the [010] track, according to modelling and calculations. Also, it implies that the diffusion stable varies lying on element dimension, hence distribution in the volume being significantly slower compared to diffusion at the nanoscale. As a result, nanoscale $LiFePO_4$ has been investigated and used Li-ion batteries [22]. Even though $LiFePO_4$ has been shown to have a lot of potential to enable new $LiCoO_2$ as a safe cathode substance used for Lithium-ion batteries, there can exist a few issues. The comparatively low $LiFePO_4$ has a particular capacity (or) energy (measured in mass) combined through some of its chemistry leaves little room at the other hand, may improve their rate performance of power. Furthermore, there are two challenges with the advent of nanoscale $LiFePO_4$, the nanomaterials have a low volumetric density, and thus preparing nanoscale $LiFePO_4$ so at a cheap price is difficult. Because nanoscale $LiFePO_4$ has a low packing density, producing high energy density batteries is problematic (by volume). Considering the significance of the interface activity of LiFePO$_4$ with the electrolyte in electrochemical performance, that is even more important also at the interfacial process involved is essential.

The underlying process involved in interfacial electrochemical reactions may be revealed by *in situ* examinations of the interface during the charge-discharge process. That knowledge can aid in the development of improved LiFePO$_4$ coatings, as well as better cycling rate as well as performance [23]. In addition to the layered LiMO$_2$ (e.g. LiCoO$_2$) as well as olivine LiMPO$_4$ (M=Fe, Mn, Ni, Co or combination, e.g., LiMn$_{1-x}$Fe$_x$PO$_4$) stated cathode substances of LiMnPO$_4$, LiCoPO$_4$ and Spinel LiMxMn4$_{-x}$O$_8$ (M=Fe, Co, Ni) are shown above, were already aggressively promoted. Materials for high-voltage cathodes may aid in increasing energy density; however, they will necessitate the use of an alternate stable electrolyte rather than a typical electrolyte.

Other new polyanoinic cathode family is Li$_2$MSiO$_2$ that have a substantially higher capacity of 330 mAh/g. Electronic conductivity, rate capability and capacity waning are all issues during cycle are all barriers to using those high-capacity Li$_2$MSiO$_2$ [24]. The highly concentrated cathode material has the potential of being mass-produced and employed in commercial Lithium-ion batteries possibly in future. In commercial Li-ion batteries, NCM (LiNi$_{1/3}$Co$_{1/3}$Mn$_{1/3}$O$_2$) cathodes are likely to play a bigger role. The solution may be sodium-ion batteries, which have made incredible strides in recent years. They could, in the near future, replace lithium-ion batteries which are now utilized in electric vehicles, cell phones and computers. Lithium and sodium are chemically very comparable and alkali metals. Although sodium does not have the same energy density as lithium, it is widely and inexpensively available.

3.3 Sodium-Ion Batteries

Sodium-ion batteries (SIB) were first researched in the 1970s; however, a quick growth, as well as the commercialization Lithium-ion batteries (LIBs), redirected attention away from SIBs. Despite the reality that SIBs and LIBs have comparable energy storage mechanisms, there are some variations between them. The discrepancy into ionic radii of Na$^+$ ions (1.02 Å) as well as Li$^+$ ions (0.76Å), for example, makes it difficult for LIBs electrodes to be employed in SIBs. Furthermore, the lower energy density is due to the larger molar mass as well as the standard electrode potential of Na (23g mol^{-1} and -2.7V Vs SHE) relative Li (6.9g mol^{-1} and -302 V

Vs SHE). The charge storage capacity, on the other hand, is mostly dictated because of the structural characteristics of materials used in the electrodes, with Na, as well as Li, accounting for just a little portion of the weight of the electrode. As a result, the power density differential between LIBs and SIBs has no bearing [25].

Sodium-ion batteries are rechargeable batteries that make use of sodium ions as charge carriers. In comparison to other battery types, the sodium-ion battery is relatively new. Sodium salts for batteries are much cheaper and more plentiful than lithium salts. Ford Motor Company (USA) made the first successful effort at a sodium battery in 1967 with the sodium-sulphur battery [26]. Because of these three factors: cost, availability, and size, SIBs are particularly appealing when intended for large-scale grid storage applications. The sodium-ion battery, developed by Williams Advanced Engineering in the United Kingdom, is significantly less expensive and safer than typical lithium-ion batteries. Sodium-ion batteries appear to be a viable alternative because of their comparable functioning principle to LIBs and plentiful Na supplies. SIB is typically made up of a carbonaceous anode with a sodium-rich cathode. Although it is often used for half-cell experiments, using a shiny metal it is strictly forbidden to use a cathode due to its extremely high reactivity and dendritic matter. Through the first charge step of a SIB, the cathode experiences oxidation and Na deintercalation, while the carbonaceous anode undergoes reduction and Na-intercalation at the same time [27].

Transition metal oxides are divided into two groups based on their structural differences: layered oxides and tunnel-oxides. With huge S-shaped, tunnel type oxides get an orthorhombic structure while including tunnels and tiny pentagon tunnels the Na-ions in the giant S-shaped tunnels can remove reversibility; whereas throughout the small tunnel, the SIBs are also not electrochemically active [28]. Na_xMeO_2 layered oxides are made up of Na ions are sandwiched between repeating layers of MeO_6. The layered oxides can be divided into two groups, according to Delmas' classification: O_3 and P_2 types, wherein Na^+ ions are found correspondingly for octahedral and prismatic sites [29]. $Na_{0.44}MnO_2$ has an orthorhombic structure and is a typical tunnel-type oxide. Inside a polymer electrolyte battery, sodium storage efficiency for $Na_{0.44}MnO_2$ was first reported by Doef *et al.* [28] the electrode had a large capacity that can be reversed 180 mAhg^{-1}, the cycling performance was poor. Layered oxides among the broad composition of Na_xMeO_2 (Me = transition metal, like Cu, Cr, Fe, Co, Ni, V, Mn and so on) be a member of the vast family with a wide range of phases and compositions. An effective structural regulation strategy for improving the layered oxides structural stability seems to be the cation replacement for transition metals using Li$^+$, Mg^{2+}, Cu^{2+}, Al^{3+}, Ti^{4+}, and some other cations, that can assist minimize layered oxide phase change upon sodium insertion/extraction. The typical crystal structures are shown in Figure 3.4.

During the charge/discharge mechanism in SIBs, Na ions transit among both the anode and cathode compartments. The selection of anode material has a considerable impact on SIBs efficiency as it can slow down electrochemical reactions. Owing to dendrite development as well as the low melting point of Na, an imperfect selection of anode materials be accessible used for SIBs unlike cathode materials [25]. The cathode, either in LIBs (or) SIBs, has been a most costly component of some of these two batteries [30]. Since Al will alloy through Li if discharged to a low potential, Cu would be employed also as active material instead of Al, for LIBs. SIB cathode materials are much less costly than LIBs cathode materials because of the lower cost of sodium. SIBs can still save around 12.6% in total costs when compared to LIBs. The energy density, on the other hand, of SIBs, LIBs have a lower rate of return, resulting in a unit price of $0.11 per Wh for LIBs and $0.14 per Wh for NIBs, depending on current technology. The distinction between these two technologies appears to be minor [31]. Although cathode materials (carbon materials and silicon materials) are more costly than anode materials, anode materials are usually frequently overrepresented in batteries to fully exploit the cathode materials' performance. As a result, cathode material performance is critical for SIBs, capacity, as well as voltage, is current indications for increasing energy density. SIBs must therefore create new cathode structures with excellent and high voltage. Sodium layered oxide cathodes have a high gravimetric energy density that is because of their low molar mass as well as extensive two-dimensional alkali-ion diffusion channels. Sodium layered oxide cathode materials including P$_2$ type and O$_3$ type materials are regarded as excellent candidates for sodium-ion complete cells. The P$_2$ type layered

Tunnel structure
(a)

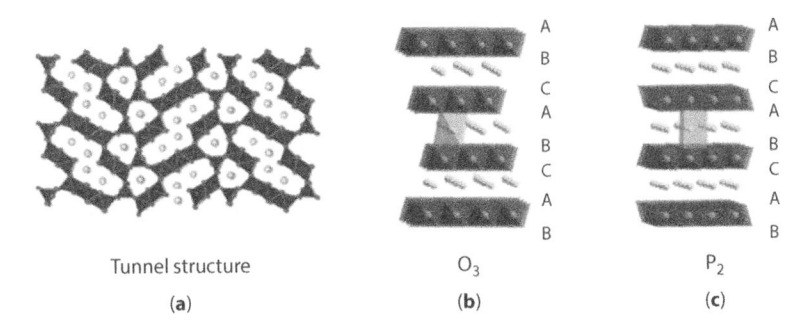

O$_3$
(b)

P$_2$
(c)

Figure 3.4 Typical crystal structures.

frameworks; however, unlike O_3 type structure, lack interstitial tetrahedral sites, providing an open passage in favour of Na ions by such a lesser the O_3 type structure has a lower diffusion barrier. As a result, P_2 type layered materials have higher ionic conductivity than O_3 type layered materials, and P_2 type layered cathodes have a high specific capacity as well as good power capability [32]. Cathode materials with such a regular shape and high tap density, for example, can considerably increase long cycling, quality and power density that will be advantageous to economic applications in the future. Furthermore, SIBs rate performance is linked to Na^+ kinetics, which is determined via electron conductivity and ion diffusion. To date, elemental doping and delicate structural design have been viable options for improving rate performance [33]. SIBs have low electrochemical performance because sodium ions have a greater diameter, consequently, it is important to design appropriate electrode materials with sodium-ions intercalation that is quick and reversible reactions. As a result, a huge effort has lately been focused on cathode materials, which are regarded as a fundamental component of SIBs. Considerable efforts have been made over the last few decades to investigate Na_xMO_2 type materials as SIBs cathodes [34, 35]. A manganese-based transition metal oxide has long been thought to be potential cathode materials, with Parant et al. introducing these in 1971. A layered structure, a tunnel structure, as well as a mixture (tunnel as well as layered) structure are created for Na_xMnO_2 $0.22 \leq x \leq 0.44$ V, $0.66 < x \leq 1$ V and $0.44 < x \leq 0.66$ V approximately, as per this report. Depending on this position occupied by alkali-ions layered cathode materials are being categorized into numerous categories, including P_2, P_3, O_2 as well as O_3 structures. The letters P and O indicate to Na ions prefer sites that are octahedral as well as prismatic, respectively. While the numbers (2 and 3) indicate the amount of transition metal oxide layers and in crystal structure that are occupied. In comparison of O_3 type of oxides, P_2 type of oxides frequently experience many phase transitions, as well as the reversible capacity may scarcely reach 120 mAhg^{-1}, as per Komaba et al. [36] findings. Due to that P_2 type oxides have a larger capacity; their cycling stability is typically poor owing to phase changes (P_2 to O_2) throughout electrochemical processes. Along with bigger Na ions, charge ordering, as well as the packing order involving Na as well as vacancies on different Na content, when Na ions are removed from alkali-ion layers, layered 3d-transition metal oxides (LTMOs) display greater complex phase transitions. Na_xFeO_2 has also been investigated as just a SIBs cathode material having the Fe^{3+}/Fe^{4+} redox pair [37, 38]. Na_xFeO_2 with such a P_2 structure has considerably better-reversed capacity retention than Na_xFeO_2 with an O_3 structure, although it has a poor specific capacity. The voltage range for charging that is directly in relation to Fe^{2+}/Fe^{3+} and Fe^{3+}/Fe^{4+} redox couples influences the iron-based electrochemical performance for Na_xFeO_2 through O_3 structure [35]. Furthermore, Na_xFeO_2 may when combined with water to form FeOOH as well as NaOH, resulting in loss of capacity and irreversible structural damage alternations. $NaCrO_2$ facilitated facile sodium ion intercalation/extraction because of the wide interpolate distance, which is in contrast to the electrochemically inert $LiCrO_2$. The $NaCrO_2$ electrode could also offer an ability that can be reversed of around 110 mAhg^{-1} with such a voltage platform of around 3V, because of owing to the redox reaction of Cr^{3+}/Cr^{4+} [39]. Along with comparing to P_2-type cathode materials, they have a higher energy density as well as a greater cut-off voltage (2.2-4.5V), Materials for O_3-type cathodes are being widely explored, and nonetheless, they get a low working potential. Although the ratio of Na to M (transition metal) in materials for O_3-type might to be kept at 1:1, the Na will be employed up to 0.8 in some circumstances. The Na and M ions occupy separate octahedral positions in O_3-type materials, which have a rack-salt layered structure.

Komaba and colleagues [36] investigated a coated O_3- $NaNi_{0.5}Mn_{0.5}O_2$ type material for SIBs in 2012, achieving a reversible capacity of 185 mAhg^{-1} throughout a potential range of 2.24-4.5V. X-ray diffraction *ex-situ* research indicated to massive discharge capacity loss was detected while cycling due to multiple phase transitions also including O3, P_3, P'_3 & P''_3. Following that, a range of O_3-type oxides was created; however, despite numerous attempts to modify the synthesis processes, the NiO phase is typically found O_3-type compounds comprising Ni as well as Li [40]. Following cathodes of the aluminium-doped O_3 type were developed with a goal for improving battery as well as structural stability performance. An example, the O_3- $NaNi_{0.8}CoO_{0.15}Al_{0.05}O_2$ cathode [41] is played a reversible capacity of 154 mAhg^{-1} but has insufficient air stability. Furthermore, a novel sort of tin-based alloy O_3- $Na_{0.9}(Ni_{0.45x/2}Mn_xSn_{0.55x/2})O_2$ cathode material was created, which gas a great operational voltage of roughly 3.45V along with maybe improved electrochemically by adding a little amount of Mn. Also, *in situ* XRD confirmed when Na insertion, as well as extraction, a reversible phase change occurs and only W-1 percent volume change, occurs during the process [42]. P_2-type materials seem to be of particular relevance in SIBs as a cathode as they have a greater off-stoichiometric potential for work region of Na (0.67<x<0) and, whereas O_3-type oxides, typically have cycling has only one phase (P_2 to O_2). Furthermore, because of the existence of an unclosed channel having a low-energy diffusion barrier allowing sodium migration, Oxides of P_2 type have superior Sodium kinetics of diffusion than the O_3-phase. As just a consequence,

P_2-type oxides have strong rate capabilities and steady cycle performance [43]. Zang et al. [44] reported a P_2-type $Na_{2/3}Mn_{2/3}Ni_{1/3}O_2$ cathode substance using a water-soluble biopolymer like gear gum, sodium alginate, as well as xanthan gum as well as the materials have been electrochemically examined, as opposed towards industrial polyvinylidene fluoride (PVDF) binder. In general, water-soluble binders have a high concentration of –OH as well as –COO- functional groups, the promote particle conductor attachment. As a result, conductivity and electro chemistry were enhanced. Xanthan gum binder, among the aforementioned water-soluble biopolymers, has a high capacity as well as excellent capacity retention of 78 per cent at 40mAhg^{-1}. The P_3-type crystalline phase seems to have the identical straight Na diffusion channel as the P_2 and O_3 phases as well as invariably converts in the direction of P_3 during the cycle progression. P_3 oxide, on the other hand, has a different oxygen stacking sequence than P_2 and O_3 oxides [45] which could lead to different electrochemical characteristics. Risthaus et al. [46] recently published experimental and theoretical work on a P_3-type $Na_{0.9}Ni_{0.5}Mn_{0.5}O_2$ cathode made through solid-state reaction. Oxide of P_3-type has a squat Na ion diffusion barrier with high capacities of 141 as well as 102 mAgh^{-1} at 10 and 100 mAgh^{-1}, etc. Furthermore, Ni decrease in the P_3 type oxide helps to boost capacity by stabilizing the anionic redox effect at greater potentials. The voltage and capacity of Mg substituted P_3 oxides containing improved over time [47]. For example, the P3-$Na_{2/3}Ni_{1/4}Mg_{1/2}Mn_{2/3}O_2$ cathode has a reasonable operating voltage of 3.6V and improved cycle stability. They have developed a full-cell battery with a target voltage of 3.45V, a specific capacity of W 120mAgh^{-1} at 0.1 C, and an appealing energy density of 412WhKg [48]. Materials with only one phase (P_2, P_3, and O_3) should be used employed when used with full-cell batteries because of their benefits and downsides. As a result, a strategy for improvement depends on integrating as well as bi-phase (or) multi-phase material having integrated benefits to synergistically achieving significant electrochemical performances. While lithium-ion batteries could be their best option for the most advanced battery technology, scientists are concerned about the practicality of a long-term lithium supply. Lithium is not a plentiful element, and also its natural resources are utilized inequitably dispersed, despite its widespread distribution in the earth's crust. Prices will rise as insist for Li commonly chemicals in transport applications grows, along with the rise of the electric vehicles industry. If Li batteries are used for large-scale energy storage, supplies will be much more limited. As a result, Lithium-based energy storage devices on a large scale might be beneficial forced to be limited [49]. Sodium-ion batteries can work as a potential in addition to Lithium-ion batteries, which have been around for a long time. Natural sodium resources are abundant around the world; sodium has been the fourth most plentiful element on the planet. In the periodic table, sodium (Na) sits below lithium (Li), and they have many chemical properties in common. The essential concepts of the SIBs, as well as LIBs, are the same: a voltage is created on the cell by the variation in chemical potential of an alkali (Li or Na) among two electrodes during charge and discharge. There are a variety of reasons to look into Na-ion batteries. High energy density becomes less important as the use of batteries expands to large-scale storage, like electrical vehicles or stationary storage coupled to renewable energy production [4]. The similarities between comparable synthesis, characterization, as well as analysis of Na-ion & Li-ion batteries methodologies, make Na-ion battery development basics easier to grasp. Moreover, when compared to the standard hydrogen electrode (SHE), Na has a redox potential of E^0 (Na^+/Na) of 2.71V, but Li has a redox potential of E^0 (Li^+/Li) of 3.0V [50]. Palomers et al. reported the brief characteristic information about Li and Na materials is listed in Table 3.1 [51].

Even though $LiCoO_2$ and $NaCoO_2$ share identical CoO_6 edge-sharing octahedral frameworks, they have different redox potentials and phase evolution during cycling. The potential difference between the two compounds varies between 0.4V and 1.0V, depending on the degree of intercalation. During cycling, the complicated voltage profile of $NaCoO_2$ suggests that many phase transitions and order events occur. Many studies have shown that $LiCoO_2$ keeps the O_3 structure until the Li level falls below 0.5. When the Na content is less than 0.5, the structure will switch from 0_3 to P_3, which is not desirable for obtaining reversibility [29]. As per Sanders, the cathode is the most expensive part for a Li-ion battery, accounting

Table 3.1 Main characteristics of Li and Na materials.

Characteristics	Li	Na
Price	4.11-4.49£Kg^{-1}	0.07-0.37£Kg^{-1}
Ionic radius	0.69Å	0.98Å
Melting point	180.5°C	97.7°
Capacity density	3.86Ahg^{-1}	1.16 Ahg^{-1}
Relative atomic mass	9.94	23.00
Coordination Preference	Octahedral and tetrahedral	Octahedral and prismatic

for around a quarter of the entire cost. The cathode material's composition is the primary distinction among Li-ion as well as Na-ion batteries, according to an assessment of their components. Since this cost of preparing for both Na-ion and Li-ion technologies, the cathode made with fresh materials is nearly identical; raw materials for Na-ion batteries, this is the primary source of cost decrease. The main benefit for Na-ion batteries was their long-term viability, and that is critical in a world aiming to eliminate carbon-based energy sources. Future work will focus on developing new cathode as well as anode material as Na-ion batteries through greater specific capacity also with voltages, such that Na-ion batteries are a feasible option, and specific energies exceeding 200 Wh/kg can be produced. Improved Na-ion battery reviews at high charge/discharge rates are enabled by electrolytes across a broad range of temperatures and demonstrating that the extended cycle life as well as shelf-life necessary, such as applications requiring large-scale energy storage, should also be developed.

3.4 Conclusion

Sodium-ion rechargeable batteries may soon be a much more expensive as well as resource-efficient alternative to present lithium-ion batteries. A revolution appears to be on the way, due to powerful prototypes and ground-breaking insights in basic research. The alkali metals lithium, as well as sodium, is chemically extremely close. Though sodium lacks the energy density of comparably uncommon lithium, it is extensively as well as inexpensively attainable. The solution may be sodium-ion batteries, which have made recently incredible development. They may, in the near future, replace lithium-ion batteries, which are now utilized in electric vehicles, cell phones, and computers. We are all in the early stages of the Li-ion era, and therefore we must act anyway to avoid repeating the polymer era, which peaked in the 1900s and has become a significant worldwide planet problem of plastic pollution in the 2000s. To avoid making the same error, the long-awaited but yet-to-be-realized Na-ion technology stands as an appealing alternative to greener and more sustainable batteries, which is the only way to electrify the planet without adding to the world environment problem.

References

1. Dunn, B.; Kamath, H.; Tarascon, J. M. Electrical Energy Storage for the Grid: A Battery of Choices. *Science (80-.).* **2011**, *334* (6058), 928–935. https://doi.org/10.1126/science.1212741.
2. Pan, H.; Hu, Y. S.; Chen, L. Room-Temperature Stationary Sodium-Ion Batteries for Large-Scale Electric Energy Storage. *Energy Environ. Sci.* **2013**, *6* (8), 2338–2360. https://doi.org/10.1039/c3ee40847g.
3. Palacín, M. R. Recent Advances in Rechargeable Battery Materials: A Chemist's Perspective. *Chem. Soc. Rev.* **2009**, *38* (9), 2565–2575. https://doi.org/10.1039/b820555h.
4. Kim, S. W.; Seo, D. H.; Ma, X.; Ceder, G.; Kang, K. Electrode Materials for Rechargeable Sodium-Ion Batteries: Potential Alternatives to Current Lithium-Ion Batteries. *Adv. Energy Mater.* **2012**, *2* (7), 710–721. https://doi.org/10.1002/aenm.201200026.
5. Choi, J. U.; Jo, J. H.; Park, Y. J.; Lee, K. S.; Myung, S. T. Mn-Rich P′2-Na0.67[Ni0.1Fe0.1Mn0.8]O2 as High-Energy-Density and Long-Life Cathode Material for Sodium-Ion Batteries. *Adv. Energy Mater.* **2020**, *10* (27), 1–10. https://doi.org/10.1002/aenm.202001346.
6. Ali, Z.; Zhang, T.; Asif, M.; Zhao, L.; Yu, Y.; Hou, Y. Transition Metal Chalcogenide Anodes for Sodium Storage. *Mater. Today* **2020**, *35* (May), 131–167. https://doi.org/10.1016/j.mattod.2019.11.008.
7. Xu, J.; Dou, S.; Liu, H.; Dai, L. Cathode Materials for next Generation Lithium Ion Batteries. *Nano Energy* **2013**, *2* (4), 439–442. https://doi.org/10.1016/j.nanoen.2013.05.013.
8. Dixit, A. Cathode Materials for Lithium Ion Batteries (LIBs): A Review on Materials Related Aspects towards High Energy Density LIBs. **2020**.
9. Fujita, T.; Toda, K. Microdisplacement Measurement Using a Liquid-Delay-Line Oscillator. *Japanese J. Appl. Physics, Part 1 Regul. Pap. Short Notes Rev. Pap.* **2003**, *42* (9 B), 6131–6134. https://doi.org/10.1143/jjap.42.6131.
10. Thackeray, M. M.; David, W. I. F.; Bruce, P. G.; Goodenough, J. B. Lithium Insertion into Manganese Spinels. *Mater. Res. Bull.* **1983**, *18* (4), 461–472. https://doi.org/10.1016/0025-5408(83)90138-1.
11. Moos. No Title Детская Неврология. *Екр* **1984**, *13* (3), 576.
12. Batteries, R. L. Fosfooliwin\\. *J. Electrochem. Soc.* **1997**, *144* (4).
13. Ohzuku, T.; Makimura, Y. Layered Lithium Insertion Material of LiCo1/3Ni1/3Mn1/3O2 for Lithium-Ion Batteries. *Chem. Lett.* **2001**, No. 7, 642–643. https://doi.org/10.1246/cl.2001.642.
14. Zaghib, K.; Mauger, A.; Groult, H.; Goodenough, J. B.; Julien, C. M. Advanced Electrodes for High Power Li-Ion Batteries. *Materials (Basel).* **2013**, *6* (3), 1028–1049. https://doi.org/10.3390/ma6031028.
15. Julien, C. M.; Mauger, A.; Zaghib, K.; Groult, H. Comparative Issues of Cathode Materials for Li-Ion Batteries. *Inorganics* **2014**, *2* (1), 132–154. https://doi.org/10.3390/inorganics2010132.

16. Antolini, E. LiCoO2: Formation, Structure, Lithium and Oxygen Nonstoichiometry, Electrochemical Behaviour and Transport Properties. *Solid State Ionics* **2004**, *170* (3–4), 159–171. https://doi.org/10.1016/j.ssi.2004.04.003.
17. Nitta, N.; Wu, F.; Lee, J. T.; Yushin, G. Li-Ion Battery Materials: Present and Future. *Mater. Today* **2015**, *18* (5), 252–264. https://doi.org/10.1016/j.mattod.2014.10.040.
18. Xu, B.; Qian, D.; Wang, Z.; Meng, Y. S. Recent Progress in Cathode Materials Research for Advanced Lithium Ion Batteries. *Mater. Sci. Eng. R Reports* **2012**, *73* (5–6), 51–65. https://doi.org/10.1016/j.mser.2012.05.003.
19. Hassel, B. A. Van; Montross, C. S. Oxide Fuel Cells, S. C. Singhal and H. Iwahara, 9. **1996**, *143* (4), 1168–1175.
20. Arai, H.; Okada, S.; Sakurai, Y.; Yamaki, J. I. Thermal Behavior of Li1-YNiO2 and the Decomposition Mechanism. *Solid State Ionics* **1998**, *109* (3–4), 295–302. https://doi.org/10.1016/s0167-2738(98)00075-7.
21. Chung, S. Y.; Bloking, J. T.; Chiang, Y. M. Electronically Conductive Phospho-Olivines as Lithium Storage Electrodes. *Nat. Mater.* **2002**, *1* (2), 123–128. https://doi.org/10.1038/nmat732.
22. Malik, R.; Burch, D.; Bazant, M.; Ceder, G. Particle Size Dependence of the Ionic Diffusivity. *Nano Lett.* **2010**, *10* (10), 4123–4127. https://doi.org/10.1021/nl1023595.
23. Gummow, R. J.; He, Y. Recent Progress in the Development of Li2MnSiO4 Cathode Materials. *J. Power Sources* **2014**, *253*, 315–331. https://doi.org/10.1016/j.jpowsour.2013.11.082.
24. Deng, D. Li-Ion Batteries: Basics, Progress, and Challenges. *Energy Sci. Eng.* **2015**, *3* (5), 385–418. https://doi.org/10.1002/ese3.95.
25. Zhu, J.; Roscow, J.; Chandrasekaran, S.; Deng, L.; Zhang, P.; He, T.; Wang, K.; Huang, L. Biomass-Derived Carbons for Sodium-Ion Batteries and Sodium-Ion Capacitors. *ChemSusChem* **2020**, *13* (6), 1275–1295. https://doi.org/10.1002/cssc.201902685.
26. Fan, X.; Liu, B.; Liu, J.; Ding, J.; Han, X.; Deng, Y.; Lv, X.; Xie, Y.; Chen, B.; Hu, W.; Zhong, C. Battery Technologies for Grid-Level Large-Scale Electrical Energy Storage. *Trans. Tianjin Univ.* **2020**, *26* (2), 92–103. https://doi.org/10.1007/s12209-019-00231-w.
27. Zhang, Y.; Gui, J.; Li, T.; Chen, Z.; Cao, S. an; Xu, F. A Novel Mg/Na Hybrid Battery Based on Na2VTi(PO4)3 Cathode: Enlightening the Na-Intercalation Cathodes by a Metallic Mg Anode and a Dual-Ion Mg2+/Na+ Electrolyte. *Chem. Eng. J.* **2020**, *399* (June), 125689. https://doi.org/10.1016/j.cej.2020.125689.
28. Doeff, M. M.; Peng, M. Y.; Ma, Y.; De Jonghe, L. C. Orthorhombic Nax MnO2 as a Cathode Material for Secondary Sodium and Lithium Polymer Batteries. *J. Electrochem. Soc.* **1994**, *141* (11), L145–L147. https://doi.org/10.1149/1.2059323.

29. Berthelot, R.; Carlier, D.; Delmas, C. Electrochemical Investigation of the P2-NaxCoO2 Phase Diagram. *Nat. Mater.* **2011**, *10* (1), 74–80. https://doi.org/10.1038/nmat2920.
30. Kim, Y.; Ha, K. H.; Oh, S. M.; Lee, K. T. High-Capacity Anode Materials for Sodium-Ion Batteries. *Chem. - A Eur. J.* **2014**, *20* (38), 11980–11992. https://doi.org/10.1002/chem.201402511.
31. Ortiz-Vitoriano, N.; Drewett, N. E.; Gonzalo, E.; Rojo, T. High Performance Manganese-Based Layered Oxide Cathodes: Overcoming the Challenges of Sodium Ion Batteries. *Energy Environ. Sci.* **2017**, *10* (5), 1051–1074. https://doi.org/10.1039/c7ee00566k.
32. Sun, Y. K. Direction for Commercialization of O3-Type Layered Cathodes for Sodium-Ion Batteries. *ACS Energy Lett.* **2020**, *5* (4), 1278–1280. https://doi.org/10.1021/acsenergylett.0c00597.
33. Yabuuchi, N.; Kubota, K.; Dahbi, M.; Komaba, S. Research Development on Sodium-Ion Batteries. *Chem. Rev.* **2014**, *114* (23), 11636–11682. https://doi.org/10.1021/cr500192f.
34. You, Y.; Kim, S. O.; Manthiram, A. A Honeycomb-Layered Oxide Cathode for Sodium-Ion Batteries with Suppressed P3–O1 Phase Transition. *Adv. Energy Mater.* **2017**, *7* (5), 1–7. https://doi.org/10.1002/aenm.201601698.
35. Wang, P.; You, Y.; Yin, Y.; Guo, Y. Layered Oxide Cathodes for Sodium-Ion Batteries : Phase Transition , Air Stability , and Performance. **2017**, *1701912*, 1–23. https://doi.org/10.1002/aenm.201701912.
36. Komaba, S.; Yabuuchi, N.; Nakayama, T.; Ogata, A.; Ishikawa, T.; Nakai, I. Study on the Reversible Electrode Reaction of NaNi Mn O2 for a Rechargeable Sodium Ion Battery. *Inorg. Chem.* **2012**, *51*, 6211.
37. Li, Y.; Gao, Y.; Wang, X.; Shen, X.; Kong, Q.; Yu, R.; Lu, G.; Wang, Z.; Chen, L. The Abundant Resource of Sodium (Na) Makes the Na-Ion Batteries a Promising Alternate to the Li-Ion Batteries in Electric Energy Storage, Especially When Iron (Fe)-Based Oxide Electrodes Are Used. Layer-Structured Na. **2018**, *2*.
38. Lee, E.; Brown, D. E.; Alp, E. E.; Ren, Y.; Lu, J.; Woo, J. J.; Johnson, C. S. New Insights into the Performance Degradation of Fe-Based Layered Oxides in Sodium-Ion Batteries: Instability of Fe3+/Fe4+ Redox in α-NaFeO2. *Chem. Mater.* **2015**, *27* (19), 6755–6764. https://doi.org/10.1021/acs.chemmater.5b02918.
39. Yu, C.; Park, J.; Jung, H.; Chung, K.; Aurbach, D.; Sun, Y.; Myung, S. Environmental Science. *Energy Environ. Sci.* **2015**, *8*, 2019–2026. https://doi.org/10.1039/C5EE00695C.
40. Mao, Q.; Gao, R.; Li, Q.; Ning, D.; Zhou, D.; Schuck, G.; Schumacher, G.; Hao, Y.; Liu, X. O3-Type NaNi0.5Mn0.5O2 Hollow Microbars with Exposed {0 1 0} Facets as High Performance Cathode Materials for Sodium-Ion Batteries. *Chem. Eng. J.* **2020**, *382* (September), 122978. https://doi.org/10.1016/j.cej.2019.122978.

41. Zhou, X.; Li, X.; Liu, Y.; Li, R.; Jiang, K.; Xia, J. Investigation of Benzo(1,2-b:4,5-B')Dithiophene as a Spacer in Organic Dyes for High Efficient Dye-Sensitized Solar Cell. *Org. Electron.* **2015**, *25*, 245–253. https://doi.org/10.1016/j.orgel.2015.06.033.
42. Yang, K.; Gao, F.; Huang, X.; Chen, L.; Hu, Y. A New Tin-Based O3-Na0.9[Ni0.45x/2MnxSn0.55x/2O2 as Sodium-Ion Battery Cathode. *J. Energy Chem.* **2018**, 5–10. https://doi.org/10.1016/j.jechem.2018.05.019.
43. Nkosi, F. P.; Soc, J. E. Insights into the Synergistic Roles of Microwave and Fluorination Treatments towards Enhancing the Cycling Stability of P2-Type Insights into the Synergistic Roles of Microwave and Fluorination Treatments towards Enhancing the Cycling Stability of P2-Type. **2017**. https://doi.org/10.1149/2.1721713jes.
44. Zhang, Y.; Zhang, S.; Li, J.; Wang, K.; Zhang, Y.; Liu, Q.; Xie, R.; Pei, Y.; Huang, L.; Sun, S. Improvement of Electrochemical Properties of P2-Type Na2/3Mn2/3Ni1/3O2 Sodium Ion Battery Cathode Material by Water-Soluble Binders. *Electrochim. Acta* **2019**. https://doi.org/10.1016/j.electacta.2018.12.089.
45. Online, V. A. Materials Chemistry A. **2018**. https://doi.org/10.1039/c8ta09422e.
46. Risthaus, T.; Chen, L.; Wang, J.; Li, J.; Zhou, D.; Zhang, L.; Ning, D.; Cao, X.; Zhang, X.; Schumacher, G.; Winter, M.; Paillard, E.; Li, J. P3 $Na_{0.9}Ni_{0.5}Mn_{0.5}O_2$ Cathode Material for Sodium Ion Batteries. *Chem. Mater.* **2019**, *31*, 5376–5383. https://doi.org/10.1021/acs.chemmater.8b03270.
47. Kim, E. J.; Ma, L. A.; Duda, L. C.; Pickup, D. M.; Chadwick, A. V.; Younesi, R.; Irvine, J. T. S.; Robert Armstrong, A. Oxygen Redox Activity through a Reductive Coupling Mechanism in the P3-Type Nickel-Doped Sodium Manganese Oxide. *ACS Appl. Energy Mater.* **2020**, *3* (1), 184–191. https://doi.org/10.1021/acsaem.9b02171.
48. Zhou, Y.; Wang, P.; Zhang, X.; Huang, L.; Wang, W.; Yin, Y.; Xu, S.; Guo, Y. Air-Stable and High-Voltage Layered P3-Type Cathode for Sodium- Ion Full Battery. *ACS Appl. Mater. Interfaces* **2019**, *11*, 24184–24191. https://doi.org/10.1021/acsami.9b07299.
49. De La Llave, E.; Borgel, V.; Park, K. J.; Hwang, J. Y.; Sun, Y. K.; Hartmann, P.; Chesneau, F. F.; Aurbach, D. Comparison between Na-Ion and Li-Ion Cells: Understanding the Critical Role of the Cathodes Stability and the Anodes Pretreatment on the Cells Behavior. *ACS Appl. Mater. Interfaces* **2016**, *8* (3), 1867–1875. https://doi.org/10.1021/acsami.5b09835.
50. Ellis, B. L.; Nazar, L. F. Sodium and Sodium-Ion Energy Storage Batteries. *Curr. Opin. Solid State Mater. Sci.* **2012**, *16* (4), 168–177. https://doi.org/10.1016/j.cossms.2012.04.002.
51. Palomares, V.; Serras, P.; Villaluenga, I.; Hueso, K. B.; Carretero-González, J.; Rojo, T. Na-Ion Batteries, Recent Advances and Present Challenges to Become Low Cost Energy Storage Systems. *Energy Environ. Sci.* **2012**, *5* (3), 5884–5901. https://doi.org/10.1039/c2ee02781j.

4

Hydrogen as a Fuel Cell

R. Rajapriya and Milind Shrinivas Dangate*

Chemistry Division, School of Advanced Sciences, Vellore Institute of Technology, Chennai, India

Abstract

Hydrogen seems to be the universe's more essential and significant element. It is constantly paired with something else. Although hydrogen has a lot of energy, an engine to run on it produces nearly rejection pollutants. Since the 1970s, NASA has used liquid hydrogen to launch into orbit the space shuttle as well as other rockets. Hydrogen has gained popularity in recent years, widely used as a source of energy, particularly in fuel cells. Fuel cells are devices that made things simpler to capture hydrogen and also turn it into useable energy. Fuel cells surprisingly found their way into a broader range of applications, including backup power, power for remote places, and distributed power generation, as well as electricity cogeneration. As in context, hydrogen is likely to take part in a big role shortly in providing a clean, low-cost, and environmentally friendly alternative to fossil fuels. The importance of hydrogen in various types of fuel cells, as well as the operating principle, advantages and disadvantages will be discussed in this chapter.

Keywords: Fuel cell, alternative energy, electricity

4.1 Introduction

Fuel cell technology was one of many technologies that have been produced and improved in many places throughout the world. Fuel cell systems have been recognized given their several advantages over traditional power sources, including excellent performance, reasonable durability, reduced time-to-market, and even environmental friendliness. In addition, because

---

*Corresponding author: milind.shrinivas@vit.ac.in

this technology is evolving at a rapid pace throughout the world, mainly in industrialized countries, many countries are attempting to replace traditional power sources by using fuel cells regularly. The fuel cell is an electrochemical cell that converts the chemical energy of a standard fuel quickly and reliably to electricity in the form of energy [1]. The action of the fuel cell is extremely similar to the Leclanche and mercury-oxide-zinc dry cells, which we are all familiar with [2]. Francis Bacon devised a hydrogen and oxygen-based fuel cell that was used in an advertisement. In 1959, a team led by Harry Ihrigh at Alli-Chalmers built a 15kw fuel cell, which was displayed at state fairs across the United States. Sir William Grove invented the first fuel cells in 1838. Fuel cells were first commercially used more than a century after Francis Bacon discovered the hydrogen-oxygen fuel cell in 1932 [3].

Chemical energy for the fuel cells is produced through hydrogen or several other fuels. This is known to produce efficient electricity. In the case of the hydrogen fuel cell, electricity, water, and heat are the products produced. Numerous potential applications segregate the type of fuel cells based on which they can provide power to the systems in huge as well as minimal amounts. Systems such as power stations require a large amount of electricity whereas a laptop computer utilizes small measures of electricity, both of which can be satisfied by a fuel cell [4]. Fuel cells are applied in many sources such as transportation, industrial/commercial residential buildings and they can store energy for the grid in reversible systems. Over conventional combustion technologies, a fuel cell operates at higher efficacy and produces electrical energy directly from the chemical energy with more than 60% productivity [5]. They do not produce emission or contain low emission and only water will be emitted, unlike conventional combustion that results in carbon-di-oxide emission. This resolves the critical climate changes around the environment. It also accounts for the cases of smog and air pollutants as well as noise pollution [3]. Like batteries, fuel cells do not run down or need a battery. The electricity and heat are produced until the fuel are present. It contains two electrodes, anode and cathode which are placed like layers; between them electrolyte is positioned; it is like a sandwich. In a hydrogen fuel cell, hydrogen is supplied in the direction of the anode as well as air is passed through the cathode. On the anode, hydrogen, molecules are separated keen on protons and electrons, and then they pass on towards the cathode which passes through an external circuit to produce electricity. The protons migrate along with the electrolyte towards the cathode and collaborate to oxygen after that electron to given water as well as heat. Alkali fuel cells are sometimes known as the Bacon fuel cell, after their creator. Since the 1960s, they have been utilized in space by NASA missions to power spacecraft as well as space capsules. The hydrogen fuel cell was regarded as an exclusive-costly technology at the time due to its utilization in high-end applications such as NASA space ships.

One of the most important discoveries and initiatives in attaining a low-carbon economy is the usage of fuel cells and hydrogen technology [6]. Hydrogen seems to be an energy carrier, and that can be confined in such a variety of ways and (Figure 4.1) transferred in a variety of modes. Hydrogen linked utilizing the fuel cell allows a continuous with a carbon-free way to synthesize power; it is adaptable as well as centralized in multiple

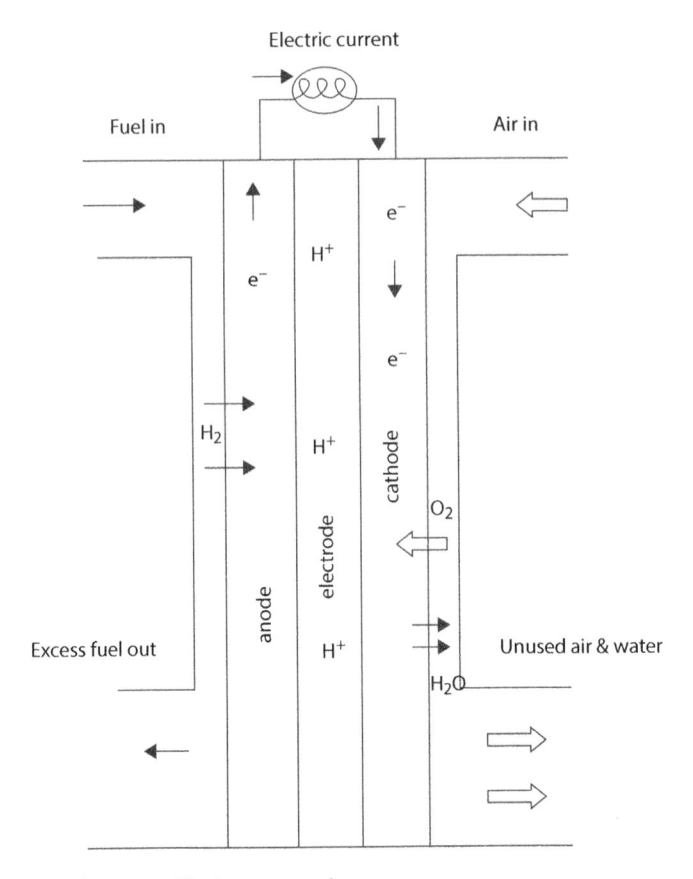

Figure 4.1 Applications of hydrogen as today.

zero-emission applications. Due to this, fuel cells must be employed in a variety of applications. Fuel cells are utilized as primary as well as backup control in commercial, industrial, and residential structures, as well as in remote (or) inaccessible places. The power cells also are found in forklifts, vehicles, taxis, ships, and submarines. Several states have recently directed their focus to utilize hydrogen fuel technology to help them in their decarbonization efforts. The benefits of fuel cell technology over traditional power sources are well known, including high efficiency, high dependability, a procedure that takes less time and is more ecologically friendly. Many governments are attempting to replace traditional power sources with fuel cells regularly, since technologies are rapidly evolving all over the world, particularly in urban areas [7]. Japan, Germany, the United States, Denmark, and China are at the forefront for hydrogen fuel cell research and development in their respective areas, as well as the implementation of this technology. The outline of fuel cell operation is given below.

4.2 Operating Principle

All fuel cells work with the same principle. At the anode, hydrogen or hydrogen-rich fuel containing positive ions (protons) are passed that is separated by the anode-coated catalyst. At the cathode, electrons and sometimes protons or water molecules are allowed to react giving hydrated water or ions. The electronics can pass only through the external electrical circuit, thus determining the electric current [8, 9]. The fuel cell works by converting chemical energy into electrical energy. Here, no intermediates are formed between the initial and final transfer of ions. This phenomenon is known as direct conversion, whereas the systems involving many transformations as well as the formation of mechanical and thermal energy are processed, are known as indirect energy conversion. The direct energy conversion produces high efficiency independent of thermal systems. This energy phenomenon breaks down the thermal or mechanical link. The basic mechanism lies in obtaining electricity through the interaction of oxygen and hydrogen [10, 11]. Chemical energy is made up of natural or manmade fuel components utilized as the sources of fuel which led the reactions of reduction and oxidation inside the fuel cell. Oxidation takes place at the anode and then the reduction proceeds at the cathode [12]. Thus, the reaction of hydrogen at the anode is;

$$\text{Acid electrolyte: } H_2 \rightarrow 2H^+ + 2e^-$$

Alkaline electrolyte: $H_2 + 2OH^- \rightarrow 2H_2 + 2e^-$

And reduction for oxygen is:

Acid electrolyte: $\frac{1}{2} O_2 + 2H^+ + 2e^- \rightarrow H_2O$

Alkaline electrolyte: $\frac{1}{2} O_2 + H_2O + 2e^- \rightarrow 2OH^-$

Where the anode adsorbs the hydrogen molecule to reduce it to hydroxyl ion and the cathode undergo (oxygen) adsorption and proceeds with reduction reacting with a water molecule. The membrane, an electrolyte, or an ionic conductor to is present in the fuel cell and will isolate the reactants from electrons. The energy produced in the form of heat due to the oxidation reaction that took place at the anode is transformed into electricity. The efficiency of such a fuel cell mechanism is higher than the conventional method because the electron between the oxidant and fuel is directly changed into electricity without any energy loss. Thus, fuel cells involve a redox reaction where the fuel is oxidized and the oxidant is reduced passing an electron by the external circuit. One such example is galvanic fuel cell where the oxidation occurs at anode and reduction at cathode producing electrons which were separated by the partial reactions resulting in the production of electricity [13].

4.2.1 Types of Fuel Cells

Fuel cells are differentiated based on their fuel kind, electrolyte own, fuel consumption and most importantly taking into account their operating temperature. It can vary from 20°C – 1500°C. The fuel cells found operating from 20°C – 100°C are known as low-temperature fuel cells, and likewise that are found working under 200-300°C and 600-1500°C are known as hot fuel cells and high-temperature fuel cells, respectively [8]. Based on the above elucidation, the types of fuel cells are:

1. Alkaline fuel cells (AFCs) – 70°C
2. Direct methanol fuel cells (DMCFs) – 60-130°C
3. Molten carbon fuel cells (MCFCs) – 650°C
4. Phosphoric acid fuel cells (PAFCs) – 180-200°C
5. Proton membrane fuel cells (PEMFCs) – 100°C and also at 1500°C -200°C
6. Solid oxide fuel cells (SOFCs) – 80°C – 1000°C

7. Zinc-air fuel cells (ZAFCs) – that utilizes zinc anode
8. Proton exchanger fuel cells (PCFCs) and comparably (RFC) regenerative fuel cells are superior types of fuel cells that will produce efficient amounts of hydrogen and oxygen via electrolytes. In the regenerative fuel cell, oxygen, as well as hydrogen, is produced more efficiently through electricity. The sun acts as fuel in RFC producing electricity, water and heat and then it is recycled for electrolysis [14].

4.3 Why Hydrogen as a Fuel Cell?

Science has demonstrated that there have been two options for assuring global energy contributes: renewable energies but also fuel cells (hydrogen-based energy), which could complement each other. The study issue as in area would aid the practical implementation of green as well as clean power systems through promoting its usage of hydrogen-based energy solutions in industrial applications at the regional, commercial as well as retail industries as renewable energy technologies [5].

The commencement of fuel cell development can be traced back to the turn of the nineteenth century. Humphry Davy evidenced a central concept which underpins the operation of a fuel cell in 1801; afterwards, in 1839, lawyer as well as amateur scientist William Grove discovered the concept for such fuel cell by accident while he disengaged the battery from an electrolysis device as well as tried to touch two electrodes when in an electrolysis experiment. This cell, which comprised platinum electrodes inserted in tubes carrying hydrogen gas but also oxygen, respectively, tubes submerged in dilute sulfuric acid was dubbed the "gas battery" by Grove. The voltage generated is roughly 1 volt. Grove later connected many of these "gas batteries" into combination using the power supply produced to power the electrolyzer which isolates hydrogen from oxygen. Grove's fuel cell failed to yield real applications owing to electrode corrosion issues as well as the volatility of such materials used. It was noticed also that the sensitivity for platinum black on contact therewith an electrolyte diminishes over time, therefore putting the electrolyte in a non-conducting porous medium extends a fuel cell's life.

Langer & Mond developed Grove's invention, coining the term "fuel cell" for the first time in 1889 after noticing that the reactivity of platinum black in the communication of electrolytes lessens over time, as well as the fuel cell's life span, is persistent, thereby enclosing the electrolyte with a porous non-conducting substance [15, 16]. Following it, the research was advanced significantly, and the 1960s-1970s saw the first fuel cells to be used in practical applications for a long time employed by NASA in spacecraft. Around the same time, General Electric was creating proton exchange membrane fuel cells, be also the foundation and fuel cells shown in the Gemini programs missions as well as the Apollo space plan to generate electricity. The fuel cells of various capacity that should be used in stationary applications have been created since 1990, to implement the institutional application of fuel cell technology.

The usage of phosphoric acid fuel cells (PAFCs) for co-generation application for its extensive use has been encouraged by advances in membrane durability as well as increased performances of fuel cell components in terms of energy. Being developed for this field is proton exchange membrane fuel cell (PEMFC) as well as solid oxide fuel cell (SOFC) technologies, particularly for tiny stationary application. The year 2000 also saw important advancements when it comes to portable devices, with the widespread adoption of the use of fuel cell systems using direct methanol (DMFC) in this sort of application. There has been a tremendous development in the use of fuel cells in the previous two decades, spanning a broad range in the portable, mobile as well as stationary sectors; there are a variety of applications [17]. This rise seems to be the result of technological advancements on the one hand, and global worries regarding energy security, efficiency, as well as durability on the other, as well as efforts to reduce last and not least, greenhouse emissions and decrease reliance on fossil fuels. Fuel cells have been studied more and more these days, as they are altering the way energy is created. They employ hydrogen as a fuel, thereby providing the ability to generate clean energy, thereby ensuring environmental protection and even enhancement. A fuel cell by definition seems to be a type of electric cell that, like battery cells, they can indeed be, continually fed by fuel, allowing it electric current generated by the electric cell's occurrence to be maintained indefinitely [18]. As a result of the fuel cell converts hydrogen or hydrogen-based fuels converted into electricity as well as heat through an electrochemical process between hydrogen as well as oxygen. At the fuel cell level, the reaction is indeed the opposite of electrolysis [8, 15].

$$2H_2 + O_2 \rightarrow H_2O + \text{Energy}$$

The chemical energy of a fuel (hydrogen), as well as the oxidant (oxygen), be converted to constant current, warmth, as well as water as reaction products [19]. Because hydrogen and oxygen gases were converted to water through an electrochemical reaction in such a fuel cell, it provides several benefits over thermal engines. Increased efficiency and extremely quiet operation, as well as the absence of pollutant emissions are all advantages for which the fuel can be even hydrogen, but if hydrogen is manufactured utilizing renewable energy sources, electrical power produced is also sustainable [12].

4.3.1 Electrolyte

It is vital to understand the elements that make up the whole, in addition, to evaluate the impact of some of the phenomenon together in the fuel cells. In a fuel cell, to act as that of an electrolyte, a substance should meet several criteria, including high chemical stability to regard to the selected electrodes in hopes of avoiding reacting High melting and boiling points are associated with these besides high-temperature cells; as well as predominantly ionic conductivity with no electronic conductivity. Liquid electrolytes (its most popular were indeed solid electrolytes both ion-exchange membranes as well as solid substance electrolytes are employed), melting electrolytes, as well as liquid electrolytes containing incorporated fuel too are used in acidic or basic solutions including an ion transport phenomena similar to that of water electrolysis are the four main types of electrolytes [15].

4.3.2 Catalyst Layer (At the Cathode & Anode)

In fuel cells, all electrochemical methods utilize to place in an area of catalyst layers. Its electrodes (catalyst layers) include catalysts molecules which accelerate the reaction in the cell rate. An electrode of the fuel cell would be a porous carbon base upon which a catalyst is placed. An electrode's thickness is normally among 5–15 m, as well as the catalyst's charge was generally involving 0.1 and 0.3 mg/cm^2 [15].

4.3.3 Bipolar Plate (Cathode & Anode)

Bipolar plates serve two purposes as in fuel cells, guiding reactant gases to a surface of the electrolytic exchange as well as driving the generated electric current. Bipolar plates require materials that are both highly conductive and gas-tight. It should also be resistant to corrosion and chemically inert. Graphite or steel, as well as composite materials, can be employed to address these issues. Channels for flow rate are "engraved" as in bipolar

```
                              2e⁻
        Anode          ┌──────────┐          Cathode
                       │          ↓
   ┌──────────┐ ┌──┐ ┌──┐ ┌──┐ ┌──────────┐
   │          │ │  │ │  │ │  │ │          │
   │ H2 → 2H⁺ + 2e⁻ │  │ │  │ │  │ │ ½ O2 + 2H⁺ + 2e⁻ → H2O │
   │          │ │  │ │  │ │  │ │          │
   └──────────┘ └──┘ └──┘ └──┘ └──────────┘
```

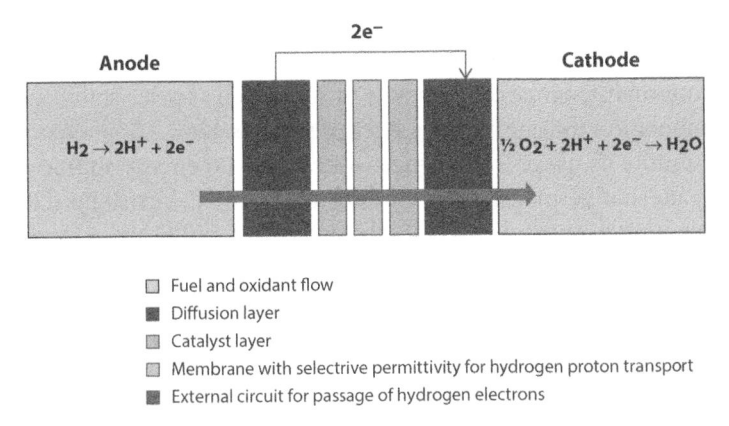

☐ Fuel and oxidant flow
■ Diffusion layer
☐ Catalyst layer
☐ Membrane with selectrive permittivity for hydrogen proton transport
■ External circuit for passage of hydrogen electrons

Figure 4.2 Applications of hydrogen as tomorrow.

plates that must generally also be thin as likely in the order it minimize the battery's mass as well as size.

The design of such flow channels affects the flow rate of reactants as well as mass transfer, and so has a decisive impact upon this fuel cell performance, making it vital to maximizing the surface of the flow to maximize the reaction surface. These components represented in Figure 4.2 [15].

4.4 Hydrogen as an Energy-Vector in a Long-Term Fuel Cell

Currently, hydrogen is regarded as a source of non-polluting energy, since, when generated through renewable sources, it cannot cause global warming [20]. Furthermore, hydrogen seems to be the only secondary energy carrier mostly on the market that's also suited for quite a broad array of applications. The truth of hydrogen can also be created by a broad range of primary energy that has been at the forefront of discussion [21]. It is an appropriately broad range for application, including transportation and portable as well as fixed use [22]. Furthermore, hydrogen may be utilized in distributed systems without releasing carbon dioxide [23]. Although hydrogen has already been used in the chemical sector, it is always used as energy. Source using technology like fuel cells [24]. Hydrogen would become an energy hub similar to how electricity is now because it can be created from a broad range of source energies being used in a wider range of applications [25].

Hydrogen has the advantage of being able to be kept as a medium for a longer time and apposed electricity. As a result, an energy carrier aids throughout maintenance on energy security as well as price stability resulting in opposition among various energy sources [26]. Some characteristics to promote its usage for hydrogen-like a backup energy source created utilizing alternative methods, according to Emre A. Veziroglu, editor of the *International Journal of Hydrogen Energy*, a publication devoted to hydrogen technology and energy; hydrogen seems to be a highly concentrated source of energy that may be made readily commercially available. It allows for the conversion of different types of good conversion efficiencies techniques of energy. When hydrogen is generated electrolytically from water, it is an infinite supply; hydrogen consumption and production is a closed cycle in which the basis of manufacturing. Water is held at a constant as well as constitutes a conventional recirculation cycle for this kind of raw material. Is the simplest and also cleanest fuel to use; the combustion for hydrogen produces nearly no polluting pollutants. When compared to other fuels, it also has substantially higher gravimetric energy density. Hydrogen could be kept in a variety of forms, including as a gas at higher than normal pressure, solid hydride or hydrogen, liquid. This could be carried across vast distance, stored in its natural state, or at one of above-mentioned modalities. Fuel cells based on hydrogen have up to 60% efficiency [5, 27].

Hydrogen has increasingly being regarded as a real future fuel by experts. Hydrogen was concentrated to 252.77 degree Celsius and has a specific weight 71grams per liter, giving it the all fuels as well as energy carrier have a higher power densities per unit of mass; 1 Kilogram for hydrogen has the same amount of energy as 2.1 Kilograms of natural gas (or) 2.8 Kilograms of oil. This feature of it makes hydrogen a fuel utilized in spaceship energy production and propulsion. When utilized in fuel cells hydrogen, it is renewable as well as non-toxic, unlike some other fossil fuels like oil, natural gas, and also coal. As an ecofriendly fuel hydrogen does have a lot of potential moreover for lowering energy imports. Though the usage of hydrogen is now expensive, due to advancements, we may be able to assist in the establishment of an economy on hydrogen in the not too distant future. The term hydrogen is derived from two Greek words that imply, "to make water". Hydrogen is a primary fuel in the supply of so-called "green energy" as it is made using several sorts of alternative energy and non-fossil resource as well as raw materials (Geothermal, Solar, Wind, Hydro Power, Biomass and so on). As a result, hydrogen-fueled systems are the best option to sustain as well as accelerate global energy stability. Hydrogen is predicted to play a significant role in global energy situations in the future.

The following are some of the benefits that support it, having the advantage of being an energy vector over other types of energy: Hydrogen could be transferred safely over long distances via pipelines. Hydrogen is indeed a non-toxic, high-specific-energy-per-mass-unit energy carrier (for example, 9.5Kg of hydrogen produces the same amount of energy as 25Kg of petrol). Hydrogen is produced using a variety of energy sources, especially renewable energy sources. Hydrogen, unlike current or heat, may be stored for extended durations [8, 9].

4.5 Application

Battery electric vehicles (BEVs) as well as Hydrogen fuel cell vehicles (HFCVs) both provide zero-emissions transportation if the power used to recharge the battery and the electricity often used to create the hydrogen are both zero-emission, i.e., renewables, nuclear, or fossil fuel without carbon and storage; this is a viable solution. BEV looks to have considerably higher transition energy than HFCVs at first glance. A battery's short-term storage, round-trip energy efficiency is important and estimated to be in the range of 80% [17]. The electrolyser's energy efficiency in a hydrogen fuel cell system is usually seen in the range of 90% (based on higher heating value), 95% for storage and 50% for the fuel cell (Higher Heating Value) resulting in the round trip energy efficiency of just 43%. Because of the HCFV's inferior round-trip efficiency some people are dismissing it in favour of BEVs. If BEVs are not used for an extended length of the batteries will self-discharge over time, and the round-trip efficiency will steadily fall, eventually reaching zero over several months. Furthermore, if the supply is primarily made up of variable renewables, the grid-connected BEV system will necessitate some long-term storage ensure to give all over the year [28]. When the storage system's energy loss, could eventually are based on hydrogen, factored in, the whole BEV system's energy efficiency on a round-trip basis will converge to the one of HFCV way. The relative gravimetric and volumetric energy and methods will play a vital part in the future hydrogen as well as battery energy storage has important roles to play storage in transportation applications. The energy numbers were transformed towards the equivalent electrical energy in the case of hydrogen, expecting 50% fuel cell efficiency. However, the fuel cell's mass, as well as volume accompanying equipment, are not included in the appropriate energy densities of various hydrogen storage types [29]. Similarly, the majority of energy densities for batteries ignore the rest of the system. Compressed hydrogen tanks are getting close to the vehicle energy densities are essential, but if exceptionally high pressures are employed [30].

As a result, several roadblocks to its development have emerged including potentially costly refuelling stations, public scepticism concerning its safekeeping hazards regarding the loss of energy when compressing hydrogen [31]. Liquid hydrogen, for example, may already achieve the US Department of Energy's gravimetric energy target with a specific energy density of 10-1.3 KWhe/Kg. However, it falls short of volumetric objectives and in testing, cryo-compressed liquid hydrogen was shown to have energy densities that are both gravimetric and volumetric that were higher than the US Energy Department's energy target. For hydrogen liquefaction, however, cryogenic cooling to 22k is required, which can waste 30-40% of the energy consumed for the amount of energy stored in hydrogen. The cost of cryogenic storage tanks is likewise quite high. As a result, whereas liquid hydrogen is expected to be also the most cost-effective as an alternate fuel for jets and other aircraft it is virtually probably too costly as well as workable for the common car as well as transportation purposes [32].

When opposed to compressed gas, solid-state benefits and often requires far lower pressures. Metal hydrides were also developed that may provide gravimetric densities of 0.30-0.47 KWhe/Kg (1.5-2.4 wt%), which is around 35% and 50% of the 0.89KWhe/Kg target. Several chemical hybrids were reported with slightly greater gravimetric densities, 0.55-065 KWhe/Kg and greater thermal densities, 0.45-0.51 KWhe/Kg. As a result, although metal and chemical hydrides still do not equal the gravimetric density for 350 and 700 bar hydrogen storage cylinders, they all have achieved volumetric density higher than 350 bar and comparing to 700 bar storage [33]. Furthermore, the applications of today, tomorrow and the future are shown in the diagram below.

4.6 Conclusion

Fuel cells also express electricity from such a variety of household fuels, includes hydrogen and renewable, as well as could be used in a broad range of applications, like cars, buses, and commercial buildings. A Fuel Cell Technologies Office (FCTO) (Figure 4.3) concentrates on fundamental research, development and invention to enhance hydrogen and fuel cells for transportation and a variety of applications in emerging technologies, allowing energy securities, resiliency and even a strong domestic economy. Traditional combustion engines as well as power plants pollute the environment and seem to be inefficient compared to hydrogen fuel cells. Automobiles

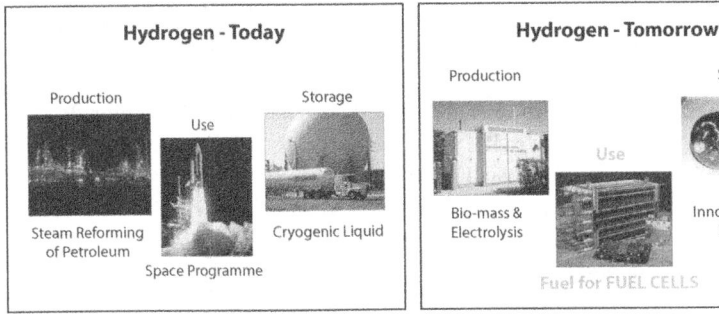

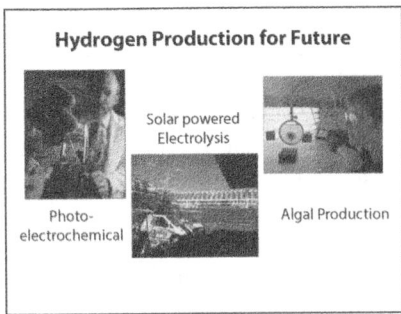

Figure 4.3 Applications of hydrogen as future.

and mobile power packs in mobile applications can both be powered by hydrogen and fuel cells. Some of the benefits of fuel cells include the reduction of Greenhouse gas emissions. Other topics discussed in this chapter are, Why hydrogen as a fuel cell? Operating principle, Fuel cell types, Hydrogen as an Energy-vector in a Long-Term Fuel cell as well as Applications.

References

1. Andújar, J. M.; Segura, F. Fuel Cells: History and Updating. A Walk along Two Centuries. *Renew. Sustain. Energy Rev.* **2009**, *13* (9), 2309–2322. https://doi.org/10.1016/j.rser.2009.03.015.
2. Douglas, D. L.; Liebhafsky, H. A. Fuel Cells: History, Operation, and Applications. *Phys. Today* **1960**, *13* (6), 26–30. https://doi.org/10.1063/1.3056989.
3. Mazur, C.; Offer, G. J.; Contestabile, M.; Brandon, N. B. Comparing the Effects of Vehicle Automation, Policy-Making and Changed User Preferences on the Uptake of Electric Cars and Emissions from Transport. *Sustain.* **2018**, *10* (3), 4–6. https://doi.org/10.3390/su10030676.
4. Burke, K. *Current Perspective on Hydrogen and Fuel Cells*; Elsevier Ltd., 2012; Vol. 4. https://doi.org/10.1016/B978-0-08-087872-0.00402-9.

5. Veziroğlu, T. N.; Şahin, S. 21st Century's Energy: Hydrogen Energy System. *Energy Convers.Manag.* **2008**, *49* (7), 1820–1831. https://doi.org/10.1016/j.enconman.2007.08.015.
6. Manoharan, Y.; Hosseini, S. E.; Butler, B.; Alzhahrani, H.; Senior, B. T. F.; Ashuri, T.; Krohn, J. Hydrogen Fuel Cell Vehicles; Current Status and Future Prospect. *Appl. Sci.* **2019**, *9* (11). https://doi.org/10.3390/app9112296.
7. Hosseini, S. E.; Andwari, A. M.; Wahid, M. A.; Bagheri, G. A Review on Green Energy Potentials in Iran. *Renew. Sustain. Energy Rev.* **2013**, *27*, 533–545. https://doi.org/10.1016/j.rser.2013.07.015.
8. Hall, J. L. Cell Components. *Phytochemistry* **1987**, *26* (4), 1235–1236. https://doi.org/10.1016/s0031-9422(00)82398-5.
9. Larminie, J.; Dicks, A. *Fuel Cell Systems Explained*; 2003. https://doi.org/10.1002/9781118878330.
10. Şoimoşan, T. M.; Moga, L. M.; Danku, G.; Căzilă, A.; Manea, D. L. Assessing the Energy Performance of Solar Thermal Energy for Heat Production in Urban Areas: A Case Study. *Energies* **2019**, *12* (6). https://doi.org/10.3390/en12061088.
11. Seyed Mahmoudi, S. M.; Sarabchi, N.; Yari, M.; Rosen, M. A. Exergy and Exergoeconomic Analyses of a Combined Power Producing System Including a Proton Exchange Membrane Fuel Cell and an Organic Rankine Cycle. *Sustainability* **2019**, *11* (12), 3264. https://doi.org/10.3390/su11123264.
12. Felseghi, R. A.; Carcadea, E.; Raboaca, M. S.; Trufin, C. N.; Filote, C. Hydrogen Fuel Cell Technology for the Sustainable Future of Stationary Applications. *Energies* **2019**, *12* (23). https://doi.org/10.3390/en12234593.
13. Badea, G.; Naghiu, G. S.; Giurca, I.; Aşchilean, I.; Megyesi, E. Hydrogen Production Using Solar Energy - Technical Analysis. *Energy Procedia* **2017**, *112* (October 2016), 418–425. https://doi.org/10.1016/j.egypro.2017.03.1097.
14. Website, H. F. P. S. European Hydrogen and Fuel Cell Technology Platform: Strategic Research Agenda. **2015**, 130.
15. T, C. Ă. T. Ă. R. I. G. R. U. S.; Rus, L. F. Experimental Study on Power Characteristic Curves of a Portable Pem Fuel Cell Stack in the Same Environmental Conditions. **2012**, *3* (4), 2–7.
16. Bockris, J. O. M. The Hydrogen Economy: Its History. *Int. J. Hydrogen Energy* **2013**, *38* (6), 2579–2588. https://doi.org/10.1016/j.ijhydene.2012.12.026.
17. Andrews, J.; Shabani, B. Where Does Hydrogen Fit in a Sustainable Energy Economy? *Procedia Eng.* **2012**, *49*, 15–25. https://doi.org/10.1016/j.proeng.2012.10.107.
18. Ball, M.; Weeda, M. The Hydrogen Economy - Vision or Reality? *Int. J. Hydrogen Energy* **2015**, *40* (25), 7903–7919. https://doi.org/10.1016/j.ijhydene.2015.04.032.
19. Samimi, F.; Rahimpour, M. R. *Direct Methanol Fuel Cell*; Elsevier B.V., 2018. https://doi.org/10.1016/B978-0-444-63903-5.00014-5.

20. Afgan, N.; Veziroglu, A. Sustainable Resilience of Hydrogen Energy System. *Int. J. Hydrogen Energy* **2012**, *37* (7), 5461–5467. https://doi.org/10.1016/j.ijhydene.2011.04.201.
21. Dincer, I. Green Methods for Hydrogen Production. *Int. J. Hydrogen Energy* **2012**, *37* (2), 1954–1971. https://doi.org/10.1016/j.ijhydene.2011.03.173.
22. Jeon, H.; Kim, S.; Yoon, K. Fuel Cell Application for Investigating the Quality of Electricity from Ship Hybrid Power Sources. *J. Mar. Sci. Eng.* **2019**, *7* (8). https://doi.org/10.3390/jmse7080241.
23. İnci, M.; Türksoy, Ö. Review of Fuel Cells to Grid Interface: Configurations, Technical Challenges and Trends. *J. Clean. Prod.* **2019**, *213*, 1353–1370. https://doi.org/10.1016/j.jclepro.2018.12.281.
24. Sharaf, O. Z.; Orhan, M. F. An Overview of Fuel Cell Technology: Fundamentals and Applications. *Renew. Sustain. Energy Rev.* **2014**, *32*, 810–853. https://doi.org/10.1016/j.rser.2014.01.012.
25. Töpler, J.; Lehmann, J. Hydrogen and Fuel Cell: Technologies and Market Perspectives. *Hydrog. Fuel Cell Technol. Mark. Perspect.* **2015**, 1–281. https://doi.org/10.1007/978-3-662-44972-1.
26. Pötzinger, C.; Preißinger, M.; Brüggemann, D. Influence of Hydrogen-Based Storage Systems on Self-Consumption and Self-Sufficiency of Residential Photovoltaic Systems. *Energies* **2015**, *8* (8), 8887–8907. https://doi.org/10.3390/en8088887.
27. Momirlan, M.; Veziroglu, T. N. The Properties of Hydrogen as Fuel Tomorrow in Sustainable Energy System for a Cleaner Planet. *Int. J. Hydrogen Energy* **2005**, *30* (7), 795–802. https://doi.org/10.1016/j.ijhydene.2004.10.011.
28. Giorgi, L.; Leccese, F. Send Orders of Reprints at Reprints@benthamscience. Net Fuel Cells: Technologies and Applications. **2013**, 1–20.
29. Shabani, B.; Andrews, J.; Watkins, S. Energy and Cost Analysis of a Solar-Hydrogen Combined Heat and Power System for Remote Power Supply Using a Computer Simulation. *Sol. Energy* **2010**, *84* (1), 144–155. https://doi.org/10.1016/j.solener.2009.10.020.
30. Kuwahata, R.; Monroy, C. R. Market Stimulation of Renewable-Based Power Generation in Australia. *Renew. Sustain. Energy Rev.* **2011**, *15* (1), 534–543. https://doi.org/10.1016/j.rser.2010.08.020.
31. Shabani, B.; Andrews, J. An Experimental Investigation of a PEM Fuel Cell to Supply Both Heat and Power in a Solar-Hydrogen RAPS System. *Int. J. Hydrogen Energy* **2011**, *36* (9), 5442–5452. https://doi.org/10.1016/j.ijhydene.2011.02.003.
32. Doddathimmaiah, A.; Andrews, J. Theory, Modelling and Performance Measurement of Unitised Regenerative Fuel Cells. *Int. J. Hydrogen Energy* **2009**, *34* (19), 8157–8170. https://doi.org/10.1016/j.ijhydene.2009.07.116.
33. MacKay, R. S. Sustainable Energy Storage–with Hot Air, or Cold Air or Liquid Air. *Contemp. Phys.* **2020**, *61* (1), 1–11. https://doi.org/10.1080/00107514.2020.1794380.

5

IoT and Machine Learning–Based Energy-Efficient Smart Buildings

Aaron Biju[1], Gautum Subhash V.P.[1], Menon Adarsh Sivadas[1], Thejus R. Krishnan[1], Abhijith R. Nair[1], Anantha Krishnan V.[1] and O.V. Gnana Swathika[2*]

[1]School of Electrical Engineering, Vellore Institute of Technology, Chennai, India
[2]Centre for Smart Grid Technologies, School of Electrical Engineering, Vellore Institute of Technology, Chennai, India

Abstract

Every decade has seen humanity make huge leaps in technology. With the passage of time, the number of gadgets and appliances at our disposal have increased and so has the energy consumed by it. Human negligence has resulted in the wastage of energy. Often, appliances are left on when they are not in use. In such a scenario, we are proposing a system that will automatically control the power supply to these appliances and thus save a large amount of money in the long run. The proposed system uses cameras powered by computer vision to detect the presence of people in a room and control the power of appliances like lights. The system can also work on appliances like television sets such that it turns off when the user moves away or falls asleep. The system is flexible and can also be used in office scenarios. The system can be used to send alerts to users when someone enters a room after a specified time, thus securing the premises. The whole system is monitored in real time, without storing the data anywhere and so protecting the privacy of the user.

Keywords: Computer vision, energy management, autonomous, face detection, NodeMCU, firebase, IoT

5.1 Introduction

The use of IoT and Machine Learning allows us to create systems that can help in reducing the power wasted unnecessarily in homes. Several works have already been done in the field of home automation, security and

---

Corresponding author: gnanaswathika.ov@vit.ac.in

power-saving using a combination of IoT and machine learning. However, most of them operate either to provide security or to turn off or on pre-set appliances when detecting a person. One system proposed that accounts for energy-saving reads information from energy meters and compares it with predicted usage based on occupancy and time using a combination of IoT and machine learning [1]. The number of people in a room is counted using image processing techniques and the energy use is predicted. Not all papers emphasize energy consumption. Some of the work has gone into predicting the movements of the people inside a home using the LSTM model [2]. The predicted parameters are used to control the subsequent appliances via a Raspberry Pi. However, this possesses some practical limitations as it requires the users to carry a phone with them at all times. Another system proposed incorporates the use of a person's emotions to control preset appliances [3]. This system also records the user pattern using different sensors to predict future needs. The controls of the appliances can also be accessed manually via Wi-Fi-enabled devices. Bluetooth technology can also be used to control the various appliances at home [4]. Electrical appliances are connected to an ARM7 LPC2148 using Bluetooth modules and function as a remote. This has various benefits, especially for the older population who would now require less physical work to control appliances.

Another method that uses wireless control involves the use of an android app to record hand gestures [5]. This is especially useful for disabled people. Face recognition has also become a trend with the advent of cheaper and efficient cameras and Machine Learning algorithms. The use of a camera with face recognition allows the system to recognize faces already saved in a database and then use it to turn on specified appliances [6]. A similar method can be used to identify the user's face and then open the doors to the house [7]. Another work describes a similar face recognition system working on passive infrared sensors [8]. This also incorporates a dongle to send video data to the owner in case the face is not recognized by the system. Another such proposed system used similar face recognition, but with a combination of PIR sensors and ultrasonic sensors [9]. The system recognizes the face of the registered user and opens corresponding doors. Convolutional Neural Networks are used to identify the face of the registered user and then open and lock doors using Raspberry Pi–controlled locks [10]. A similar system can also be implemented near door locks and alert users if needed [11]. It would open the door for the faces it recognizes; however, in cases in which it does not recognize the face, an email or notification can be sent to the owner. The system operates using a stepper motor controlled by a Raspberry Pi. All these works rely on identifying people to turn on appliances or give access to specific areas of the house. Another paper uses focus-of-attention strategy to reduce the required computation. Graphic user interface allows users to perform tasks by using controls like switches, which are created using MATLAB [12]. Bluetooth module connected to the Arduino to Rx and Tx pin which sends the information to the microcontroller is used to read the information and to send to the relay drivers turning appliances on and off [13].

Implementation of CNN for security has also been done in combination with robotics [14]. Here, a robot with a camera monitors the environment for any abnormalities. If detected, an alert is sent via connected apps. Electrical Appliances are connected to the relay which is then controlled by the NodeMCU based on the data available [15, 16].

Though all the previously mentioned works control appliances using a variety of technology, none of them is focusing on reducing the energy consumed. Hence, there is a need to create a system that incorporates many of the above-mentioned features but will primarily focus on saving energy which otherwise would be wasted due to human negligence.

5.2 Methodology

Figure 5.1 shows the block diagram of the proposed system. The proposed system will consist of a camera that has to be situated above the TV which will cover the entire room so that the system can detect the person as well as check the drowsiness of the person sitting in the front of the TV. The camera will capture the video and Raspberry Pi will take this video as input with the help of a USB port. Raspberry Pi consists of a CNN model trained for detecting people as well as their drowsiness. The Raspberry Pi will send unique messages to firebase according to what it detects. The NodeMCU

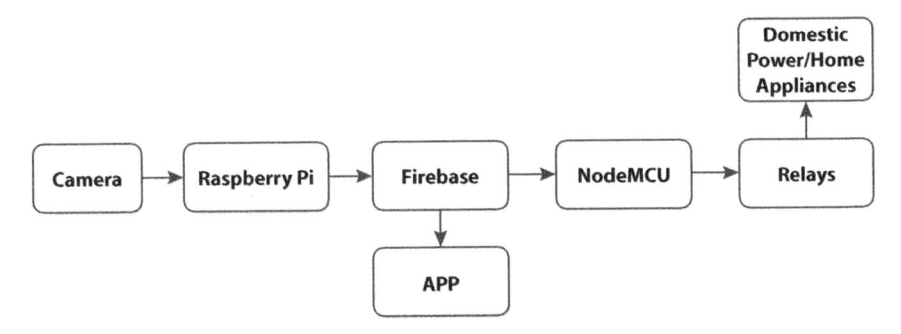

Figure 5.1 Block diagram of the proposed system.

will take these messages from firebase and actuate the respective relay according to the message. In this way, if drowsiness is detected, appliances such as TV, speaker system and lights will be turned off and if no person is detected then the whole power supply of the room will be turned off. The status of appliances is sent from NodeMCU to the firebase and with the help of an app, the status will be displayed from the firebase database and users can monitor and turn on/off the appliances remotely. This app also consists of a lockdown mode for security. The user has to activate this mode in the app, before leaving the building and should deactivate this mode when the user reaches back to the building. During lockdown mode,

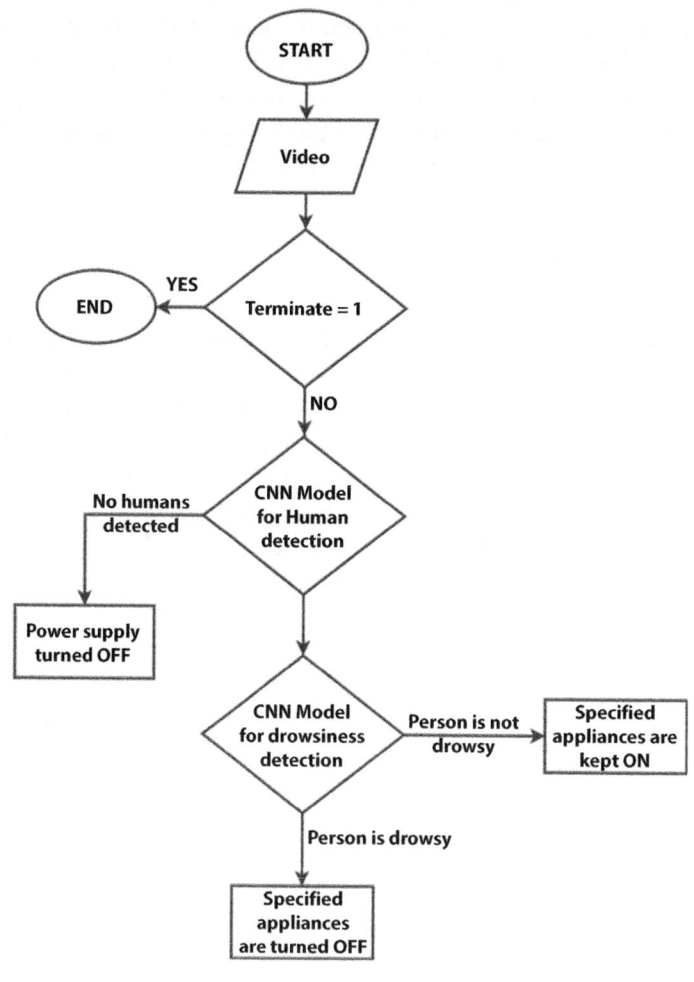

Figure 5.2 Algorithm.

the app will send alert messages to the user when an intruder is detected inside the building. The Raspberry Pi will also capture a 10-second video of that instance and store it in firebase cloud storage which the user can access later to identify the intruder.

The electrical switch of the respective appliances and the whole power supply of the room should be turned on for the entire duration of working for the proposed system. The power supply of each specified appliance can be controlled with the app. The power supply can still be controlled manually and the proposed system is an add-on for the traditional switches. Thus, in case of emergencies, supply can be turned off using physical switches.

Figure 5.2 shows the algorithm used to actuate the relays. The video is taken as input and it is converted into a real-time image. A variable, Terminate is kept as 0 by default. The use of Android app allows the user to shut down the proposed IoT system (Terminate set as 1). When Terminate = 0, the image is then run through the CNN model which is used to detect people. If a person is detected then the image is run through another CNN model for drowsiness detection and if drowsiness is detected then specific appliances such as TV and speaker system will be turned off; otherwise there will be no change in power supply. If no one is detected within the room by the CNN algorithm, then the whole power supply of the room will be turned off. The system will continue to search for humans in the room from the video input, as long as it is given permission from the user application via the Terminate variable.

5.3 Design Specifications

5.3.1 NodeMCU

NodeMCU is an open-source IoT-based firmware and its hardware is based on an ESP-12 module. It is based on the Lua Scripting language. A standard NodeMCU module has 12 general-purpose input/output pins. The network capabilities of the NodeMCU allow it to change values corresponding to the detection of humans and faces in the Firebase database. NodeMCU is also capable of controlling the relays according to the values in Firebase.

5.3.2 Relay

Relay is an electrical switch that controls or switches (on & off) a high-voltage circuit using a low-voltage source. For this project, we are using a 5 Channel, 5V, Optically Isolated Relay Module. It is equipped with high-current relays that can work under AC-250V/10A or DC-30V/10A. It can be interfaced directly with a microcontroller (NodeMCU). This relay module optically isolates the high-voltage circuit from the low-voltage circuit. The relay output consists of a Normally Closed (NC), Normally Open (NO) and a Common (COM) terminal. When the relay is not powered the NC terminal gets connected to the COM terminal and the NO terminal remains disconnected. But, when the relay is powered, the COM terminal gets connected to the NO terminal and the NC terminal gets disconnected.

5.3.3 Firebase

Google's Firebase is used as the real-time database in this project. The Raspberry Pi sends unique messages to firebase according to what the CNN model detects and NodeMCU responds according to these messages. The messages are received instantaneously so changes can be done in the control of relays without much lag.

5.3.4 Raspberry Pi

Raspberry Pi is a versatile and powerful single-board computer developed by the Raspberry Pi Foundation. It has multiple GPIO pins along with regular USB ports. These allow it to be used flexibly in any project and hence it is a favourite of enthusiasts around the world. Due to its compact size and capable processing power, it can be used easily in our application. It can be installed near the cameras placed on the television without being too distracting. The USB port allows the use of normal webcams. The data from the webcam can be easily processed to do the required detections from the live video. Wireless network capabilities allow it to send data to Google's firebase.

5.3.5 Camera

Any regular camera can be used for this purpose. Even a webcam can be used if it has a USB port to connect to the Raspberry Pi.

5.4 Results

A camera is used to identify the presence of a person in a room. In case a person is not detected in the camera, the Raspberry Pi will send 0 to the firebase and this value is further sent to NodeMCU which activates the relay to turn off the lights and other specified appliances in the room. If a person is detected then Raspberry Pi will send 1 to firebase and none of the relays will be activated by the NodeMCU, all electrical appliances will remain "ON". Figure 5.3 shows the CNN model detecting a person.

The camera is installed on top of the television so it can detect if the user is asleep. The Raspberry Pi detects the drowsiness of the person using CNN model. If the person is asleep, then Raspberry Pi will send 1 to firebase and this value is further sent to NodeMCU which will activate the relays of the TV and speaker system which then turns off these specified systems. Figure 5.4 shows the working of CNN model for drowsiness detection. When the face is detected, it will identify the eyes and then check the

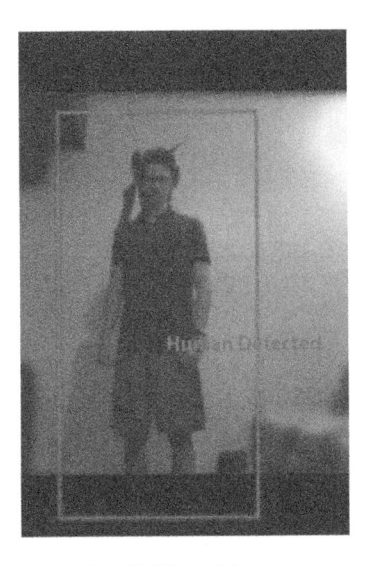

Figure 5.3 Detection of human using CNN model.

Figure 5.4 Detection of drowsiness using CNN model for drowsiness.

pixels in that area. When the model detects the pixel values corresponding to closed eyes, it will count for a preset amount of time, after which it will send 1, indicating the viewer is asleep. Figure 5.5 shows the working of the firebase. The value 0 in both the cases indicate that no one is detected and the value 1 in both the cases indicate that person is detected as well as their drowsiness.

Figure 5.6 shows the NodeMCU circuit diagram of the proposed system. The 5-channel relay is connected to the digital pins of NodeMCU which gives high and low output values. The NC terminal and COM terminal of the relay channel are connected to the wires of respective appliances. A 5V battery is used for supplying the power to NodeMCU. The status

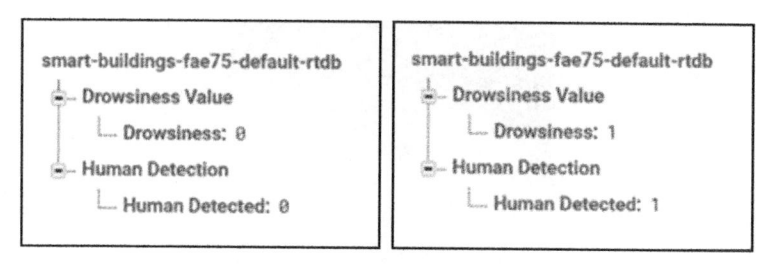

Figure 5.5 Working of firebase.

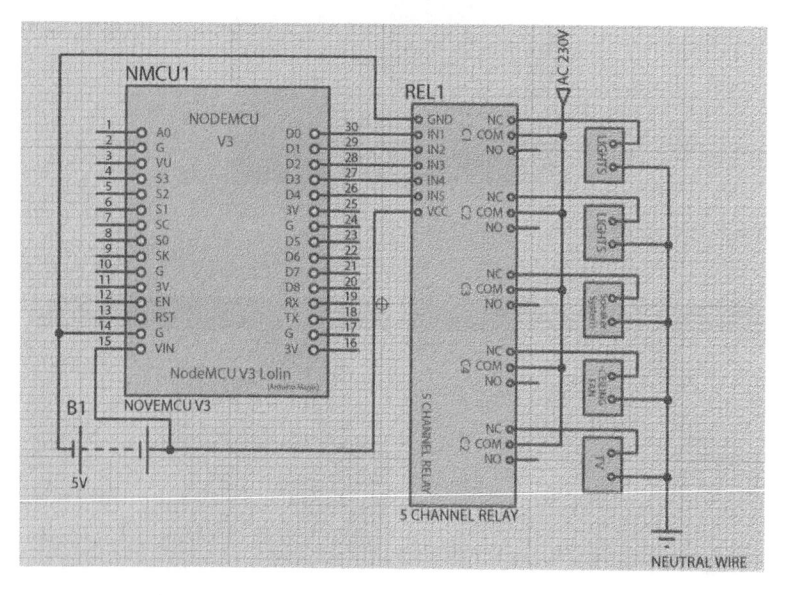

Figure 5.6 Circuit diagram.

of different appliances is also uploaded into the firebase so that users can monitor and control the devices at their convenience via a mobile app.

5.5 Conclusion

The proposed system will turn off the complete power supply of the room if no one is found in the room. Hence, it will turn off the electrical appliances if a user forgets to do so while leaving the room. The system can be further improved by using machine learning for predicting the activities of the people and then automatically turning off the appliances that are not required. The real-time video is taken as input and is not stored, thus protecting the privacy of the user. A PCB and a PCB housing can be made for the proposed system which will save space and make the installation process much easier. The system can turn off preset appliances like television, if the user has a tendency to sleep while it is on. The one disadvantage is that the system cannot completely replace physical switches and is dependent on it. The shift to fully software switches is a risky move, especially in the case of emergencies. Additionally, the lockdown feature helps the owner to get alert messages on his mobile app whenever their house is breached. The system consisting of Raspberry Pi and camera, Relay, and NodeMCU would cost around 7000 INR or approximately 95$.

References

1. Mahato, D. K., Yadav, S., Saxena, G. J., Pundir, A., & Mukherjee, R. (2018, February). Image Processing and IoT Based Innovative Energy Conservation Technique. In *2018 4th International Conference on Computational Intelligence & Communication Technology (CICT)*, pp. 1-5. IEEE.
2. Manu, R. D., Kumar, S., Snehashish, S., & Rekha, K. S. (2019). Smart Home Automation using IoT and Deep Learning.
3. Jaihar, J., Lingayat, N., Vijaybhai, P. S., Venkatesh, G., & Upla, K. P. (2020, June). Smart home automation using machine learning algorithms. In *2020 International Conference for Emerging Technology (INCET)*, pp. 1-4. IEEE.
4. Naresh, D., Chakradhar, B., & Krishnaveni, S. (2013). Bluetooth based home automation and security system using ARM9. *International Journal of Engineering Trends and Technology (IJETT)*, Volume, 4, 4052.
5. Kheratkar, N., Bhavani, S., Jarali, A., Pathak, A., & Kumbhar, S. (2020, May). Gesture Controlled Home Automation using CNN. In *2020 4th International Conference on Intelligent Computing and Control Systems (ICICCS)*, pp. 620-626. IEEE.

6. Haria, H., & Shaikh Mohammad, B. N. (2020, April). Home Automation System Using IoT and Machine Learning Techniques. In *Proceedings of the 3rd International Conference on Advances in Science & Technology (ICAST)*.
7. Ghafoor, S., Khan, K. B., Tahir, M. R., & Mustafa, M. (2019, November). Home Automation Security System Based on Face Detection and Recognition Using IoT. In *International Conference on Intelligent Technologies and Applications*, pp. 67-78. Springer, Singapore.
8. Sowmiya, U., & Mansoor, J. S. (2015). Raspberry Pi based home door security through 3g dongle. *Int. J. Eng. Res. Gener. Sci, 3*, 138-144.
9. Gupta, P., & Rajoriya, M. (2020). Face recognition based home security system using IoT. *Journal of Critical Reviews, 7*(10), 1001-1006.
10. Irjanto, N.S., Surantha, N. (2020). Home Security System with Face Recognition based on Convolutional Neural Network. *International Journal of Advanced Computer Science and Applications (IJACSA)*, Vol. 11, Issue 11.
11. Bhatia, P., Rajput, S., Pathak, S., & Prasad, S. (2018, November). IOT based facial recognition system for home security using LBPH algorithm. In *2018 3rd International Conference on Inventive Computation Technologies (ICICT)*, pp. 191-193. IEEE.
12. Bahajathul F., Lakshmi, A.D., Ravali, B., Dokku, R.R, 2018, Real Time Face Detection Using Matlab, *International Journal of Engineering Research & Technology (IJERT)* Volume 07, Issue 02 (February 2018).
13. Malav, V., Bhagat, R.K., Saini, R., Mamodiya, U. (2019). Research Paper On Home Automation Using Arduino. *Journal of Advancements in Robotics (JoARB)*, Vol. 4, Issue 10.
14. Shin, M., Paik, W., Kim, B., & Hwang, S. (2019). An IoT platform with monitoring robot applying CNN-based context-aware learning. *Sensors, 19*(11), 2525.
15. Gaikwad, R., & Joglekar, A. (2017). Energy Saving Through Home Automation and Data Mining Technique. *International Journal of Engineering Research in Computer Science and Engineering (IJERCSE)*, Vol. 4, Issue 10.
16. Vishwakarma, S., Upadhyaya, P., Kumari, B., & Mishra, A. (2019). Smart Energy Efficient Home Automation System Using IoT. *2019 4th International Conference on Internet of Things: Smart Innovation and Usages (IoT-SIU)*, 1-4.

6

IOT-Based Smart Metering

Parth Bhargav[1], Umar Ansari[1], Fahad Nishat[1] and O.V. Gnana Swathika[2]*

[1]*School of Electrical Engineering, Vellore Institute of Technology, Chennai, India*
[2]*Centre for Smart Grid Technologies, School of Electrical Engineering, Vellore Institute of Technology, Chennai, India*

Abstract

Conventional metering demands a scenario where every month a person comes from the electricity department to the consumer's premises to take the readings of the electricity meter and further generate a due bill for the previous month. This is a tiresome process because the person visits each and every house in that particular area and collects the readings. It takes a lot of time and demands a huge manpower requirement. Many times the end user also face problems like extra bill amount, notifications from the electricity department even though the bill has already been paid. To solve the aforementioned problems, a hybrid Internet of Things–based energy management system which eliminates the third party between the end user and the energy management companies is proposed in this paper. Nowadays, smart energy meters are manufactured by the companies that are internally connected to different sensors and constructed to measure electricity voltage, current and power consumption. The data obtained from the smart meter is transmitted to the Arduino microcontroller through wire and both are physically connected. The Wi-Fi module is attached to the Arduino so that it is connected to the internet in the hotspot-enabled area. When the Arduino is connected to the internet it starts sending data to the cloud server (in this case the cloud server of the energy management company) and perform cloud processing of the data received. The end user or consumer logs in with a unique user ID and password and is able to see their energy consumption live, and if they wish they can make payment also by the gateway service.

Keywords: Smart metering, IOT, Arduino, ThingSpeak, Wi-Fi, quantitative analytics, missing data prediction

---

Corresponding author: gnanaswathika.ov@vit.ac.in

Milind Shrinivas Dangate, W.S. Sampath, O.V. Gnana Swathika and P. Sanjeevikumar (eds.) Integrated Green Energy Solutions Volume 1, (71–92) © 2023 Scrivener Publishing LLC

Abbreviations and Nomenclature

IOT	Internet of Things
API	Application Programming Interface
kWh	Kilowatt-hour
HTTP	Hypertext Transfer Protocol
MQTT	Messaging Protocol
GSM	Global system for mobile communication
x	The number of blinks of an LED
y	Wide variety of gadgets of strength
z	Cost of intake

6.1 Introduction

Smart metering usually involves the deployment of a smart meter at the client's location that performs the normal procedure of studying, processing, and giving the feedback of intake facts to the consumer [1–3]. Those smart meters are so called since the introduction of static meters, due to the fact that they covered at the very least one microprocessor [4–7]. For over 15 years, smart meters were used with clients like energy flowers, huge corporations and so forth; however, lately they are used with mass utilities together with families and small enterprises. Nevertheless, it is still a gradually growing field, particularly due to the fee of smart meters, and moreover because of the lack of the infrastructure that manages both the meters community and also the big quantity of expertise from meters; moreover, as offer contrivances for sundry cease-utilizer packages [8–13]. This approach allows the energy agencies to take the electrical energy meter readings month-to-month without sending a person to each residence. This may be done by way of Arduino microcontroller unit that continually monitors the power meter [14–19]. This undertaking in particular offers a smart electricity meter, which makes use of the functions of embedded structures, i.e., coalescence of hardware and software programs with a view to putting into effect favored functionality. This device invariably facts the studying and additionally the stay meter studying is exhibited on web-web page [20–27].

6.1.1 Motivation

At the end of the month, someone from the power board visits people's place of residence to read the electricity meter and prepare a bill, which the client then pays. The main downside of this tool is that someone has to go to every house to read the meter and then the bills are prepared. Normally there may be errors like greater bill quantity, or feedback ping from power board, albeit the payments are paid counted [31–34]. The antidote for most of these quandaries is to hold music of the consumer's load on a timely substructure, which holds a guarantee of particular billing. To conquer this disadvantage a concept that brings transparency of energy usage between consumer and lodging issuer, for that the smart meter utilizing IOT and Arduino are brought [35–37]. In this technique, Arduino is utilized due to electricity efficiency, i.e., it uses less power, and is most expeditious. In this paper, electric energy meters that are mounted at our homes for domestic use are not outmoded; however, a minuscule amendment to the already mounted meters transmutes them into smart meters [38]. This device continually facts the studying and the meter analyzing may be exhibited on the website to the patron on request then the power department and the terminus utilizer of residence [39, 40].

6.1.2 Objectives

- To design and broaden IOT-based smart metering for domestic programs.
- To resolve the trouble of power theft that is occurring at a massive level.
- To reduce the power fee given with the aid of a quit-user.

6.2 Methodology

6.2.1 Advent of Smart Meter

Conventional techniques of metering are cumbersome and it also takes lot of time and manpower. Also it has huge impact on the profits of the energy management companies as they have to pay salaries to the hired manpower and spend some money in the maintenance of the reading device. Many times the end user also faces problems like extra bill amount, notifications from electricity department even though the bill has already been paid. To solve this problem using technology we are proposing a hybrid Internet of Things–based energy management system which eradicates the third party in between the end user and the energy management companies. We are using Arduino because of its energy efficiency and less power consumption. Nowadays, smart energy meters are manufactured by the companies that internally have different sensors and are constructed to measure electricity voltage, current and power consumption. The data obtained from this smart meter is transmitted to the Arduino through wire as both are physically connected. The Wi-Fi module is attached to the Arduino and the Arduino is further connected to the internet in the hotspot-enabled area. When the Arduino is connected to the internet it starts sending data to the cloud server (in this case, the cloud server of the energy management company) and perform cloud processing of the data received. The end user or consumer has to log in with a unique user ID and password and they see their live energy consumption and if they wish they also make payment by the gateway service.

The concept of IOT lets human beings attach primary gadgets that we use in our daily lifestyles to link with each other via the internet. Devices related and governed through the IOT protocols may be remotely administered and analyzed. Net of factors concept offers fundamental infrastructure and opportunities to set up a bridge between the real international and computing structures [8]. With the speedy growth of an increasing number of wireless devices on the market, this concept becomes increasingly crucial. It links hardware peripheral to one another via internet. The Manufacturer Espressif Systems 8266 Wireless module used within the system provides a connection to the internet in the system. Today, the demand for electricity of the whole population is increasing at a constant rate for various domains like agriculture, industry, home use/domestic use, hospitals, etc. As a result, handling power maintenance and demand becomes more and more complex; also, there is a pressing need to save as much power as possible [32]. As the need for electricity from the new generation of people continues to increase, corresponding technological improvements are required. The proposed system provides a technical improvement for ordinary electricity meters using IOT technology. In addition, we must also solve other problems, such as power larceny and meter tampering, which in turn causes profitability losses for the country. Monitoring and optimizing power consumption and minimizing power waste are the main goals for achieving a better power structure and network. For Billing, an individual human is required for accessing each customer's electricity meter and engendering a bill by taking unit readings from the electricity meter [28–31]. In order to eliminate this time-consuming process and solve all the given restrictions, we have created a system from scratch contingent on IOT technology.

Design of smart meters based on Wi-Fi systems is based on three main goals:

- Provide instant automatic meter reading function basis.
- Use electricity in the best way.
- To cut down the power wastage.

Correspondingly, the system should additionally be a subsidiary of the server. Consequently, the concrete system is fundamentally divided into two:

- End User/Back End.
- Front End.

The data from the back end system is exhibited on a web page, accessible to both consumers and accommodation providers. This system is depicted on the Arduino micro controller [2]. It is divided into three components in structure: controller, larceny detection circuit, and Wireless fidelity unit. The controller executes rudimentary calculations and acts upon the information. Larceny detection circuit provides details about any meter interference, and the wireless fidelity unit plays the most paramount role to send information from the controller via Internet. Universal messaging platform. The Arduino controller is Programming on the Arduino software IDE is a prerequisite for running on the Arduino board. The Wireless Fidelity module is the important and chief component utilized in the complete Internet of Things operations. The core of the Arduino board is to provide connections between the different components of the depicted system. The microcontroller board is predicated on the AT mega central mainframe processor. That is the core of the gadget and is obligatory for rudimentary operations that must be performed (for example, automatic electricity bills and tamper detection input for tampering circuits) [7]. The load represents the equipment that needs to be energized. The AC supply of power is connected to the system through a transformer for supplying power to the system [16]. Then read the reading from the meter Process and update via Wireless Fidelity via Espressif Systems 8266 Wireless Fidelity module.

If any tampering is encountered by the system, it updates the following condition and transmits the packets over the cloud on the required web page and web page used to exhibit energy readings. Every 5 seconds the pulse it sent from the energy meter to the cloud for processing and also it updates energy reading timeline on the web page every 5 seconds. The energy reading is exhibited on the LCD exhibit [14]. If any interference has been done in the energy meter, the buzzer issues a loud sound.

All statistics and particulars from the apparatus is facile. It is found on the website Thingspeak.com [19].

Advantage

- Reduce energy waste.
- Prevent power shortages during the dry season.
- Make each customer a protector of their own interests Power supply.
- Authentic-time bill monitoring.
- Reduced receiving timeline.

Application field

- Residential and commercial buildings in public energy Supply system.
- Government Energy plant.

Among the suggested techniques, the customers handle their power by understanding the use of its power frequently or once a while [11]. This scheme provides many paths including the interchange linking the two Utilities and buyers additionally provide different Customers' facility to ignore payment power; Vibrant supply cuts off in terms of utilities. When the user pays the bill the front end user supply also reconnects. In integration to the subsisting innovation system, it includes sending a vigilant message to the utilizer, every 15 days, Perpetual alert messages, including payment details and payment sanctions until the payment is made [26]. Evade further Energy consumption, we set limits for each energy consumption Family, if the constraint is exceeded, the minimization method is adopted Put down the appliance according to the user's accommodation Automatic and manual. If the electronic watch is faulty, a notification is additionally sent to the utilizer or user [21].

Internet of Things server ThingSpeak is utilized as a cloud solution provider. ThingSpeak is the first live web-based solution provider/implement for engendering IOT projects [18]. The current value and voltage value are perpetually stockpile in the server. Alarms are lined up on the node web server. Most of the lodged substructure plays a role in three modes.

1. Unmanned mode
2. Physical mode
3. Problem scrutiny

Unmanned mode: In this one, it crosses the inhibition The contrivance automatically switches OFF. Contrivance culled through utilizer accommodation.

Physical mode: In this mode, the switch is manually turned ON. In manual mode, the customer consumes as much preset as possible.

Problem scrutiny: The power panel is used to it in a physical procedure but is prone to human error. This is a concern to customer as they have to obtain adjustment from vitality supply board.

6.2.2 Modules

A meter is connected to a load (Electrical Bulb) and readings are recorded for a period of 10 hours. Every hour, the reading is sent to the ThingSpeak server and data is updated on cloud through Arduino UNO and Wi-Fi module. The live data is downloaded into an excel sheet from ThingSpeak server with date and timestamp. If the system stops working for some duration then the Statistical Analytics is employed to predict the missing reading until the system starts working again.

The smart meter consists of the Hardware (Front End) which comprises Energy meter, Arduino, Wi-Fi module and loads & Software (Back End) as in Figure 6.1.

6.2.3 Energy Meter

Energy meter is connected to the load through wires. It is ON for a total period of 10 hours and energy consumption is recorded for this period. In the fresh meter, reading of 2 units is already calibrated. After using for 10 hours, the meter reading changes depending on the load that is applied.

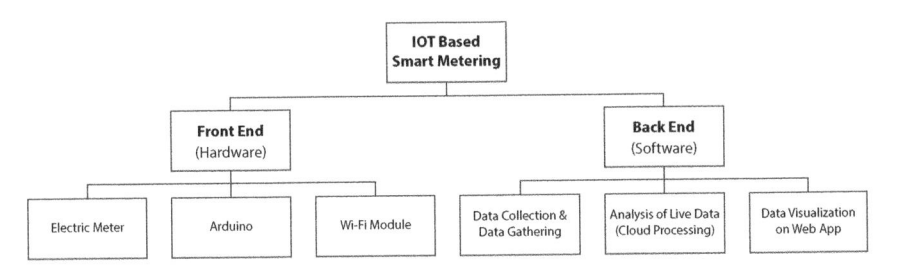

Figure 6.1 Flow chart.

6.2.4 Wi-Fi Module

Wi-Fi module is used for providing net connection to the System through mobile Hotspot. Due to Wi-Fi sensitivity, voltage overflow may occur which is curbed in the hardware setup where the voltage regulator is used for the safety of component.

6.2.5 Arduino UNO

Arduino UNO is used to read input from the hardware. The code is written in the Arduino IDE for counting the ON and OFF time of the LED.

6.2.6 Back End

After connecting all the components, the data gathering/data collection technique is applied and these data is uploaded to the cloud server. ThingSpeak API is used to show data analytic and further all the data is captured in the Excel sheet with date and timestamp. The electricity consumption data is available in ThingSpeak server and it is accessible globally for the user.

Implementation: The module is further divided into two components, namely physical part that includes hardware prototype, and the second part is the web page.
Physical part: It is composed of Arduino board and Espressif Systems 8266 Wireless Fidelity module, LCD exhibit, alarm and power transmission. Arduino Uno enhancement board Arduino is a Microcontroller board, which is predicated on AT mega 328P. It is powered by the power supply at the potency outlet. And utilizes AC to DC adapter or with a battery [29].
Wireless module: Espressif structures manufacture Wi-Fi constancy module Low-cost additives made by way of manufacturers Microcontroller module capable of wireless networking. Espressif structures 8266 the Wireless Fidelity section is a gadget in a chipset section with the subsequent aptness: 2.4GHz variety. It makes use of a 32-bit RISC CPU strolling at eighty it is far predicated on TCP/IP (Transmission manipulate Protocol). It is by far the most consequential element inside the gadget that performs IOT operations. It has sixty-four KB boot ROM, sixty-four KB Injunctive sanction RAM, 96 KB statistics RAM. Wireless constancy unit execution internet of things operation with the aid of sending power meter records to a web page.

LCD Exhibit - LCD (liquid crystal exhibit) screen is an Electronic exhibit module, found in a lot of applications. 16 * 2 denotes 16 exhibits in each line of characters, and there are 2 such lines. In this LCD each character is exhibited in a matrix of 5 * 7 pixels. The 13 and 14 staple of the exhibit to be utilized as particulars staple Arduino interface that is used to exhibit wattage.

Working of E-Meter - The vitality and purport of the evaluation circuit board are known as a vitality meter. Vitality is aggregation the power consumed and utilized by the heap in a categorical duration. It is utilized as a component of housing and machinery Alternating Current driven circuits, used to estimate potential utilization. The electric energy meter is more economical and gives more delicate reading. The fundamental component of power is in watts. When utilizing 1 kilowatt at 60 degrees Minutes, it is considered a unit that uses energy. These meters estimate instantaneous voltage and flow. This one Integrate forces in a period of vitality Utilized in that era and age.

Arudino AT-MEGA - It is an AVR-compliant low-powered 8-bit CMOS micro controller Upgraded minimized injunctive authorization. It has less potential consumption and minute acreage [5]. Recollection withal has a devoted chip Expunge mode; you could expunge the complete chip within 10 ms. in comparison to Expunge EPROM, you could expunge and reprogram the contrivance within the circuit. However, EEPROM is at the most luxurious and least dense ROM [10].

Regulated power supply - Power plan includes rugged power transformer (additionally disseverment between information and benefits) and dissipate the layout of the controller circuit. Controller. The circuit contains a single Zener diode or three. The terminal directly arranges the controller to distribute the required resultant voltage [18].

Liquid crystal exhibit (LCD) - Supplementally it has 512 bits of figure engenderer Random Access Memory. The recollection is Characters avail oneself of for customer characteristics [11].

RTC - Genuine Time Clock (RTC) is utilized for data resetting and store in internet based and disconnected mode. Relay loop-as long as forward and invert Connect the ground to the congruous terminal Switch cycle. The most diminutive involute shape Includes a loop utilized as an open circuit for the solenoid, and Close the switch contact [15].

The Internet of Things server - ThingSpeak is utilized as a web-based computing solutions service provider. ThingSpeak is the first live cloud service provider implemented for engendering IOT projects. The current and voltage data are perpetually saved in the back end web computing device. Alarms also can be programmed on the server [22].

Wireless-Fidelity module - Espressif systems 8266 is constancy section Felicitous for integrating wireless fidelity constancy usability to the contemporary version. Take a microcontroller adventure via GM noncontemporary transmitter-receiver serial sodality. This section may even be resurrected to clear up As an unrestrained contrivance cognate to wireless fidelity [27].

Voltage electric eye - DC kineticism from AC frame as a way to make a bequest to the micro circuit, we are capitalizing on the voltage detection circuit. The loop offers a particular era for fabricating this DC logo. Detection potential with the aid of utilizing a voltage transformer, the denotement received is modified within the predominant op-amp stage, and within the 2nd operational amplifier is organized [16].

Current sensor - By detecting the current, Current transformer, and rectify in the main operation Amplifier stage and booster for the second operational amplifier arrangement [12].

The reason for the design of electricity meters predicated on the Internet of Things is to minimize indoor power consumption. It eschews Human intervention, truncates costs and preserves manpower. Both automatic and manual are available. The table sent Pay first hand to your mobile phone before the due date Human meddling. The computerization minimizes the total work cost and makes the framework more efficacious and precise. This system is mainly utilized in smart cities with public places Wireless Fidelity hotspot areas. The work is predicated in the cyber world The concept of things. Designed to supersede old energy Instruments with advanced implementation. It uses Automatic power reading, through which its power is optimized thereby truncating power waste. of Upload the meter reading to Thingspeak.com Channels for concrete energy utilization Both the server and the server views the meter customer. Conclusion In the era of smart city development, this work fixates on connection and networking Factors of the Internet of Things. Energy consumption calculation is predicted on the calibration pulse count is Use PIC16F * & A MCU design and implementation Embedded system domain. In the proposed paper, Internet of Things and PLC-predicated meter reading system aim to The energy electricity meter readings are perpetually monitored. If the customer does not pay the monthly bill, the accommodation provider disconnects the power supply, eliminating manual intervention, providing efficacious meter readings, and averting billing errors.

The paper achieves the following goals:

1. It is facile to access consumer information from electricity meters through the Internet of Things.
2. Real-time larceny detection on the utilizer side.
3. LCD exhibit energy squander unit and temperature.
4. Internet of Things electricity meter uses Wi-Fi module to access consumption. The system performance is enhanced in the following ways: Connect all home appliances to the Internet of Things. Consequently, the following goals are achieved in the future to conserve costs, power and prevent larceny:
   i. An IOT system is built for users to monitor energy consumption and pay bills online.
   ii. A system is built where users enable short messaging service in their mobile phone when they want or when they exceed the doorway of electricity consumption.

6.3 Design of IOT-Based Smart Meter

6.3.1 Energy Meter

An electric meter or power meter is the tool that estimates the proportion of electricity used by a residential, transaction, or electrical appliance as shown in Figure 6.2. Table 6.1 provides the design specification of an energy meter. Electrical appliances use electricity meters installed on customers' premises for payment and monitoring purposes. The most commonly used

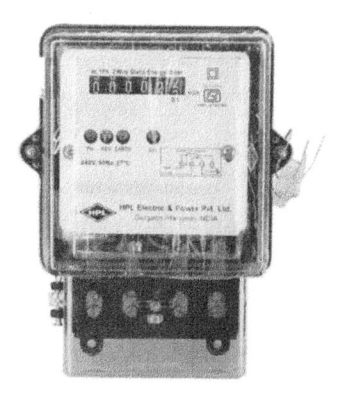

Figure 6.2 Energy meter.

billing unit is in kilowatt hours (kWh). They are usually read once each payment period. When occasional power saving is required, some meters may measure demand, higher power consumption sometimes. The "Time of the Day" meter calculation allows electricity prices to be converted during the day, recording usage at the highest possible time and lowest, and time costs. Also, in some areas meters have a transfer of demand response in demand during peak load.

Design specification:

Table 6.1 Design specification of energy meter.

Voltage rating	240 V AC
Current rating	10 A
Frequency	50Hz +/-5%
Power factor	+1 to -1
Rated power	1 KiloWatt-Hour (KWH)
Phase	AC 1 phase
Wire Type	2 wire static energy meter

6.3.2 Arduino UNO

Arduino UNO microcontroller board is based on ATmega328P as shown in Figure 6.3. It has 14 digital input/input ports, 6 analog inputs, 16MHZ-crystal oscillator, Arduino responds to 5v offers provided by the optocoupler and continues to calculate supply and data energy consumption

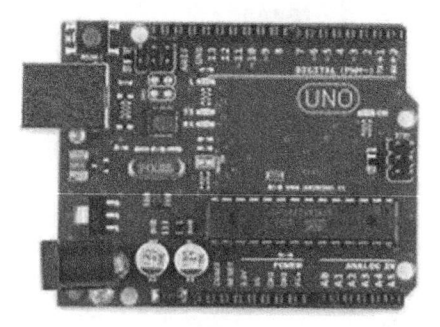

Figure 6.3 Arduino UNO.

and cost, continuously maintained on a web page. Table 6.2 provides the design specification of the Arudino.

Design specification:

Table 6.2 Design specification of Arduino UNO.

Microcontroller	ATmega328P – 8 bit AVR family micro-controller
Operating Voltage	5V
Recommended Input Voltage	7-12V
Frequency (Clock Speed)	16 MHz

6.3.3 Wi-Fi Module

The ESP8266 Wi-Fi Module is a SOC containing an integrated Transmission Control Protocol/Internet Protocol (TCP/IP) that provides any microcontroller access to your Wi-Fi network cloud as in Figure 6.4. Table 6.3 provides the design specification of Arudino.

Figure 6.4 Wi-Fi module.

Design specification:

Table 6.3 Design specification of Wi-Fi module.

Power Supply	+3.3V only
Current Consumption	100mA
I/O Voltage	3.6V (max)
Flash Memory	512kB

6.3.4 Calculations

This set up is very easy reckoning.
3200 blinks = 1 unit depends on manufacturer.
In these case 3200 blinks of LED is 1 unit.
Basically,
No. of units (q) =(x/3200)
Assumption: Cost of 1 unit of electricity to end user = Rs 10.
Therefore;

$$f=q^* \ 5 \ rs$$

6.3.5 Units

Normally, basic unit of electricity is Kilowatt hour (KWH).
1KWH = 1000 watt for 1 hour.
Example,
One 200watt bulbs used for 1 hour gives 0.2 KWH.

6.4 Results and Discussion

6.4.1 Working

The hardware setup is as shown in Figure 6.5. The electric bulb of 200 W is connected to the Electric Meter as a load. The bulb consumes around 0.2 units per hour in Electric Meter. The meter records the reading for 10 hours. The meter has 2 units on it when received from the manufacturer. The meter includes LED inside it which blinks 3,200 times to make 1 unit. The LED produces square wave and it is going to count ON and OFF time of LED through Arduino code. Arduino UNO and Wi-Fi Module are used to send this data to the cloud. Wi-Fi is used for providing web connection. Every hour Arduino is updating the live data in the ThingSpeak API server. From the server the user is able to download the live data in an Excel sheet from ThingSpeak server with date and timestamp.

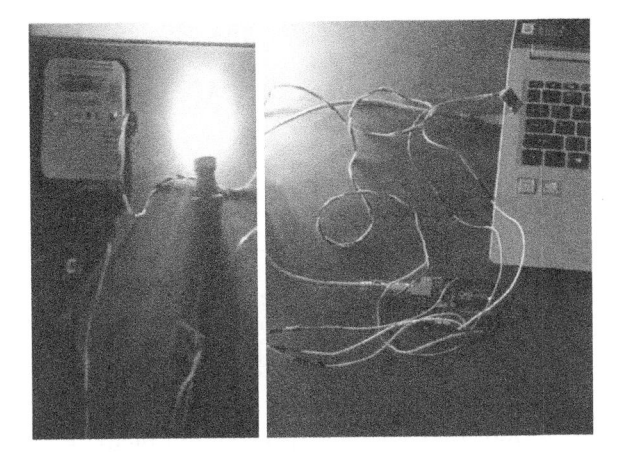

Figure 6.5 Working model.

6.4.2 Readings Captured in the Excel Sheet

The meter reading is downloaded into the Excel sheet as shown in Figure 6.6 so that the consumer tracks consumption of units by downloading the live data from ThingSpeak server with date and timestamp.

Figure 6.6 Electrical units consumed in 1 hour.

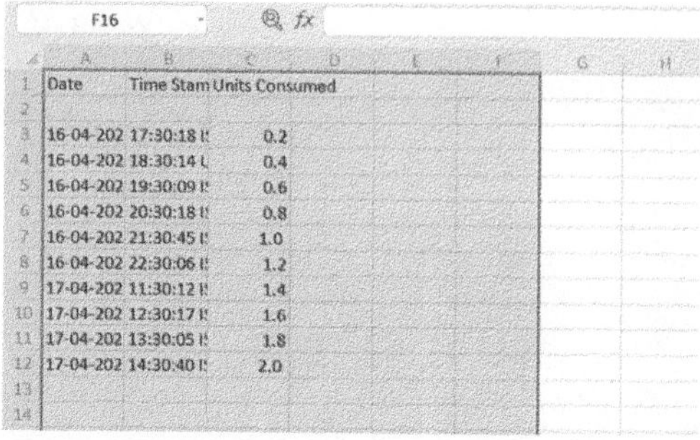

	A	B	C
1	Date	Time Stam	Units Consumed
2			
3	16-04-202	17:30:18	0.2
4	16-04-202	18:30:14	0.4
5	16-04-202	19:30:09	0.6
6	16-04-202	20:30:18	0.8
7	16-04-202	21:30:45	1.0
8	16-04-202	22:30:06	1.2
9	17-04-202	11:30:12	1.4
10	17-04-202	12:30:17	1.6
11	17-04-202	13:30:05	1.8
12	17-04-202	14:30:40	2.0

Figure 6.7 Excel sheet of electrical units consumed in 10 hours.

In 10 hours, 2 units are consumed as shown in Figure 6.7. 200W Bulb connected as a load consume 0.2 unit every hour.

6.4.3 Predication Using Statistical Analytics

The fault occurs through any abnormal condition. If the fault occurs and the electrical system stops working, then it gets difficult to identify the unit consumed.

We have applied Statistical Analytics to find the missing data. The standard deviation is obtained as in Figure 6.8.

6.4.4 Quantitative Analytics

Statistical evaluation is the technology of gathering statistics and uncovering styles and traits. After gathering statistics you could examine it to summarize the facts.

$$\sigma = \sqrt{\frac{\sum (x_i - \mu)^2}{N}}$$

σ = population standard deviation
N = the size of the population
x_i = each value from the population
μ = the population mean

Figure 6.8 Formula for standard deviation.

6.4.5 Predication of Missing Data

The data is predicted as in Figure 6.9.

6.4.6 Hardware Output

ELECTRIC METER AFTER 1 HOUR:

This meter has already 2 units of consumed electricity since we cannot modify it as it is sealed with glass, so 2 unit is treated as reference point in Figure 6.10.

```
IDLE Shell 3.9.2
File Edit Shell Debug Options Window Help
Python 3.9.2 (tags/v3.9.2:1a79785, Feb 19 2021, 13:44:55) [MSC v.1928 64 bit (AMD64)] on win32
Type "help", "copyright", "credits" or "license()" for more information.
>>>
===== RESTART: C:/Users/parth/AppData/Local/Programs/Python/Python39/123.py =====
0.2
0.4
0.6
0.8
1.0
1.2
1.4
1.8
2.0
Mean=1.044
Median=1
Mode=0.912
SD=0.614636
Missing data1.66
>>>
```

Figure 6.9 Output of the statistical analysis.

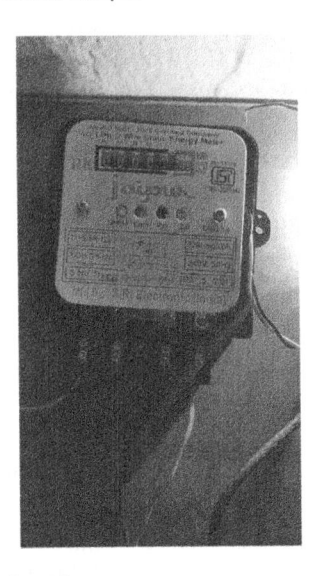

Figure 6.10 Electric meter after 1 hour.

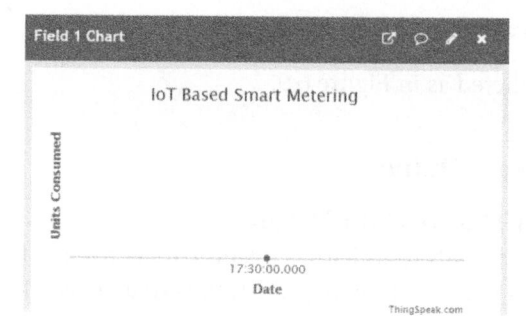

Figure 6.11 ThingSpeak website after 1 hour.

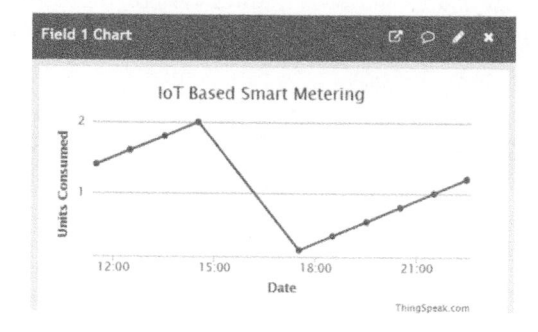

Figure 6.12 ThingSpeak website after 10 hours.

THINGSPEAK WEBSITE AFTER 1 HOUR:
The system is switched ON at 16:30:00 and the line graph model of unit consumed after 1 Hour is as shown in Figure 6.11. It shows that 0.2 units of electricity is consumed in 1 hour.

The final output when the meter is ON for 10 hours is shown in Figure 6.12.

The above line graph model shows the total electrical units consumed in 10 hours when load is applied.

6.5 Conclusion

A prototype of an IOT-based smart meter is constructed. Smart meters are soon going to replace the conventional domestic meters and hence this paper focusses on realizing a smart meter with the help of IOT, Arduino UNO and other components which is cost efficient and affordable. A whole new era of learning has arrived. Smart metering systems provide companies in the service sector with the ability to monitor media supply networks and respond well to current events. Data is obtained from hard-to-reach meters and those available over long distances. The design of the plan, which forms the ready solution for the needs of urban infrastructure development, includes concepts of building a smart city and Industry 4.0. big data and IOT.

References

1. Sasanenikita N, "IOT based energy meter billing and monitoring system", *International Research Journal of Advanced Engineering and Science* (2017).
2. Pandit S, "Smart energy meter using internet of things (IOT)", *VJER Vishwakarma Journal of Engineering Research*, Vol. 1, No. 2 (2017).
3. Maitra S, "Embedded Energy Meter - A New concept to measure the energy consumed by a consumer and to pay the bill", 2008.
4. Depuru SSSR, Wang L & Devabhaktuni V, "Electricity theft: Overview, issues, prevention and a smart meter based approach to control theft", *Energy Policy*, Vol. 39, No. 2, pp. 1007-1015.
5. Giri Prasad S, "IOT based energy meter", *International Journal of Recent Trends in Engineering & Research (IJRTER)*.
6. Sehgal VK, Panda N, Handa NR, Naval S & Goel V, "Electronic Energy Meter with instant billing", *Fourth UK Sim European Symposium on Computer Modeling and Simulation (EMS)*, (2010), pp. 27-31.
7. Yaacoub E & Abu-Dayya A, "Automatic meter reading in the smart grid using contention based random access over the free cellular spectrum", (2014).
8. Ashna K & George SN, "GSM based automatic energy meter reading system with instant billing", *Proceeding of International Multi Conference on Automation, Computing, Communication, Control and Compressed Sensing (Imac4S)*, Vol. 65, No. 72 (2013), pp. 22-23.
9. Malik NS, Kupzog F & Sonntag M, "An approach to secure mobile agents in automatic meter reading", *IEEE International Conference on Cyber Worlds, Computer Society*, pp.187-193.
10. Praveen MP, "KSEB to introduce SMS-based fault maintenance system", *Hindu News*, (2011).
11. G Ainabekova, Z Bayanbayeva, B Joldasbekova, A Zhaksylykov (2018). The author in esthetic activity and the functional text (on the basis of V. Mikhaylov's narrative ("The chronicle of the great jute"). *Opción*, Año 33. 63-80.
12. Z Yesembayeva (2018). Determination of the pedagogical conditions for forming the readiness of future primary school teachers, *Opción*, Año 33. 475-499.
13. J. Widmer, Landis," Billing metering using sampled values according lEe 61850-9-2 for substations", *IEEE* 2014.

14. Cheng Pang, Valierry Vyatkin, Yinbai Deng, Majidi Sorouri, "Virtual smart metering in automation and simulation of energy efficient lighting system", *IEEE* 2013.
15. Amit Bhimte, Rohit K.Mathew, Kumaravel S, "Development of smart energy meter in labview for power distribution systems", *IEEE Indicon*, 2015 1570186881, 2015.
16. H. Arasteh, V. Hosseinnezhad, V. Loia, A. Tommasetti, O. Troisi, M. Shafie Khan, P. Siano, "IOT Based Smart Cities: A survey" *IEEE* 978-1-5090-2320-2/1631.00,2016.
17. Clement N. Nyirendre, Irvine Nyandowe, Linda Shitumbapo, "A comparison of the collection tree protocol (CTP) and AODV routing protocol for a smart water metering", pp. 1-8, 2016.
18. Kurde, A. and Kulkarni, V., 2016. IOT based smart power metering. *International Journal of Scientific and Research Publications*, 6(9), pp.411-415.
19. Bhilare, R. and Mali, S., 2015, December. IOT based smart home with real time E-metering using E-controller. In *2015 Annual IEEE India Conference (INDICON)* (pp. 1-6). IEEE.
20. Yaghmaee, M.H. and Hejazi, H., 2018, August. Design and implementation of an Internet of Things based smart energy metering. In *2018 IEEE International Conference on Smart Energy Grid Engineering (SEGE)* (pp. 191-194). IEEE.
21. Ahmed, E., Yaqoob, I., Gani, A., Imran, M. and Guizani, M., 2016. Internet-of-things-based smart environments: state of the art, taxonomy, and open research challenges. *IEEE Wireless Communications*, 23(5), pp.10-16.
22. Barman, B.K., Yadav, S.N., Kumar, S. and Gope, S., 2018, June. IOT based smart energy meter for efficient energy utilization in smart grid. In *2018 2nd International Conference on Power, Energy and Environment: Towards Smart Technology (ICEPE)* (pp. 1-5). IEEE.
23. Rastogi, S., Sharma, M. and Varshney, P., 2016. Internet of Things based smart electricity meters. *International Journal of Computer Applications*, 133(8), pp.13-16.
24. Avancini, D.B., Rodrigues, J.J., Rabêlo, R.A., Das, A.K., Kozlov, and Solic, P., 2021. A new IOT-based smart energy meter for smart grids. *International Journal of Energy Research*, 45(1), pp.189-202.
25. Joshi, D.S.A., Kolvekar, S., Raj, Y.R. and Singh, S.S., 2016. IOT Based Smart Energy Meter. *Bonfring International Journal of Research in Communication Engineering*, 6 (Special Issue), pp. 89-91.
26. Shrouf, F., Ordieres, J. and Miragliotta, G., 2014, December. Smart factories in Industry 4.0: A review of the concept and of energy management approached in production based on the Internet of Things paradigm. In *2014 IEEE International Conference on Industrial Engineering and Engineering Management* (pp. 697-701). IEEE.

27. Mir, S.H., Ashruf, S., Bhat, Y. and Beigh, N., 2019. Review on smart electric metering system based on GSM/IOT. *Asian Journal of Electrical Sciences*, 8(1), pp. 1-6.
28. Hiwale, A.P., Gaikwad, D.S., Dongare, A.A. and Mhatre, P.C., 2018. Iot Based Smart Energy Monitoring. *International Research Journal of Engineering and Technology (IRJET)*, 5(03).
29. Dong, S., Duan, S., Yang, Q., Zhang, J., Li, G. and Tao, R., 2017. MEMS-based smart gas metering for Internet of Things. *IEEE Internet of Things Journal*, 4(5), pp. 1296-1303.
30. Singh, A. and Gupta, R., 2018. IOT based smart energy meter. *International Journal of Advance Research and Development*, 3(3), pp. 328-331.
31. Majee, A., Bhatia, M. and Swathika, O.G., 2018, March. IOT based microgrid automation for optimizing energy usage and controllability. In *2018 Second International Conference on Electronics, Communication and Aerospace Technology (ICECA)* (pp. 685-689). IEEE.
32. Odiyur Vathanam, G.S., Kalyanasundaram, K., Elavarasan, R.M., Hussain Khahro, S., Subramaniam, U., Pugazhendhi, R., Ramesh, M. and Gopalakrishnan, R.M., 2021. A Review on Effective Use of Daylight Harvesting Using Intelligent Lighting Control Systems for Sustainable Office Buildings in India. *Sustainability*, 13(9), p. 4973.
33. Mehmood, Y., Ahmad, F., Yaqoob, I., Adnane, A., Imran, M. and Guizani, S., 2017. Internet-of-things-based smart cities: Recent advances and challenges. *IEEE Communications Magazine*, 55(9), pp. 16-24.
34. Sahani, B., Ravi, T., Tamboli, A. and Pisal, R., 2017. IOT based smart energy meter. *International Research Journal of Engineering and Technology (IRJET)*, 4(04), pp.96-102.
35. Yaghmaee, M.H. and Hejazi, H., 2018, August. Design and implementation of an Internet of Things based smart energy metering. In *2018 IEEE International Conference on Smart Energy Grid Engineering (SEGE)* (pp. 191-194). IEEE.
36. Chin, W.L., Li, W. and Chen, H.H., 2017. Energy big data security threats in IOT-based smart grid communications. *IEEE Communications Magazine*, 55(10), pp. 70-75.
37. Ali, W., Dustgeer, G., Awais, M. and Shah, M.A., 2017, September. IOT based smart home: Security challenges, security requirements and solutions. In *2017 23rd International Conference on Automation and Computing (ICAC)* (pp. 1-6). IEEE.
38. Pau, M., Patti, E., Barbierato, L., Estebsari, A., Pons, E., Ponci, F. and Monti, A., 2018. A cloud-based smart metering infrastructure for distribution grid services and automation. *Sustainable Energy, Grids and Networks*, 15, pp. 14-25.
39. Song, W., Feng, N., Tian, Y. and Fong, S., 2017, June. An IOT-based smart controlling system of air conditioner for high energy efficiency. In *2017 IEEE International Conference on Internet of Things (iThings) and IEEE Green*

Computing and Communications (GreenCom) and IEEE Cyber, Physical and Social Computing (CPSCom) and IEEE Smart Data (SmartData) (pp. 442-449). IEEE.
40. Hiwale, A.P., Gaikwad, D.S., Dongare, A.A. and Mhatre, P.C., 2018. IOT based smart energy monitoring. *International Research Journal of Engineering and Technology (IRJET)*, 5(03).

7

IoT-Based Home Automation and Power Consumption Analysis

K. Trinath Raja[1], Challa Ravi Teja[1], K. Madhu Priya[1] and Berlin Hency V.[2*]

[1]School of Electronics and Communications, Department of SENSE VIT University, Chennai, India
[2]School of Electronics Engineering (SENSE), VIT University, Chennai, India

Abstract

In modern days we can observe that there is an exponential increase in the usage of the electrical equipment in our home. Often we may forget to turn off the fan or light, which increases the power consumption and results in lots of wastage of electricity. We have seen that there is an increase in gas leak accident cases these days, which causes tremendous life and collateral loss. So to provide a solution for this we have designed our model to solve the existing problems. The aim of the project is to develop an IoT-based home automation system which includes home appliances like fan, light and gas detection. We will control the fan and light through Google Assistant using voice commands. ThingSpeak is a free IoT-based cloud server which is used to store or analyse the data using MATLAB; it will be linked to IFTTT, which is an acronym for "If This Then That", which is used to create the trigger based on the events. By using Google Assistant in the IFTTT the voice commands will be given as input and webhook will be given as output. The user can give commands to Google Assistant to control home appliances like fan and lights. The commands will be sent to microcontroller which will decode and act accordingly. The microcontroller used is NodeMCU (ESP8266) and the communication will be taken care of by inbuilt Wi-Fi in NodeMCU. A special advantage of this project is that fan speed will get adjusted automatically based on the temperature and humidity in the room, which will decrease the power consumption.

Keywords: ThingSpeak, IFTTT, DHT sensor, home automation, power consumption, Internet of Things (IoT), google assistant, node MCU

---

*Corresponding author: Berlinhency.victor@vit.ac.in

Milind Shrinivas Dangate, W.S. Sampath, O.V. Gnana Swathika and P. Sanjeevikumar (eds.) Integrated Green Energy Solutions Volume 1, (93–104) © 2023 Scrivener Publishing LLC

7.1 Introduction

Home Automation is remote controlling and monitoring various electronic and electric appliances present in the home. Technology development is indeed a boon to the whole world. The main motto of the technology is to reduce effort and increase efficiency. Appliances in various areas are remotely controlled using Internet of Things (IoT). This project mainly targets disabled people and patients. The commands given by the admin serve as input to the smart mobile phone. The mobile phone recognizes the commands and performs operations based on the received command. The system remotely controls fan and light based on the room temperature. In the proposed system the admin will give the commands to the Google Assistant to control the various home appliances like fan and light. The commands will be decoded and sent to the microcontroller which controls the relay connected to it. Home is the place where people take their rest after their busy workday. People are so tired that they find it hard to move themselves to switch their lights on or off. The proposed system may help people to monitor the devices via commands using Google Assistant. Since not everyone can afford the automation systems, it is quite important to find a system which is economic and cost effective.

7.2 Literature Review

In [1] the proposed system is based on the development of the Home Automation system using IoT technology. In the proposed paper the author mainly focused on the friendly interface for the user to control the whole system without the help of the developer. By using the proposed system, we can control the applications from anywhere around the world using NodeMCU along with relays to control the switches remotely from the server which is built on Node.js. The paper [1] also enables the authentication for security concerns. The paper's proposed motto is to enable the IoT devices not the IoT homes.

In [2] the author is mainly concerned about saving energy in homes using IoT technology. The paper explains the system to reduce energy consumption by changing the lifestyle of the habitat; it detects the data about user and energy usage by the home appliances and recognizes the wastage of energy based on the user's location, so as per his location the proposed system will change the status of the home appliances. The algorithm is also optimized to decide the number of times the system has to notify the user

In [3] the proposed program focuses on the Home Performance Management Framework based on the Smart TV Set-top. This paper proposes a smart TV set-top box-based appliance framework, which enables the user to monitor and control his or her household items from anywhere in the world. Here they have installed a software module as a home gateway in the set-top box form, as the default service. These Smart TV set-up boxes are still evolving, and solutions for integrating home network and IP streaming capabilities into standard smart TV set-top boxes are constantly being tried. But nothing like the proper and effective plan has been developed yet.

In [4] the proposed system talks about the home automation system which includes a smart fan as one of the smart devices. The system is mostly useful for blind people, patients and disabled people who cannot monitor the temperature manually. Without human intervention the fan is automatically monitored based on the room temperature. By giving the set point through the mobile phone the user can be monitor the speed of the fan, which serves as the main advantage of the system.

In [5] the paper proposed the smart fan system which includes an ultrasonic sensor and wireless IoT technologies. The fan gets automatically turned OFF/ON based on the presence of a person. An LCD screen is used to display the temperature of the room, number of people present inside the room and also the speed of the motor present inside the fan. If the temperature of the room is high then the speed of the fan is high; similarly, for lower temperature the speed will be decreased. The model is implemented in almost every sector since it results in a large amount of power saving and thus improves the lifetime of the fan.

In [6] the paper has developed a system which will allow computers and smartphones to remotely control and monitor all the features and functions of home appliances using Internet of Things. The paper focuses on the remote operation of various equipment and other electronic and electric appliances using different control systems. The system uses IoT technology to create a wireless connection between the main user and the other devices. The entire network consists of a single main user which makes the system more secure as only the admin can have the authority to access the other nodes present under different users.

In [7] the proposed paper has generalised the concept of IoT and raised a point that IoT can be embedded anywhere around the world; the proposed idea clearly states that a single user admin can control the entire architecture including all the nodes. The nodes are connected to the cloud server. This whole IoT-enabled building is connected to the internet and allows mobiles and laptops to control them. The usage of IoT in every field will definitely bring a lot of change, which includes efficiency and economy.

In [8] the proposed system is developed in order to maintain living conditions at home. Various parameters including fan, light, air, temperature and humidity are controlled and automated using Internet of Things. The paper completely focuses on an IoT-based monitoring and sensing system for home automation. This paper uses the EmonCMS platform for data collection and analysis and remote controlling of home appliances and devices. This paper is highly effective and efficient and can be used across many buildings by increasing the number of sensors, measured parameters, and control devices.

7.3 IoT (Internet of Things)

In general, IoT is used to connect many devices like smart mobile phones and electric appliances to the internet. This brings communication between things and people. In many worst circumstances the internet is helpful in bringing out many solutions in connecting with nearby remote places. The concept used in the Google Assistant-controlled smart fan and light is the Internet of Things. Human-made objects have the ability in transferring the data over a network. This interaction is called as IoT. IoT sensors, gateways and many protocols are used to build a complete system. The physical objects are getting connected and controlled digitally and remotely there is a large amount of workings which acts as an additional advantage of IoT. The machines are able to communicate with each other, which leads to faster and timely output without the human intervention. The rise in Internet of Things will reform various sectors, like healthcare, automation energy, transportation, etc.

7.4 Architecture

The block diagram shown in Figure 7.1 clearly shows the whole description of the proposed idea. As shown in the figure, we can see that the proposed paper is using the Google Assistant in the mobile phone to give commands to change the status of the applications. The Google Assistant is connected to the IFTTT website which converts the commands into required script format and gives the respected output as webhook trigger. The webhook trigger is connected to the cloud that is ThingSpeak using the Thinghttp services which was provided by the ThingSpeak platform for free. In the

IoT-Based Home Automation and Power Consumption Analysis 97

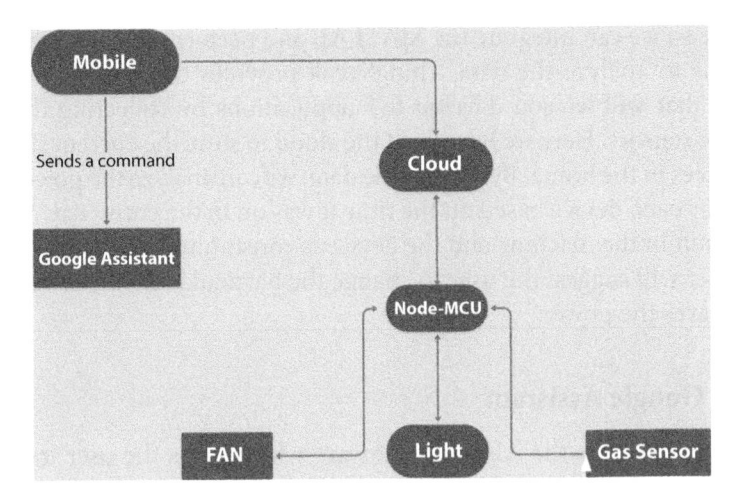

Figure 7.1 Block diagram of our proposed idea.

cloud we will be monitoring the current status of the applications and then we can give the commands to the system accordingly. So here NodeMCU which is connected to internet is connected to both the cloud and applications like light, fan and gas sensor; it sends the signal to the cloud about current status. Simultaneously it sends the Alert Notification to the Registered mobile in the IFTTT, indicating if there is any gas leakage in the house.

7.5 Software

7.5.1 IFTTT

IFTTT means "If This Then That"; it is the software platform which is used to connect the apps, services and devices of various developers to trigger one or more automations involving the apps and services. In IFTTT we can connect various services and trigger the event when a certain condition meets the threshold. This paper used the IFTTT to trigger the SMS when a leakage of the gas is detected. It means the proposed idea will take input as webhook and outputs the SMS as trigger. To use this service, we have to download the app and log into the account in which the applet has been created.

7.5.2 ThingSpeak

ThingSpeak is the IoT-based cloud platform where you can upload the data using the sbc like Arduino and Raspberry Pi. ThingSpeak is a MATLAB software so we can integrate the MATLAB and perform the necessary calculations to analyze the data. ThingSpeak provides the web service API (REST) that will let you develop IoT applications by collecting the data from the sensors. Here we have used the cloud to store the current status of the devices in the home. By storing the data, we can analyze the power consumed by each device based on the time it was on in the entire day. If there is any fault in the machine and the device is consuming heavy power, then our paper will suggest the user to change the particular device or repair it, which saves the power.

7.5.3 Google Assistant

Google Assistant is the AI-based software which helps the user to access the internet and control the apps and the devices in mobile using voice commands. This was created by one the top companies in the world, Google. This paper uses the Google Assistant to give the voice commands to control the devices in the room, such as the command "Turn off the Fan", which will turn the fan off, and the command such as "Turn on the Light" will turn the light on. By using this there is no need to touch anything. This is very easy for old people or people with disabilities to control the devices in home. We can control these devices from anywhere in the world, which saves a lot of power if we forget to turn off the devices when we go to the office.

7.6 Hardware

7.6.1 DHT Sensor

DHT sensor is used for detecting the humidity and temperature of the surrounding environment allowing us to automate the speed of the fan. This Sensor is attached to Arduino and the current room temperature is detected by the DHT sensor and sent to cloud. If the temperature is high then the speed of the fan will be increased; if the temperature is low then the fan speed will decrease, which will help in reducing power consumption.

7.6.2 Motor

DC motor is used as the Fan in this paper which has the voltage range (4.5-7)V and has maximum peak motor current of 600mA.

7.6.3 NodeMCU

NodeMCU is a hardware device which has default Wi-Fi. It is an open-source IoT platform that comprises firmware that works on Espressif Systems' ESP8266 Wi-Fi Module and hardware based on the ESP-12 module. This is the system's brain, and it connects the Hardware components to the internet.

7.6.4 Gas Sensor

We are using MQ6 gas sensor which is used to detect the LPG gases. This sensor is placed near the stove, If there is any gas leakage the sensor detects the gas and sends the signal to NodeMCU, which in turn sends the alert notification to the user. This is used to eliminate major accidents.

7.7 Implementation, Testing and Results

Our proposed design of IoT-based home automation and power consumption analysis is used for controlling the house energy and using it in efficient manner, while maintaining the minimum required comfort-of-living conditions is implemented as shown in Figure 7.2. The sensors placed inside and around the house continuously monitor Humidity and Temperature and the information should be accessible at all times and uploaded every few minutes to the cloud server for display, processing, and archiving.

Figure 7.2 shows our hardware equipment setup for the proposed idea. It contains 2 LEDs, Fam and the Arduino which are connected to the computer. It can be seen that the Arduino is connected to bread board and

Figure 7.2 Hardware system of our proposed idea.

components like lights and fans are connected to Arduino. Based on the signal we get the necessary operations will be performed; we are using the 100 watt plug box for our experiment.

Results provided in Figure 7.3 show the recorded values of temperature, humidity. NodeMCU module will send the data to the cloud and it will be stored for the next operations. Additionally, these values are recorded and displayed over time through Blynk app or ThingSpeak dashboard, as shown in Figure 7.3. Furthermore based on the temperature of the room the fan speed will get adjusted by the motor driver, and based on the voice command or through Blynk app we can control the lights and fans.

As shown in Figure 7.4, We have used Google Assistant for the controlling of the appliances through voice command. The appliance connected with the individual transfer can be turned on or off according to the client's solicitation to the Google Assistant. The microcontroller utilized

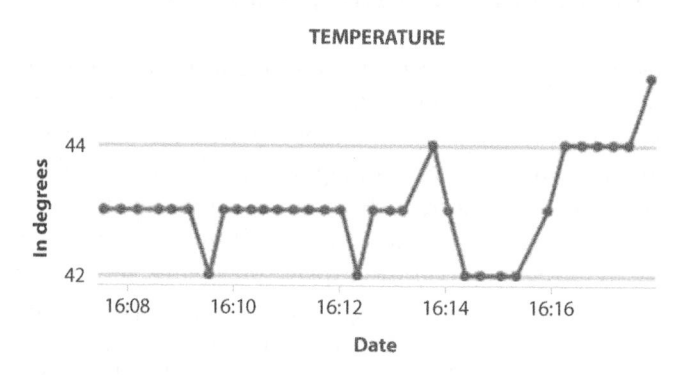

Figure 7.3 Temperature data in ThingSpeak.

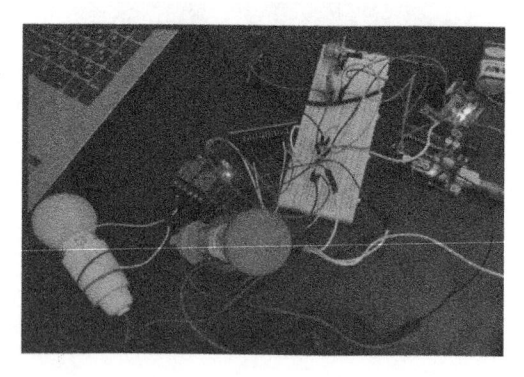

Figure 7.4 Bulb has glowed as per Google command.

is NodeMCU (ESP8266); furthermore, the correspondence between the microcontroller and the application is set up through Wi-Fi (Internet).

In Figure 7.5, our proposed idea has decreased the power consumed by the appliances, like if you see the smart fan consumes 50 watts of power in one day and the normal fan is consuming 75 watts of the power in one day, this shows our model has more advantage on the power consumption, which is very useful and reduces the consumption of natural resources.

Figure 7.6 shows the constant power consumed by the ceiling fan at every temperature. Figure 7.6 is the output of our proposed smart fan and self-adjusts based on the temperature in the room. Here by using temperature sensor we adjust the speed of the fan, which saves power and also makes the room more comfortable.

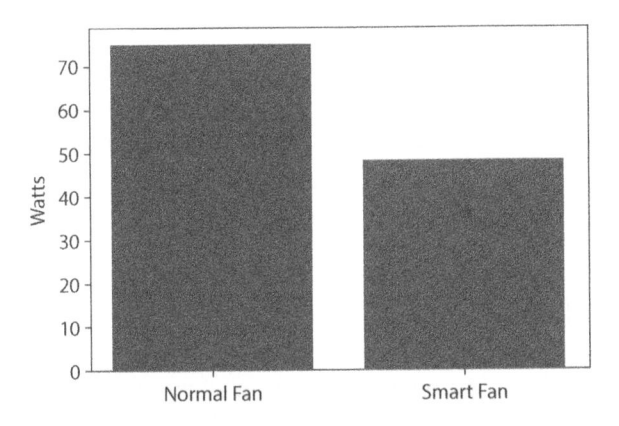

Figure 7.5 Comparing the power consumed by the normal fan and smart fan.

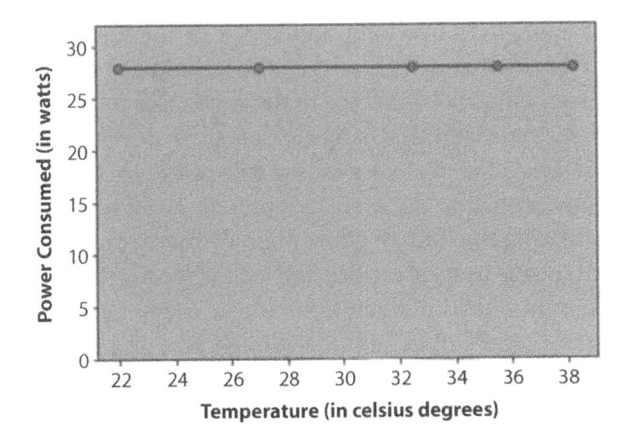

Figure 7.6 Temperature vs. power consumed generally.

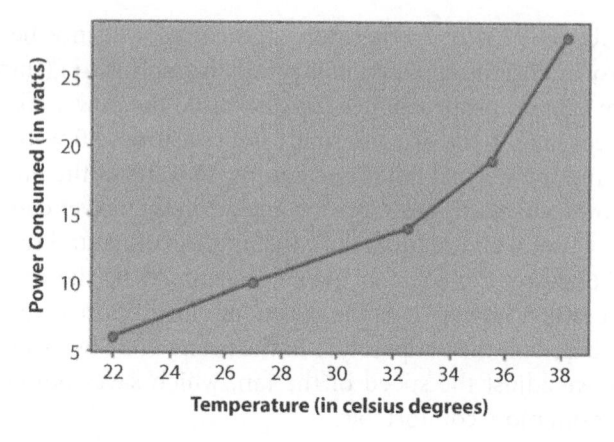

Figure 7.7 Temperature vs. power consumed by our proposed idea.

In Figure 7.7 we clearly see the difference in power consumed in manually powered appliances and smart appliances. The power consumption is not constant, it is variable and depends upon the temperature in the surrounding so if there is lower temperature, the fan works with less speed, which in turn reduces the power consumed.

7.8 Conclusion

The IoT-based Home Automation has been successfully made. The voice commands are given to Google Assistant which have been further added to IFTTT website and the ThingSpeak account. When the user gives the commands through Google Assistant the system is able to control and monitor various home appliances like light and speed of the fan. The commands given first are decoded and then sent to the nodeMCU and it control the relays. The device which is connected to the respective relay turns ON/OFF based on the user's commands. Using cost-effective devices the home automation system is developed at very reasonable price. Therefore, the project overcomes many problems like security, cost, etc. In addition, the system is capable of decreasing the human effort, thereby increasing the efficiency. It is also convenient for usage purpose and will improve the comfort of our home. We can expand this project in the future to connect more electrical equipments such as tube light, air conditioner, etc., and have control over them using our smartphone. All the functions can be integrated into one single mobile app.

References

1. H. K. Singh, S. Verma, S. Pal and K. Pandey, "A step towards Home Automation using IoT," 2019 *Twelfth International Conference on Contemporary Computing (IC3)*, 2019, pp. 1-5, doi: 10.1109/IC3.2019.8844945.
2. Junbo Wang, Z. Cheng, Lei Jing, Y. Ozawa and Yinghui Zhou, "A location-aware lifestyle improvement system to save energy in smart home," *4th International Conference on Awareness Science and Technology*, 2012, pp. 109-114, doi: 10.1109/iCAwST.2012.6469598.
3. J. Kim, E. Jung, Y. Lee and W. Ryu, "Home appliance control framework based on smart TV set-top box," in *IEEE Transactions on Consumer Electronics*, vol. 61, no. 3, pp. 279-285, Aug. 2015, doi: 10.1109/TCE.2015.7298086.
4. Mehran Ektesabi, Saman A. Gorji, Amir Moradi, Suchart Yammen, V. Mahesh K. Reddy, Sureerat Tang, "IoT-based home appliance system", pp. 37–46, 2018, DOI: 10.5121/csit.2018.81604.
5. Gokul Dev. P, Arun V. S, Mr. S. Mani, "IoT based energy conserving smart fan", e-ISSN: 2582-5208, Volume:02/Issue:09/September-2020.
6. Shopan Dey, Ayon Roy, Sandip Das, "Home Automation Using Internet of Thing", 978-1-5090-1496-2016, 2016 IEEE.
7. S. Dey, A. Roy and S. Das, "Home automation using Internet of Thing," *2016 IEEE 7th Annual Ubiquitous Computing, Electronics & Mobile Communication Conference (UEMCON)*, 2016, pp. 1-6, doi: 10.1109/UEMCON.2016.7777826.
8. Majid Al-Kuwari, Abdulrhman Ramadan, Yousef Ismael, Laith Al-Sughair, Adel Gastli, "Smart-Home Automation using IoT-based Sensing and Monitoring Platform", *2018 IEEE 12th International Conference on Compatibiity (CPE-POWERING 2018)*, 2018, pp. 1-6, doi: 10.1109/CPE.2018.8372548.

8
Advanced Technologies in Integrated Energy Systems

Maheedhar* and Deepa T.[†]

Vellore Institute of Technology, Chennai, Tamil Nadu, India

Abstract

Worldwide energy shortage problems can be overcome by integrated energy systems (IES), which maximize the efficiency of energy supply. Integrated energy systems combine the electric power technology, distributing energy technology and thermal technologies to provide heating, cooling, and humidity control by utilizing thermal wastage from power plants and furnace-based industries. This technology can be utilized in commercial and institutional buildings but in limited fashion, because in summer time not much heat is required. By introducing the Thermal Activated Technologies (TAT) device, temperature and humidity can be controlled in a building. Therefore, IES along with Combined Heat and Power (CHP) can reduce the air pollutants, carbon emission and enhance the system's efficiency. The way heating, cooling, air conditioning, lighting and power loads are treated the IES will get success. Research and Development focuses on component and equipment integration, modular machine development, machine integration with grid and buildings. Furthermore, energy analysis underpins the majority of the analysis undertaken for intermittent renewable energy and integrated energy systems. Reducing the exergy losses in components like PV surface modules, wind rotor, and boilers significantly increases quality of energy. This chapter focuses on the IES in buildings, different advanced systems used in CHP for heat and power generation and economic aspects of IES.

Keywords: Combined heat and power, exergy, integrated energy system, renewable energy, energy

---

*Corresponding author: maheedhar32@gmail.com
[†]Corresponding author: deepa.t@vit.ac.in

8.1 Introduction

Renewable energy varies based on time and local conditions. Thus, integrated energy systems, energy storage and demand side management are playing a key role in the energy sector. Consumers may benefit from efficient light gains, customer choice, and energy security with integrated energy systems. Present research and development in integrated energy systems targets the commercial buildings to reduce the fossil fuel use and emission of air pollutants, improve energy security, return on investments, and enhance electric grid quality power [1]. Because of ineffective use of thermal energy byproducts, distributed energy and CHP (combined heat and power) in commercial buildings are limited. By using the TAT (Thermal Activated Technologies) equipment along with the CHP system thermal byproducts can be used effectively in summer season to provide cooling and humidity control. CHP system alone could not serve economically in building sectors, such that IES (Integrated Energy Systems) integrated in the building sector can serve economically and effectively [2]. IES's advantages can be attained by merging operation and planning. The condition of one infrastructure can affect the total system performance and economy [2]. A disturbance in one system can have an impact on the other linked system, which in turn has an impact on the first. The capacity of IES to manage cooling, heating, ventilation, electricity, and lighting, as well as run the bulk of loads that are generated directly or indirectly by CHP output, is critical to its success. This study focuses on the possibility

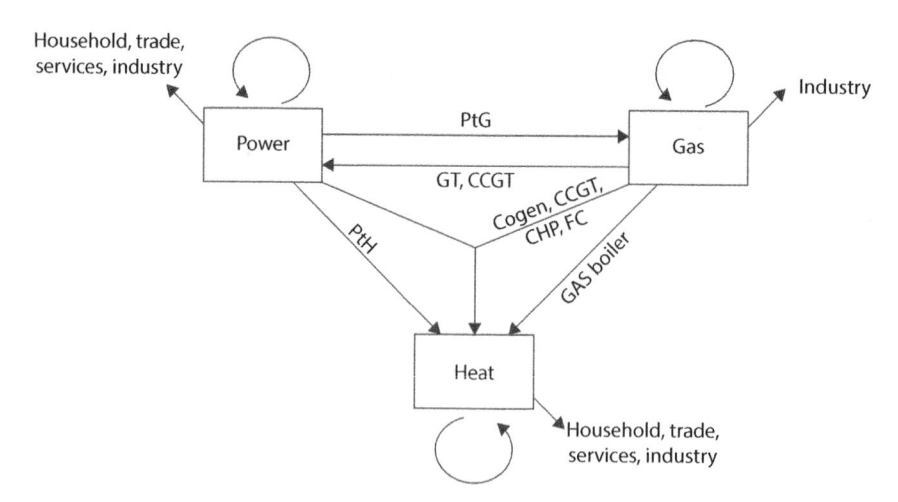

Figure 8.1 Integrated energy systems [9].

of IES integration in buildings in the aspect of economic and performance. Integrated Energy System is defined as co-generating of heat along with power. CHP technology is included in this, along with chillers and humidifier systems. The Figure 8.1 shows the schematic diagram of IES.

8.2 Combined Heat and Power

In industrial and commercial sectors Combined Heat and Power plays an important role. A number of advanced technologies are available in CHP to generate electricity and thermal energy effectively, cleanly and economically [1]. This section reviews the present technologies which are available and investigates development and performance.

8.2.1 Stirling Engines

This Stirling engine technology is an advanced technology in combined heat and power applications. This engine is an external combustion engine that allows a very good control for process of combustion [4]. Compared to external combustion engine this Stirling engine is a most prominent device for power generation and has more merits. This engine is available in three different types: α, β and δ [10]. Efficiency of High Thermal Differential (HTD) Stirling engine is 40% at a speed of 2000-4000 rpm and at a temperature range of 923-1073K [3]. Due to performance of regenerator and heat exchanger this practical Stirling engine cycle is different from ideal cycle. The Stirling engine has mainly five essential parts: compression, cooler, regenerator, heater and expansion space. Hydrogen, air, helium and nitrogen are used as a running fluid due to high amount of heat flow capacity. Helium is a noble gas that is safer to utilize. The Stirling engine was designed with swept volume of 220cc and total volume of 580cc, and it was tested with helium.

This Stirling engine is an advanced engine in combined heat and power applications and the parameters of the Stirling engine parameters are tabulated in Table 8.1.

The highest brake power attained was 96.7W with helium gas at 550°C at charge pressure of 10 bar at 700 rpm. Electricity was generated by using biomass and agricultural waste, sugarcane bagasse, wood and sawdust. Maximum power generated by the sawdust of 46W and minimum amount of power generated by trimming of tree wood waste of 21W. The ignition time of sawdust is 4 minutes minimum, and maximum time 10 minutes for trimmed wood. The below Figure 8.2 shows the electricity produced from the biomass and agricultural waste.

Table 8.1 Stirling engine specifications.

Parameter	Value
Weight	20kg
Piston diameter	70mm
Displacer Diameter	69mm
Stroke	50mm
Working gas	He
Power	100W
Rpm	600rpm
Hester Temperature	550°C
Cooler Temperature	30 - 50°

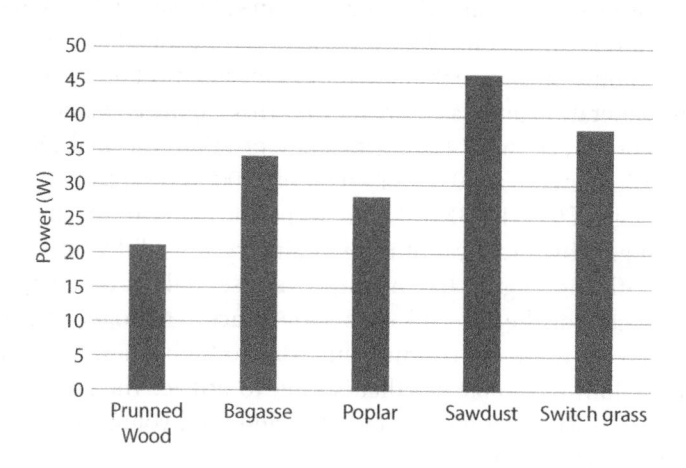

Figure 8.2 Electricity generated from biomass agricultural waste.

This single cylinder, gamma type Stirling engine will produce 1kw power, which is useful for building electrification.

8.2.2 Turbines

The Department of Mechanical, Energy, and Management Engineering (DIMEG) developed a tiny steam turbine with a novel design that allows for controlled flow over a broad range of pressures without sacrificing efficiency [5]. This turbine features a revolving channel and a tangential supply nozzle with three deflector ducts. Due to absence of blades, the structure is very simple and gives better resistance to fatigue. Compared to traditional turbines, this turbine can rotate in low speeds. This turbine can operate in low flow rates with minute variation in efficiency. To control the flow a tangential nozzle is fitted in this innovative turbine; this nozzle can be opened from zero to rotating channel's cross-sectional area without affecting efficiency. Opening fraction K_p defined as

$$K_p = \frac{Nozzle\ area}{Channel\ area}$$

At nominal condition, efficiency of the turbines is 50%. If the flow rate is 63% of nominal flow the efficiency definitely will be more than 47%. Figure 8.3 shows the efficiency curve.

The size of steam turbine is considered as two stages and the parameters of each stage was tabulated in Table 8.2. This micro steam turbine combined with the micro gas turbine and demonstrated combined micro plant and this plant is in CHP configuration, and the turbine used is a single-shaft, two-stage turbine. 91kW of power produced by this plant by the combustion of biomass along with the hot water flow rate generates 183.5 kW of heat and distributes it. This system can be used in commercial and industrial buildings for power and heat utilization either for a single building or multibuildings. This system does not require much maintenance, and it is economical with low environment impact. The primary energy saving by the system is 23.3%, thermal efficiency is 44.76%, and electrical efficiency is 22.35%.

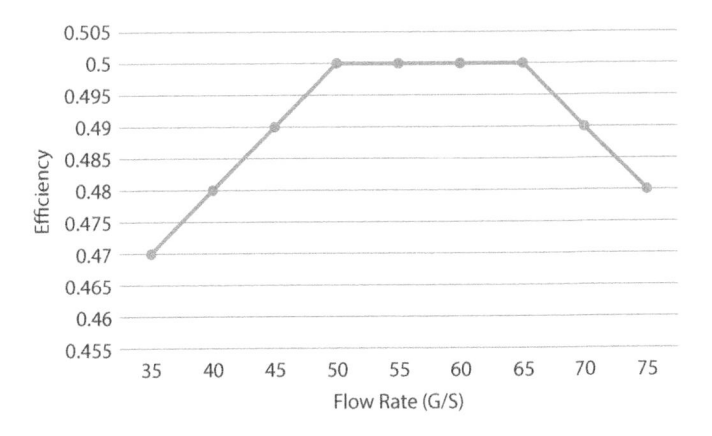

Figure 8.3 Efficiency curve.

Table 8.2 Sizing of steam turbine.

Parameter	1st stage	2nd stage
Nominal rotational speed (rpm)	12000	12000
Feeding pressure (bar)	16	4
Turbine inlet temperature (°C)	201.4	143.6
Isentropic enthalpy drops (kJ/kg)	251.8	228.9
Ratio of velocity (u/c_0)	0.1	0.2
Throat area of Nozzle (mm²)	34.86	130.68
Width of the rotating channel (mm)	5.9	11.43
Height of Deflector (mm)	11.81	22.86
Diameter of Rotor (mm)	113	215

8.2.3 Fuel Cell

Over the last few decades, fuel cell technology has been integrated in Combined Heat and Power system, and implemented in building sectors to provide heat and power together [6]. The fuel cell has many advantages. It is very clean, environmentally friendly, with high efficiency, and has the capability to produce both thermal energy and electrical energy together from hydrogen or natural gas. Anode, cathode, and electrolyte make up a fuel cell, which transforms chemical energy into electrical and thermal energy via an exothermic reaction between hydrogen and oxygen at the anode and cathode sides, respectively. The reaction speed in a fuel cell depends upon the electrolyte, and catalyst used. Depending on the electrolyte used and operating temperature, a fuel cell is classified as Phosphoric Acid fuel cell (PAFC), Alkaline fuel cell (AFC), Sulfuric Acid Fuel Cell (SAFC), Molten Carbonate Fuel Cell (MCFC), Solid Oxide Fuel Cell (SOFC), Solid Polymer Fuel Cell (SPFC), and Proton Exchange Membrane Fuel Cell (PEMFC) (PEMFC). The PEMFC's operating temperature is 80°C, which is low and SOFC operating temperature is 600°-1000°C, which is very high. The Figure 8.4 the system's architecture.

This system consists of DC/DC converter, which is helpful to make the voltage stable which is generated by the fuel cell. From DC/DC converter the power will be delivered to the building directly. The amount of heat produced through the fuel cell is determined by the enthalpy change ΔH_r,

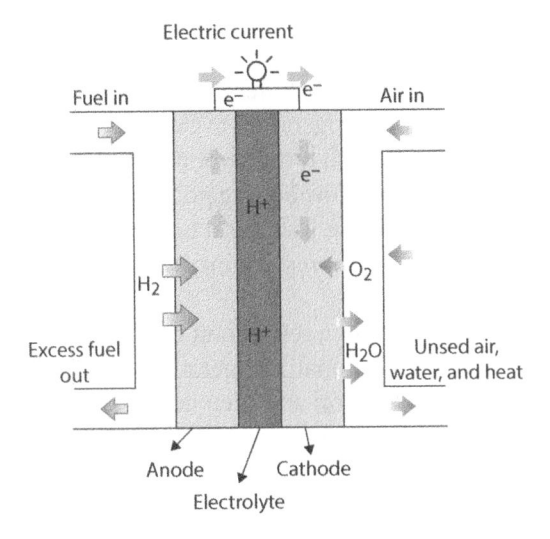

Figure 8.4 Proton exchange membrane fuel cell.

which is negative owing to exothermic reactions, and is described in terms of Gibbs free energy ΔG and entropy ΔS.

$$\Delta H_r = \Delta G + T \Delta S$$

In a fuel cell a small amount of enthalpy is useable to generate both heat and electricity. The amount of heat produced by the fuel cell is calculated by the

$$Q = m * C_p * (T_{final} - T_{initial})$$

Where m = density of the component (kg/m³)
C_p = specific heat (kJ/kg K)
T_{final}, $T_{initial}$ = final and initial temperature (°C)

Total power generated by the fuel cell is 500W but consumed power is 480W; the remaining 20W power is consumed by the fuel cell's auxiliary equipment which is helpful to operate the fuel cell. At 50% of fuel cell efficiency the quantity of heat delivered outside is 1.293 kJ, The total heat produced by a fuel cell is 365.1 kJ, which fuel cells require in order to raise the temperature. The amount of heat and power delivered by a fuel cell can be used to operate the building effectively with less cost and with high efficiency.

8.2.4 Chillers

To make occupants comfortable, chillers are used to provide cooling in buildings [7]. By means of a refrigeration cycle, heat will be removed from the liquid by using chillers, which is the same as that of a domestic refrigerator. In compression systems, a low boiling point liquid refrigerant absorbs heat from the returning building water and boils in an evaporator to produce a gas. The gas is then compressed, increasing the temperature even further.

Absorption refrigeration uses a refrigerant that boils at low temperature and pressure, similar to traditional refrigeration. The refrigerant gas, on the hand, is absorbed in a solution and then heated in a "generator," allowing it to re-evaporate at a high pressure and temperature. The gas is condensing after that, releasing its latent heat, which is subsequently released into the atmosphere.

The following methods can be used to reject heat from the chillers:

- By cycling heat via a condenser, heat was ejected into the atmosphere
- The use of water mist in the atmospheric air to improve cooling effect is known as evaporative cooling
- Water cooling is generally utilized for big systems and requires a connection to a cooling tower.

Heat recovery can be utilized to reuse chiller waste heat for room heating or hot water production. By reversing the cycle such that heat is delivered to the building rather than chilling, a heat pump may achieve the exact opposite of the refrigeration process. Some systems are reversible, meaning they can provide both heat and cooling. Chilled water is generated by chillers and directed to air handling units (also known as fan coil units) in heating, ventilation, and air conditioning systems, where it is used to cool the air that circulates throughout the structure. The re-cooling of the "warmed" water is accomplished by returning it to the chiller unit. The act of cooling ventilation air reduces humidity as well. When heating rather than cooling is necessary, air handling devices may contain a heating coil. The refrigerant (rather than chilled water) is delivered to terminal units that serve various temperature zones through a separate system. Traditional HVAC systems are more efficient, small and versatile than VRF systems, which employ a single exterior condensing unit and numerous interior evaporators. To offer high occupant comfort and improved system performance, HVAC systems require frequent maintenance.

8.2.5 PV/T System

Residential and commercial energy consumption is rising, resulting in a significant increase in greenhouse gas emissions due to growing demand for electricity and heat [8]. The fundamental goal of a building's energy regulation is to ensure internal thermal comfort while reducing energy usage [11]. In the residence, a smart energy building with an effective energy management system may save up to 23% on electricity operating costs and up to 30% on peak usage. One of the most complicated energy systems that has lately been the focus of research is renewable energy resources, domestic smart energy automation systems, and smart heat/electricity network communication. Depending on the circumstances, PVT panels can attain a total efficiency of 70%, with a thermal efficiency of 50% and an electrical efficiency of more than 20%. As sun availability and ambient temperature increase, solar incidence radiation on the PVT surface increases, lowering reliance on the auxiliary heater for hot water production and, as a result, heater power consumption. By proposing this smart electricity-heat supply system for a building's energy system, a path is laid out for bringing the entire energy system up to smart energy system standards in the future.

Figure 8.5 shows the energy utilization factor for every month. The total amount of hot water generated each year is 526.2 m^3, of which 123.4 m^3 is used to meet the whole demand of the residence and the remaining 402.8 m^3 is supplied to the building heating grid. The PVT panels produce 3647.4 kWh of yearly power, of which the heat storage tank holds 400.2 kWh. The remainder is used to satisfy the annual electrical demands of the family. PVT total energy efficiency and annual energy utilization factor are 14% and 0.61, respectively.

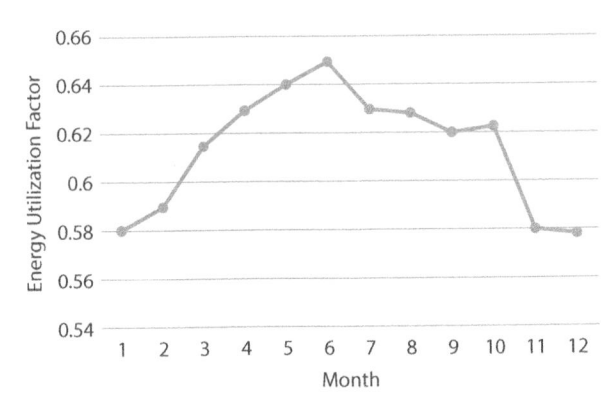

Figure 8.5 Graphical representation of energy utilization factor for every month.

8.3 Economic Aspects

This study's marketing plan focuses on the potential for IES as a structure system that assists with chilling demands, rather than just as a standard CHP [1]. The research makes use of RDC's DIStributed Power Electronic Rationale Selection (DISPERS) model. This approach is based on a spreadsheet that compares different choices to traditional equipment in order to determine IES'S economic potential. Fuel consumption, on-site power production, electric and natural gas bills, installation expenses, and return on investment are all calculated using the DISPERS model. Simple payback is being utilized as a cost-effective selection criterion, and load profiles are being used to anticipate cooling, heating, hot water, and electricity use for a variety of buildings. The IES system may be used in buildings that have Utility and distribution systems that are compatible with CHP. Figure 8.6 shows the distribution of IES market potential by cooling operating scheme.

According to the analysis, each cooling system had its own operating strategy that provided the greatest economies. It made the most sense to run the chiller to meet both baseload and peak requirements, as well as running on-peak and off-peak, for CHP using absorption chillers. The most cost-effective choice for running the EDC (Engine Driven Chiller) full-time, storage, or the EDC just during peak hours was the basic loaded plan. Even at low capacity, base load operation makes more sense since EDCs are more expensive to construct and run.

IES appears to have potential economies for other building sectors, but hurdles are blocking widespread implementation. The greatest possibility for IES in office buildings, which include over 10GW of total IES and Engine-driven coolers and CHP with absorption units has a lot of potential,

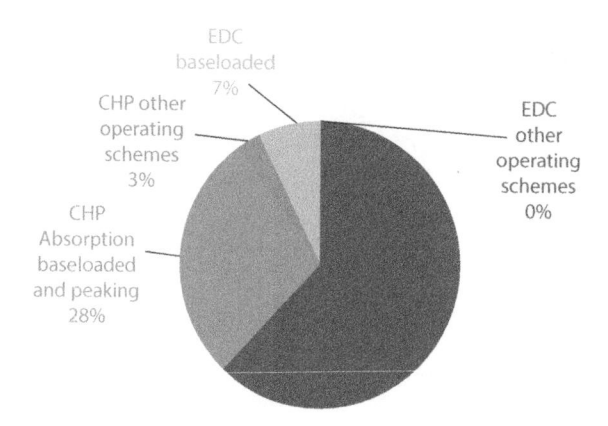

Figure 8.6 Distribution of IES market potential by cooling operating scheme.

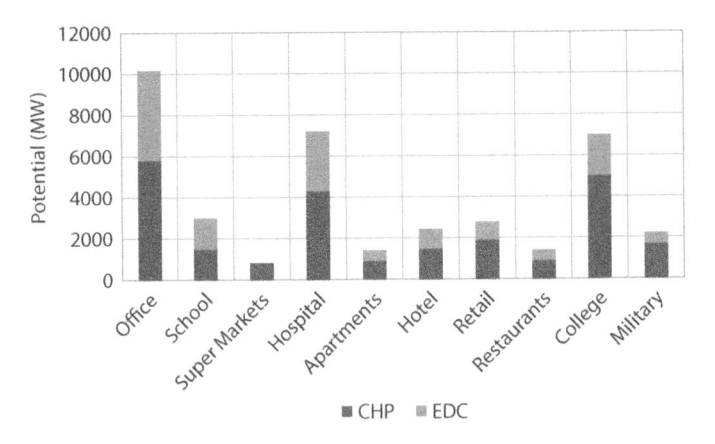

Figure 8.7 IES potential by building type.

as illustrated in the diagram below. CHP with absorbed contributes for about 3.6GW of the 4.5GW total, whereas EDCs account for 1.1GW, giving over half of the EDC potential.

IES potential based on building type was shown in Figure 8.7. Hospitals and universities each have about 7GW of IES potential, despite having a long history of using CHP. Although schools, businesses, and hotels are minor sectors, their large heating and cooling loads provide opportunities for more IES, immersion units, and EDC. The CHP with absorber or EDC's ratio is the highest in schools and residential complexes, accounting for more than half of the entire IES potential.

8.4 Conclusion

Worldwide energy shortage problems can be overcome by integrated energy systems (IES), which maximize the efficiency of energy supply. Renewable energy varies based on time and local conditions. Thus, integrated energy systems, energy storage and demand side management are playing a key role in the energy sector. This Stirling engine is the most renowned mechanism for power generation and has more qualities than the external combustion engines. Efficiency of High Thermal Differential (HTD) Stirling engine is 40% at a speed of 2000-4000 rpm and at a temperature range of 923-1073K. The Stirling engine generates electricity by using biomass and agricultural waste. The maximum power generated by the sawdust of 46W and minimum amount of power generated by trimming of tree wood waste of 21W. The micro steam turbine combined with the micro gas turbine and demonstrated combined micro plant 91kW of power produced by this plant by the combustion of biomass along with the hot water flow rate generates 183.5 kW of heat and distributes it. The advantages of the fuel cell are that it is very clean, environmentally friendly, has high efficiency, and has the capability to produce both thermal energy and electrical energy together from hydrogen or natural gas. Operating temperature of PEMFC is 80°C, which is low, and SOFC operating temperature is 600°-1000°C, which is very high. Total power generated by this fuel cell is 500W but consumed power is 480W; the remaining 20W power is consumed by the fuel cell's auxiliary equipment which is helpful to operate the fuel cell. By refrigeration cycle heat will be removed from the liquid by using chillers, which is the same as that of a domestic refrigerator. In compression systems, a low-boiling-point liquid refrigerant absorbs heat from returning building water and boils in an evaporator to produce gas. The gas condenses after that, releasing its latent heat, which is subsequently released into the atmosphere. Traditional Heating, Ventilation, and Air Conditioning (HVAC) systems are more efficient, compact, and adaptable than VRF systems, which use a single external condensing unit and several inside evaporators. An HVAC system requires regular maintenance to provide good occupant comfort and better performance of the system. In residential settings, a smart energy building with an effective energy management system may save about 23% on electricity operating costs or nearly 30% on peak demand. The PVT panels produce 3647.4 kWh of annual power, of which 400.2 kWh is kept in the heat storage tank and the surplus meets household annual electricity needs. By 2030, the potential building market for IES will be over 70 GW. The economics of IES offers promise for other building sectors, but hurdles impede widespread implementation. Office buildings offer the largest potential for IES, with over 10GW of total IES; CHP with absorption units and engine-driven chillers, for example, has a lot of potential.

References

1. Paul LeMar, "Integrated Energy Systems (IES) for Buildings: A Market Assessment," Resource Dynamics Corporation for Oak Ridge National Laboratory, 2002, 409200.
2. Jonet Dancker, Christian Klabunde, Martin Wolter, "Sensitivity factors in electricity-heating integrated energy systems," *Energy*, 229 (2021) 120600.
3. Hojjat Damirchi, Gholamhassan Najafi, Siamak Alizadehnia, Barat Ghobadiana, Talal Yusaf and Rizalman Mamat, "Design, Fabrication and

Evaluation of Gamma-Type Stirling Engine to Produce Electricity from Biomass for the micro-CHP system," *The 7th International Conference on Applied Energy – ICAE2015* (2015) pp. 137-143.
4. Mohammad Hassan Khanjanpour, Mohammad Rahnama, Akbar A. Javadi, Mohammad Akrami, Ali Reza Tavakolpour-Saleh, Masoud Iranmanesh, "An experimental study of a gamma-type MTD stirling engine," *Case Studies in Thermal Engineering*, 2021, 10087.
5. S. Barbarelli, E. Berardi, M. Amelio, N. M. Scornaienchi, "An Externally Fired Micro Combined-Cycle, with Largely Adjustable Steam Turbine, in a CHP System," *International Conference on Industry 4.0 and Smart Manufacturing (ISM 2019)*, 2020, pp. 532-537.
6. Sofia Boulmrharj, Mohammed Khaidar, Mustapha Siniti, Mohamed Bakhouya, Khalid Zine-dine, "Towards performance assessment of fuel cell integration into buildings," *6th International Conference on Energy and Environment Research, ICEER 2019*, pp. 288-293.
7. https://www.designingbuildings.co.uk/wiki/Chiller_units.
8. Amirmohammad Behzadi, Ahmad Arabkoohsar, "Feasibility study of a smart building energy system comprising solar PV/T panels and a heat storage unit," *Energy*, 2020, 118528.
9. https://www.tuhh.de/transient-ee/en/projectdescription.html.
10. Iskander Tlili, Youssef Timoumi, Sassi Ben Nasrallah, "Analysis and design consideration of mean temperature differential Stirling engine for solar application," *Renewable Energy*, 33 (2008) 1911-1921.
11. Golpira H, Khan SAR, "A multi-objective risk-based robust optimization approach to energy management in smart residential buildings under combined demand and supply uncertainty," *Energy* (2019), 170, 1113-1129.

9
A Study to Enhance the Alkaline Surfactant Polymer (ASP) Process Using Organic Base

M.J.A. Prince* and Adhithiya Venkatachalapati Thulasiraman

Department of Petroleum Engineering, AMET Deemed to be University, Kanathur, Tamil Nadu, India

Abstract

This paper depicts the utilization of another kind of natural base that replaces and enhances customary inorganic salts like sodium carbonate. The natural soluble base assessed in Alkaline surfactant polymer (ASP) slug carrying ordinarily utilized polymers and surfactants. The impact of natural soluble base on interfacial tension (IFT), viscosity and adsorption contrasted with that of a customary salt in these concentrations. The natural soluble base was seen as perfect with saline waters containing high total dissolved salts and divalent ion concentrations. They are utilized with saline waters without softening and they give preferable outcomes over a regular base in proper proportions where they can be utilized. These natural soluble bases have the benefit of enhancing dissolution with brine, polymers and surfactants. Their non-toxic characteristics make them especially reasonable recovery applications,

Keywords: Natural base, unsoften saline, capillary number and sodium carbonate

9.1 Introduction

During ASP slug preparation, the salt responds to modest quantities of esters in raw petroleum to shape surfactants to consolidate towards the infused surfactant to deliver coordinated blends at the oil/saline water

---

Corresponding author: prince466@gmail.com

interface [1, 13]. The soluble base is likewise professed to decrease the measure of surfactants adhere towards the arrangement, particularly in carbonate reservoirs. By expanding the purpose of pH onto limestone is surpassed and then it turns out to be contrarily electrified. This expands the electrostatic shock in the middle of the stone surface medium and on the contrary the negatively charged surfactants and subsequently diminishes adsorption [1, 13].

The surfactant lessens the gap in the middle of the saltwater and remaining oil and in this way builds the capillarity number. The capillarity number (Nc) is utilized to communicate the powers following up on an ensnared bead of oil inside a permeable media.

Nc is an element that belongs to the Darcy speed (v) applied with an aid of portable on the caught stage, the consistency (μ) of the versatile stage, and the IFT (σ) in the middle of the versatile and the caught oil stage [13]. Beneath is the correlation of Darcy speed, thickness and IFT to the narrow Number.

$$Nc = v\,\mu/\sigma$$

Polymer is utilized to build the consistency of the infusion liquid for proper concentration [2, 13]. The blend of surfactant, salt and polymer is reused in a procedure where lingering crude can be monetarily expelled from the store [13]. ASP was assessed in the research facility and utilized generally in this case with incredible success [13].

A field in China is most likely to use ASP on an all-inclusive application an expansion in bubble recuperation of 20% oil saturation. The use of the ASP procedure diminished the oil immersion from 35% to under 5% [3, 13]. There have been various other distributed lab assessments that affirm the positive capability of utilizing ASP to evacuate leftover oil. Be that as it may, even with these focal points and the achievement of manyfield ASP extends, the procedure is not without certain drawbacks. The solid salt effectively affects polymer execution and much of the time, an extra polymer is required to accomplish the ideal viscosity [4, 13]. A procedure that gives favourable circumstances yet takes out or decreases a portion of the current issues related to the ASP technique is required and this paper propounds such a procedure [5, 13].

For improved oil recovery IOR process, it is frequently attractive to utilize the created liquid as the infusion liquid to make the undertaking practical and conservative. Lamentably most produce liquid or accessible water which has cations having the valency of 2 [13]. For an ASP slug the divalent cation concentration should be under 15ppm to dodge the response of the salt with divalent cations that shapes insolvable scales by the following responses [13]:

a. Enhances and lessening the concentrations of polymer required.
b. Reduces the consumption and scaling issues.
c. Irradiates different issues related to taking care of salt.

9.2 Materials and Methods

The natural soluble base concentrated right now got from that point sodium ions of certain feeble polymer acids. The natural antacid is planned into the ASP slug to supplant the inorganic soluble base, for example, NaOH and Na_2CO_3, and their blend.

During ASP procedure for recuperating leftover crude, the natural salt mix with polymers, surfactants and different added substances that are commonly utilized for enhancing oil recuperation are brought into the oil concentration through at least one infusion borehole and the oil is recouped from at least one production well. The ASP procedure where traditional salt as:

$$Na_2CO_3 + Ca^{+2} = 2Na + CaCO_3$$

Supplanted with natural antacid will be assigned as NA. It is once in a while alluring to include the natural antacid to the infusion saltwater originally followed by different parts, for instance, surfactant and additionally polymer:

$$Na_2CO_3 + Mg^{+2} = 2Na^{+1} + MgCO_3$$

All in all, the water is dealt with utilizing particle trade or some other favoured H_2O relaxing method [6, 7, 13]. The direct expenditure, the activity expenditure and the slop removal would be apparent with regular turns into the plug for the ASP procedure [8, 13]. Moreover, if aqua treatment is planned, and the feasibility test demonstrate fruitlessly or the task has dropped rashly, the cost for the water treatment cannot be recuperated [9, 13]. Terminating the treatment process alongside the utilization of salt bearable surfactants makes offshore improved oil recovery (IOR) extends increasingly practical and conservative [11, 13].

Considering the above necessities for boosting the oil recovery, the targets of this investigation are to discover a substitution for alkaline and a procedure that:

a. Contributes to the potential of Hydrogen for upgrading the IFT.
b. Perfect to cations having the valency of 2 and water softening with theelimination of slug is unnecessary.
c. Reduction of the direct front venture essential to water treatment procedure and related expenses for synthetic compounds, removal, and the supportextended when ordinary inorganic antacids are utilized.
d. Reduction in the adherence of surfactants and polymers.
e. Less cost in application.
f. Enhances and decreases the usage of surfactants.

The natural salt of decision relies upon cost, accessibility and execution [10]. Among the most alluring natural soluble bases that are promptly decomposed by microorganisms and blend well with different surfactants, polymers and different added substances normally utilized in IOR [13].

9.3 Similarity Study of NA in the Saline Water Containing Cations Having a Valency of 2

A significant capacity of the natural soluble base with different valent cations might be available in the infusion saline water and the interstitial water [12, 13]. The Ca and Mg in the unrelaxed saline solution surpass 10 ppm for ideal ASP infusion. This common saltwater will by and large be softening to evacuate the Ca, Mg in saline water for ordinary ASP to forestall scale development.

The one utilizing Na_2CO_3 as the soluble base in unsoften saline solution, which responded with the Mg and Ca in the saline water and structures insoluble Calcium Di-Hydroxide and Magnesium Di-Hydroxide. The arrangements including natural antacid are both clear similar to the soft saline water test containing Na_2CO_3 [13].

Utilizing natural alkali makes the infusion procedure less complex since there is no requirement for a water softening procedure for the infusion

9.4 Results and Discussion

A capillary number of around 10^{-6} is found after fulfilment of the run of the softening water flood and this must be expanded by in any event a few sets of greatness to effectively dislodge the oil. The IFT between crude and water is observed to be 110 to 150 Nm/m. The utilization of the best possible surfactant can without much of a stretch reducing the IFT to 10^{-5} N/m or bringing about a relating increment in the narrow number by at any rate a few sets of size. The correlation of the capillary number with oil increment is shown in Figure 9.1 [13].

9.4.1 Alkalinity Contributed by NA for Intensifying the IFT Characteristics

Similarly, as an inorganic soluble base, the natural salt improves the potential of Hydrogen of the infusion saline water in this way giving alkalinity which is utilized to produce surfactant which response with esters that would be available in the ensnared raw petroleum. These intermittently produced surfactants can consolidate with the additional surfactant which performs cooperatively in expelling the remaining crude. Likewise, the natural soluble base adjusts the dynamic locales on the outside of the permeable zone to decrease the loss of polymers that may happen by adsorption.

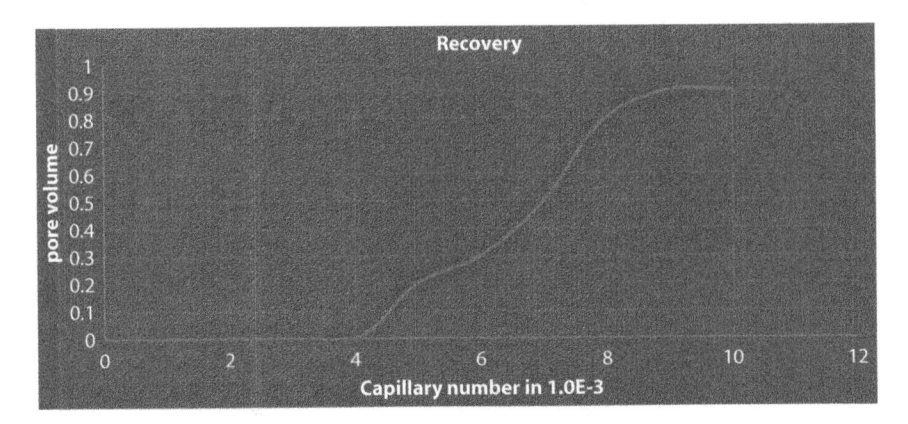

Figure 9.1 Recovery of oil as a function of capillary number.

The impact of untreated saline water with pH utilizing traditional and natural salt by utilizing 1% Na_2CO_3, the pH with untreated saltwater was around 1 pH lesser when contrasted with the 1% Na_2CO_3 in softened saline water. The pH variety is after the effect of the response of the soluble base by cations having the valency of 2 to frame Ca and Mg scales [13]. The arrangement of scales lessens the compelling measure of salt as well as may harm the development and gear.

9.4.2 Interfacial Tension Properties

A distinction in the boundary between two contrast phases strains averse to raw petroleum for salt waters carrying inorganic and natural antacid. The testing arrangement was readied utilizing 0.11 weight percent of the surfactant and 0.112 percentage of the polymer to reproduce the ASP framework. The relaxed saline solution is utilized for liquid holding the inorganic soluble base to forestall the precipitations because of the response of the inorganic salt with the cations having the valency of 2 [13]. The unsoftened saline water is utilized for liquid carrying natural antacid since it is compatible with the cations having the valency 2 present in the saltwater and in this way, aqua relaxing procedure isn't required.

9.4.3 The Similarity of NA + Polymer

The estimated consistency of 0.15% polymer in the relaxed and untreated saline solutions utilizing traditional salt and the natural soluble base was observed. Surfactant is added to the framework to reenact the concentrations of ASP. The consistency was estimated at the base opening temperature of 55C. The observation shows that polymer consistency is reduced within the sight of conventional soluble base (Na_2CO_3) in the softened saltwater. The thickness is further conversely the polymer thickness is not affected by the natural salt and the divalent cations with the unsoften saline solution [13].

Sodium carbonate contributes the least consistency in treated and unsoften brine contrasted with the natural antacid. Likewise, the untreated saline gives a low thickness in collation with the treated aqua when utilizing Na_2CO_3 as the soluble base [13].

The pH of 1% natural soluble base in the softened and unsoftened saltwater is about the equivalent.

9.4.4 Traits of Adsorption

The consequence of organic salt with the adherence of surfactant towards the surface of the reservoir has been assessed utilizing the static adsorption technique portrayed. The test was led at 60 °C utilizing the saline water arrangements that appears in Table 9.1. Relaxed saltwater was utilized where sodium carbonate was the soluble base and unmellow saline solution where natural antacid was utilized. In the two cases, 1% by weight antacid was utilized alongside 0.11% polymer. Surfactant is utilized as the surfactant. Adsorption of surfactant from salt waters holding no soluble base was found to surpass 10^{-3} g/Kg.

9.4.5 Economics

Although the natural soluble base is more expensive than customary inorganic antacids as sodium carbonate, this expense could be balanced with the reserve funds in the direct front speculation vital for aqua handling and the expense of the extra polymer is essential when utilizing inorganic salts [13].

9.4.6 Regular NA Injection Recommendation

The measure of natural antacid, surfactant and polymer required depends on the consistency of the oil, the porousness of the arrangement, the development properties, the adsorption properties, the treatment structure, and so forth. A regular infusion plan is shown in Table 9.1 [13].

Table 9.1 Injected composition with a concentration.

S. no.	Ingredient	Concentration wt%
1.	Surfactant	0.3-0.5
2.	Organic Alkali	0.5-1.2
3.	Polymer	0.5
4.	Brine	trace

9.5 Conclusions

Our examinations have discovered that saline solution inhibition, fewer adherence details for ASP could be defined by adding slats that are weak acids, from now on alluded to as natural soluble bases. The natural soluble base is utilized to give alkalinity, reduce adsorption, complex multivalent decreased with precipitation within the sight of traditional salt in the untreated saline solution. This implies extra polymer was expected to develop the thickness for the treatment cations, limit the surface gear and limit the concentration harm contrasted with comparative plans utilizing customary inorganic soluble bases.

An organic soluble base is powerful as a traditional antacid in getting the least interfacial tensions figured into ASP infusion liquids. Organic salt contributes similarly also in treated like in untreated saline solution. Organic salt does not frame accelerates with divalent cations, for example, calcium and magnesium. Organic salt does not diminish the impact of polymer in expanding the thickness of infusion liquids and improves polymer execution in untreated waters. Traditional antacids, for example, sodium carbonate, can lessen polymer effectiveness in both hard and soft waters. The characteristics of natural salt permit it to be utilized in untreated water in this manner diminishing in advance expenses of water treatment, dealing with and removal.

References

1. Touhami, Y. Mechanism for the Interaction between Acidic Oils and Surfactant-Enhanced Alkaline Solutions. *J. Coll. Inter Sci.*, 1991; **17**, 446-455.
2. S. A. Shedid. Experimental investigation of alkaline/surfactant/polymer (ASP) flooding in low permeability heterogeneous carbonate reservoirs. In: *Proceedings at the SPE North Africa Technical Conference and Exhibition, 16, Cairo, Egypt, 2015*, p. 16-20.
3. J. J. Sheng. A comprehensive review of alkaline-surfactant-polymer (ASP) flooding. *Asia Pacific Journal of Chemical Engineering*, 2014, **9**, p. 20-21.
4. Clark, S.R. Design and Application of an Alkaline-Surfactant-Polymer Recovery System to the West Kiehl Field. In: *Proceedings at the SPE Rocky Mountain Regional Meeting, Casper, WY, 2011*, p. 11-23.
5. D. Cuong. Modelling and optimization of alkaline-surfactant-polymer flooding and hybrid enhanced oil recovery processes, *Journal of Petroleum Science and Engineering*, 2018, p. 578–601.
6. Al-Hashim. Alkaline Surfactant Polymer Formulation for Saudi Arabian Carbonate Reservoirs. In: *Proceedings at the SPE/DOE 35353 prepared*

for presentation at the Tenth Symposium on Improve Oil Recovery, Tulsa, Oklahoma, 1996, p. 21-24.
7. S. S. Riswati, W. Bae, C. Park, A. K. Permadi, I. Efriza, and B. Min. Experimental analysis to design optimum phase-type and salinity gradient of alkaline surfactant polymer flooding at the low saline reservoir. *Journal of Petroleum Science and Engineering* 2019, **173**, p. 1005–1019.
8. Vargo, J. Alkaline-Surfactant/Polymer Flooding of the Cambridge Field" In *Proceedings at the SPE 55633 presented at the Rocky Mountain Regional Meeting, Gillette, Wyoming,* 1999, p. 15-18.
9. L. Kexing, J. Xueqi, H. Song, and W. Bing. Laboratory study displacement efficiency of viscoelastic surfactant solution in enhanced oil recovery. *Energy and Fuel,* 2016, **30**, p. 6-8.
10. French, T.R. Design and Operation of the Alkali-Surfactant-Polymer Project. *Pet. Eng. Intern,* 2009, **25**, p. 35-37.
11. Schramm, L. *Surfactants Fundamentals and Applications in the Petroleum Industry*, Cambridge University Press, UK, 2004, pp. 125-27.
12. Manrique, E. Alkali/Surfactant/Polymer at VLA 6/9/21 Field in Maracaibo Lake: Experimental Results and Pilot Project Design. In: *Proceedings at the SPE 59363 presented at the Improved Oil Recovery Symposium, Tulsa, Oklahoma,* 2005, p. 3-5.
13. Paul Berger, Improved ASP Process using Organic Alkali, *Proceedings of SPE/DOE Symposium on Improved Oil Recovery,* 04/2006, https://slideplayer.com/slide/14988899/.
14. Tereza Neuma de Castro Dantas, Flavia Freitas Viana, Tamyris Thaise Costa de Souza, Afonso Avelino Dantas Neto *et al.* Study of Single-phase polymer – alkaline-micro emulsion flooding for enhanced oil recovery in sandstone reservoirs, *Fuel,* 2021, 10.1016/j.fuel.2021.121176.
15. P.D. Berger, Ultra-low concentration surfactants for sandstone and limestone floods, *Proceedings of SPE/DOE Improved Oil Recovery Symposium,* 04/2002.
16. Yefei Wang, Optimized Surfactant IFT and Polymer viscosity for surfactant-polymer flooding in heterogeneous formations, *Proceedings of SPE Improved Oil Recovery Symposium,* 04/2010.

10
Flexible Metamaterials for Energy Harvesting Applications

K.A. Karthigeyan[1], E. Manikandan[2]*, E. Papanasam[2] and S. Radha[1]

[1]Department of ECE, SSN College of Engineering, Chennai, India
[2]School of Electronics Engineering, Vellore Institute of Technology, Chennai Campus, Tamil Nadu, India

Abstract

Metamaterials are manmade structures exhibiting many exotic properties like a negative index of refraction, cloaking, and so forth. These are majorly used as electromagnetic wave manipulators. Based on the structure and material properties either it will transmit, reflect or absorb the incident waves. In particular, it can be designed to absorb a particular region of electromagnetic spectrum starting from radiofrequency to ultra-violet region. The structure geometrical parameters are adjusted for obtaining the desired frequency range for a particular material. By using a suitable structure, the incident electromagnetic wave is absorbed and can be converted into useful electrical energy further. In the case of solar cells, the metamaterial is being integrated with the system for intensifying the absorption of the incident waves. Also, thermal energy harvesting is being done using metamaterial absorbers. By designing the suitable structure and using proper rectifying circuits, RF energy harvesting can also be done. This chapter starts with an introduction to energy harvesting, and continues with metamaterial, its design principle, absorber designs for various energy harvesting applications, circuit analysis, and finally the inference section.

Keywords: Metamaterial, absorber, flexible, circuits, energy

*Corresponding author: manicrescent2019@gmail.com

10.1 Introduction

In recent years, Wearable technology has been dominating the world, and it is growing faster and faster due to the advanced developments in computing technologies [1]. It is being widely utilized in different fields including healthcare, military, consumer, agriculture, and so forth. The major issue with wearable electronics is to power those for getting the functionality i.e., a need for rechargeable batteries or a kind of medium for directly providing the required power. In batteries, the chemical energy is converted into electrical energy and the efficiency conversion plays a vital role. Also, it is expected to have a miniaturized battery for integration with wearable devices. Other energy harvesting (EH) techniques are proposed and available for usage [2].

Energy harvesting defines the process of converting ambient available energy into useful electrical energy. The source of energy can be in any form, including thermal, solar, electromagnetic, and kinetic. The generalized structure of the EH system is shown in Figure 10.1. The conversion medium is responsible for producing electrical energy. Further, the generated alternate current (AC) is converted into direct current (DC) and the booster can be included in the system if necessary. The generated energy is supplied to the load for its operation. The energy conversion medium can be in any form. For example, materials can be utilized to convert the incident vibrational energy to useful electrical energy. But the problem is, there is a need for materials to resonate in the lower frequency regime. Similarly, for the other energies also the proper selection of materials is required for the conversion. This poses a great challenge in EH applications. Apart from material selection, the flexibility in integration with wearable devices is also of great importance [3–5].

Nowadays, artificial structures called metamaterials are being utilized as a medium for EH applications. The advantage is, the geometrical parameters and the structure can be optimized for absorbing incident energy based

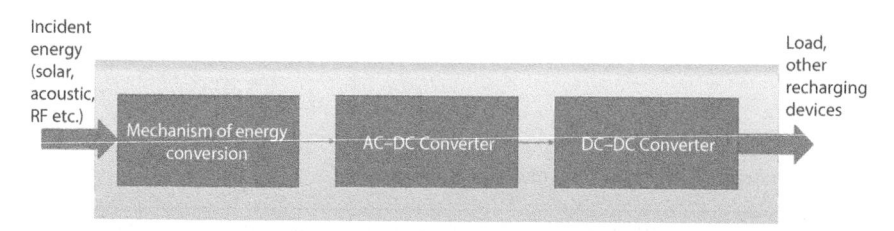

Figure 10.1 Generalized block diagram of energy harvesting system.

on their frequency response characteristics. This way, any form of energy thermal, electromagnetic shall be absorbed and converted into electrical energy with proper electric circuits. This chapter starts with the introduction of the EH system and its mechanisms, introduction to metamaterial and design aspects, metamaterials for absorbing energy, and finally, the issues in the metamaterials-based energy harvesting (MMEH) system [6].

10.2 Metamaterials

Metamaterial Design Metamaterial is an emerging field of study which provides an opportunity to develop artificial materials for enhanced optical, electrical, mechanical, or acoustic functionalities by modifying the constituent architecture from macro to nanoscale [7, 8]. Metamaterials are artificial materials made from assemblies of several elements molded from composite materials, *viz.*, metals and plastics. The property of metamaterial depends on its structure and cautiously engineered meta-atoms in contrast to the properties of the material that constituted it. Based on the properties required architecture can be built in two- or three-dimensional repeating structural arrays. They are sensibly designed to have a fascinating property that is missing in naturally occurring and chemically synthesized materials. The metamaterial is different from photonic crystal, another type of engineered material in terms of the length scale of the unit cell (meta-atom). The unit cell size of the photonic crystal is close to the wavelength of light while the size of the meta-atom is much smaller than the wavelength of the light, which makes the metamaterial an effective media for light. The development of metamaterial is expected to provide a feasible solution to the enduring issues, *viz.*, far-field, large-scale, and high-resolution imaging. Distinct electromagnetic properties of metamaterials entice significant attention from researchers from several disciplines. Early-stage research in metamaterial is focused on microwave region to implement negative refraction and superlens and the same has been swiftly progressed to terahertz and optical frequencies.

Based on the electromagnetic properties, all materials fall under four regions, which are depicted in Figure 10.2. Region 1 covers materials having positive permittivity and permeability; most of the dielectric material falls under this region. Metals, doped semiconductors, and ferroelectric materials fall under region 2 with negative permittivity and positive permeability as they exhibit negative permittivity below plasma frequency. Region 3 is the most fascinating region in material parameter space with

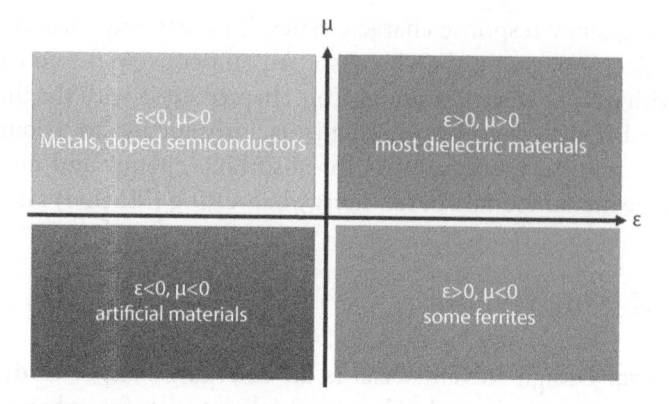

Figure 10.2 Classification of materials.

negative permittivity and permeability and no such material available in nature. The metamaterial is factually beyond natural material and is rationally engineered to make use of this remarkable property. A few ferrite materials with positive permittivity and negative permeability fall under region 4 though the magnetic response of ferrites rapidly fades away above microwave frequencies.

10.2.1 Energy Harvesting Using Metamaterials

Recently, electromagnetic metamaterials have received attention in many applications including energy absorption because of their unique properties such as negative index of reflection, cloaking, etc. It shows unique physical properties for sound, light, and elastic. Based on these, it is being utilized for EH applications. The utilization of metamaterials for absorbing solar energy, acoustic energy and radio frequency (RF) energy is discussed in the following section.

10.2.2 Solar Energy Harvesting

Solar energy is a kind of renewable that is available free of cost and also non-polluting in nature. Generally, the solar cell is used for converting incident radiation into electrical energy, i.e., based on the principle of absorption. The absorbed energy creates an electron-hole pair and the electrons are then converted into electrical current. Recently, thin-film solar cells are emerging in the market but the absorption properties might be decreased due to the thin layer of materials. In other ways, metamaterials could be utilized for absorbing the incident radiation, and with proper electrical circuits, it can be converted into electrical current. The numerical simulation of metamaterial should be carried out with proper boundary conditions for obtaining the optimal results.

10.2.2.1 Numerical Setup

The simulation will be carried out using the finite element modeling method. The unit cell of the metamaterial with periodic boundary conditions in the x-y direction will be used for the numerical analysis. The wave propagation will be perpendicular, i.e., along the z-direction. Usually, the structure has three layers, namely, a structured top metallic patch, a dielectric substrate, and a bottom metallic ground plane. The bottom ground plane will act as a perfect reflector so that the absorption characteristics can be improved further. The structure and the dielectric properties of the substrate will decide the resonant characteristics [9–11].

The designed metamaterial unit cell with proper boundary conditions for analysis is shown in Figure 10.3. The obtained corresponding absorption spectra in the solar regime are shown in Figure 10.4. It is observed that the structure has obtained near-perfect absorption in the visible region. In addition, some absorption in ultra-violet (UV) and infrared (IR) is also noticed. Further optimizing the structure, it is possible to harvest the thermal energy. i.e., IR region also. Many designs have been proposed for improving the absorption characteristics [12].

Similarly, a solar thermal absorber using metamaterial was proposed and experimentally demonstrated. The structure was metallic trenches, those able to provide selective absorption with higher selectivity. The Graphene material used in the structure showed better thermal conductivity characteristics also. The fabricated structure and the corresponding results in the region (0-1) μm are shown in Figure 10.5. The observation is, it has achieved the near perfect absorption in the designed band [13].

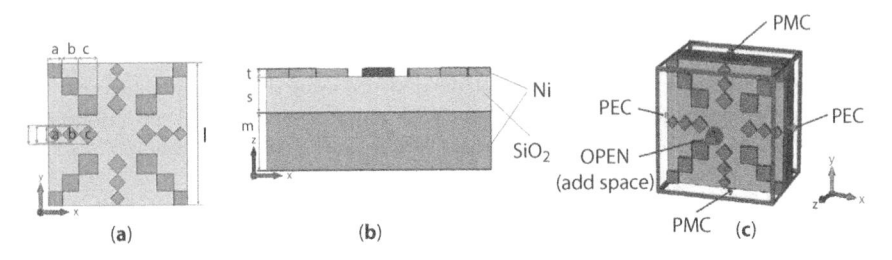

Figure 10.3 Designed metamaterial unit cell (a) top view (b) side view and (c) boundary conditions PEC-Perfect Electric Conductor, PMC – Perfect Magnetic Conductor. Reprinted with permission from [12].

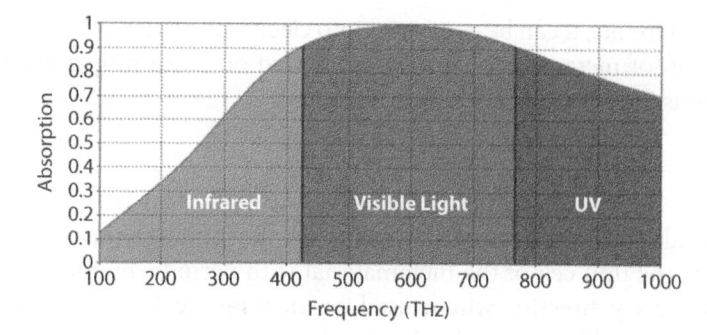

Figure 10.4 Obtained simulated absorption characteristics of the designed structure in the entire sun regime. Reprinted with permission from [12].

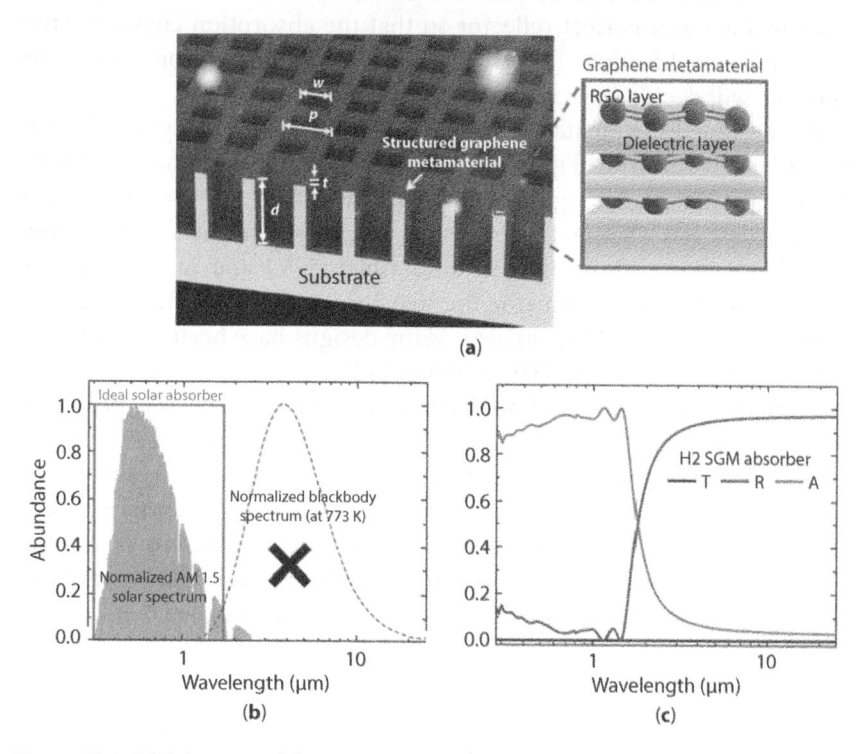

Figure 10.5 (a) Schematic of the metamaterial designed using structured graphene (b) Solar absorber result – Ideal case (c) Simulated transmission, reflection and absorption characteristics of the designed metamaterial. Reprinted with permission from [13].

An attempt to absorb the region from IR to UV was done and the corresponding absorption spectra is shown in Figure 10.6. It is not covering the entire spectrum of the sun but it is noticed that some absorption from IR to UV [10].

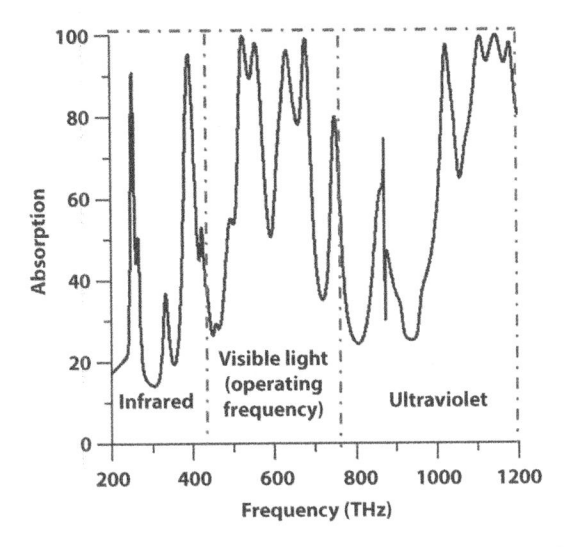

Figure 10.6 Obtained absorption characteristics for the proposed design from 200THz to 1200THz. Reprinted with permission from [10].

10.2.3 Acoustic Energy Harvesting

Like metamaterials, phononic crystals can also be utilized for absorbing vibroacoustic energy. In this, acoustic waves are being treated as elastic waves. Like photonic crystal, phononic crystal has a forbidden gap in the structure which does not allow the propagation of certain frequencies. This way, those frequencies can become localized energy and converted into useful electric energy [14]. The energy harvester based on phononic crystal with integrated piezoelectric energy harvester is shown in Figures 10.7 and 10.8. The corresponding simulated localization of wave energy is shown in Figure 10.7b.

The other way periodic arrangement metals (i.e., aluminum) in two-dimensional shall be treated as a metamaterial. Thus the metamaterial shows the possibility of using it as an acoustic energy harvester [15]. The fabricated structure and the corresponding frequency response is shown in Figure 10.9. In this structure, the defect in the arrangement is used as a source for localization, and the two resonances 35Hz, 63Hz are raising from them. These two examples show the use of metamaterial for acoustic energy harvesting.

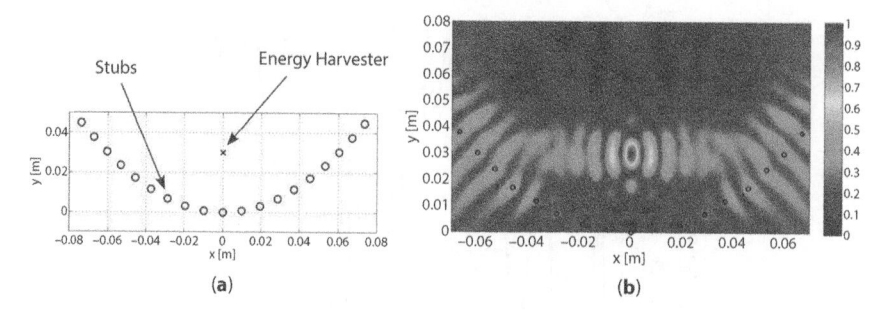

Figure 10.7 (a) Schematic representation of PAM configuration (b) Simulated RMS displacement PAM – Parabolic Acoustic Mirror, RMS – Root Mean Square. Reprinted with permission from [15].

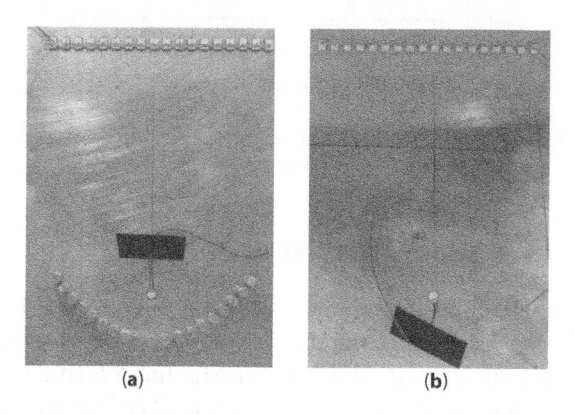

Figure 10.8 (a) Energy harvester in PAM (b) In the absence of PAM. Reprinted with permission from [15].

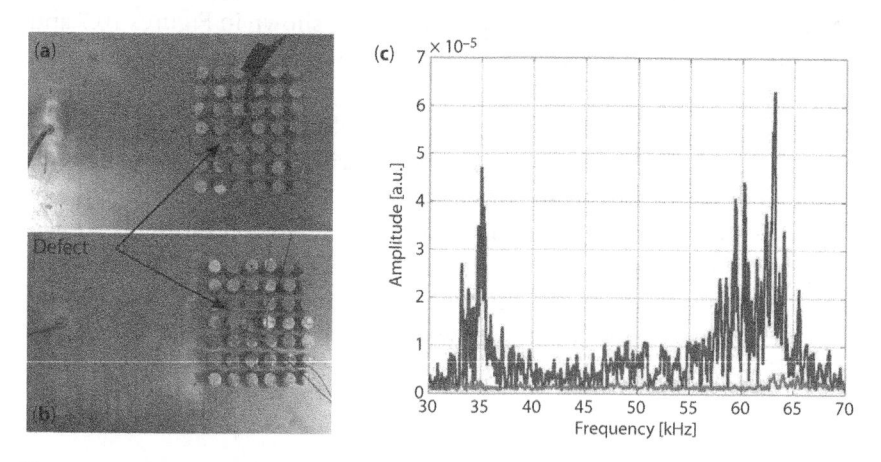

Figure 10.9 (a) 2D lattice structure made of Al stubs with defects circular form (b) two rectangular form harvester (c) frequency response for the circular, blue-due to defect, red-due to stub. Reprinted with permission from [15].

10.2.4 RF Energy Harvesting

The importance of metamaterial for RF energy harvesting is not a new method. Indeed, it has existed for the last two decades. In addition to antennae, metamaterials were incorporated to improve the absorption characteristics. The numerical evaluation setup remains the same as discussed in the solar energy harvesting section. The difference could be in the resonance which is due to the geometrical parameters [16–21]. A simple structure based on a modified square resonator for absorbing RF energy is shown in Figure 10.10. The harvester frequency vs. efficiency for an array of 9*9 structure for different polarization angles at normal incidence is shown in Figure 10.11. At the resonant frequency, it has achieved nearly 100% efficiency [22].

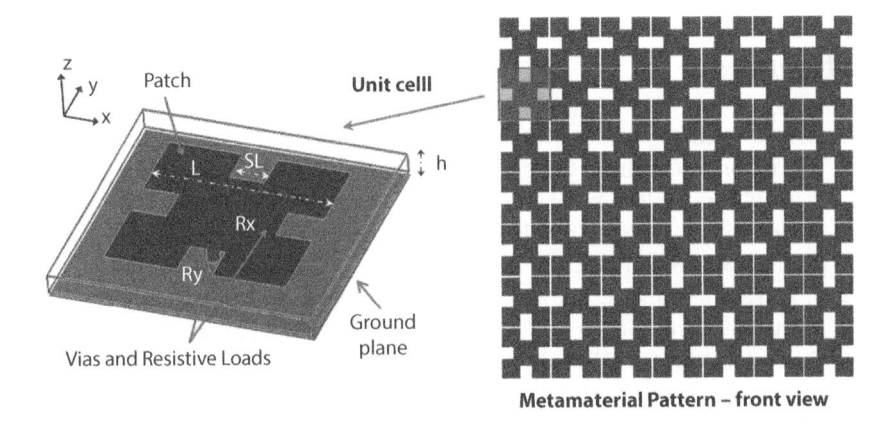

Figure 10.10 Designed metamaterial unit cell for energy harvesting. Reprinted with permission from [22].

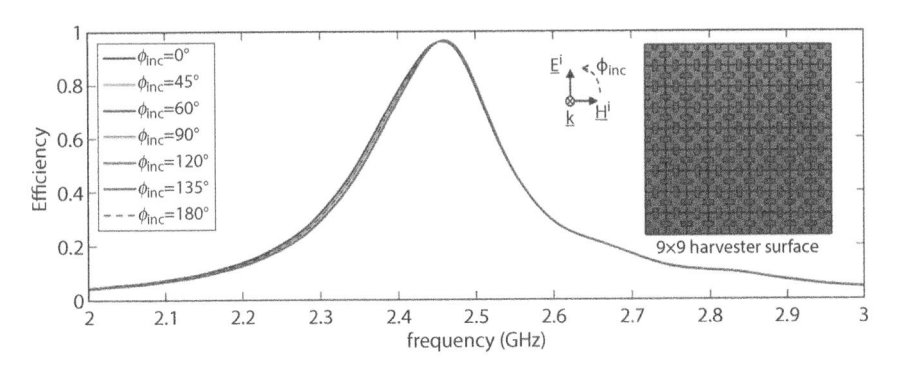

Figure 10.11 Harvester frequency vs. efficiency for an array of 9*9 structure for different polarization angles at normal incidence. Reprinted with permission from [22].

10.3 Summary and Challenges

Metamaterial-based absorbers can be utilized for absorbing any form of electromagnetic energy. Unlike other techniques, the simple and planar structure makes it easier in terms of fabrication and also integration with other devices. The major issue with these structures is the conversion of the incident radiation into electrical current. Suitable electrical circuits have to be designed and properly matched with the absorber structure. This imposes a challenge in the design aspect. Apart from this, the numerical evaluation should be carried out in some proper constraints [23]. A focus on these issues and proper designs will make it more efficient. Further, it is expected that the metamaterial-based absorbers will dominate the EH application in the near future.

References

1. Serhat Burmaoglu, Vladimir Trajkovik, Tatjana Loncar Tutukalo, Haydar Yalcin, Brian Caulfield, Chapter 14 - Evolution Map of Wearable Technology Patents for Healthcare Field, Editor(s): Raymond Kai-Yu Tong, *Wearable Technology in Medicine and Health Care*, Academic Press, 2018, pp. 275-290, ISBN 9780128118108,
2. J. A. Paradiso and T. Starner, "Energy scavenging for mobile and wireless electronics," in *IEEE Pervasive Computing*, vol. **4**, no. 1, pp. 18-27, Jan.-March 2005, doi: 10.1109/MPRV.2005.9.
3. Govindaraman, L. T., Arjunan, A., Baroutaji, A., Robinson, J. & Olabi, A.-G. Metamaterials for Energy Harvesting. *Ref. Modul. Mater. Sci. Mater. Eng.* **0–13** (2021) doi:10.1016/b978-0-12-815732-9.00127-3.
4. Manikandan, E., Padmalaya, G., Sreeja, B. S. & Radha, S. Solar power: More than what we see. *Water Energy Int.* **59RNI**, (2017).
5. Rawat, V. & Kale, S. N. Metamaterials for Energy-Harvesting Applications : A Review. *Nanotech Insights* **6**, 2–8 (2015).
6. Chen, Z., Guo, B., Yang, Y. & Cheng, C. Metamaterials-based enhanced energy harvesting: A review. *Phys. B Condens. Matter* **438**, 1–8 (2014).
7. Liu, Y. & Zhang, X. Metamaterials: A new frontier of science and technology. *Chem. Soc. Rev.* **40**, 2494–2507 (2011).
8. Sun, J. & Litchinitser, N. M. *Metamaterials. Fundamentals and Applications of Nanophotonics* (2016). doi:10.1016/B978-1-78242-464-2.00009-9.
9. Kumar, R. Perfect Selective Metamaterial Absorber With Thin-Film of GaAs Layer In The Visible Region for Solar Cell Applications. *Res. Sq.*

10. Hossain, M. J., Faruque, M. R. I. & Islam, M. T. Perfect metamaterial absorber with high fractional bandwidth for solar energy harvesting. *PLoS One* **13**, 1–15 (2018).
11. Bağmancı, M. et al. Broad-band polarization-independent metamaterial absorber for solar energy harvesting applications. *Phys. E Low-Dimensional Syst. Nanostructures* **90**, 1–6 (2017).
12. Bagmanci, M. et al. Solar energy harvesting with ultra-broadband metamaterial absorber. *Int. J. Mod. Phys. B* **33**, (2019).
13. Lin, K. Te, Lin, H., Yang, T. & Jia, B. Structured graphene metamaterial selective absorbers for high efficiency and omnidirectional solar thermal energy conversion. *Nat. Commun.* **11**, 1–10 (2020).
14. Qi, S., Oudich, M., Li, Y. & Assouar, B. Acoustic energy harvesting based on a planar acoustic metamaterial. *Appl. Phys. Lett.* **108**, 1–6 (2016).
15. Carrara, M. et al. Metamaterial-inspired structures and concepts for elasto-acoustic wave energy harvesting. *Smart Mater. Struct.* **22**, (2013).
16. Bakir, M., Karaaslan, M., Dincer, F., Delihacioglu, K. & Sabah, C. Perfect metamaterial absorber-based energy harvesting and sensor applications in the industrial, scientific, and medical band. *Opt. Eng.* **54**, 097102 (2015)
17. Abdulmjeed, A., Elwi, T. A. & Kurnaz, S. Metamaterial vivaldi printed circuit antenna based solar panel for self-powered wireless systems. *Prog. Electromagn. Res. M* **102**, 181–192 (2021).
18. Dincer, F. Electromagnetic energy harvesting application based on tunable perfect metamaterial absorber. *J. Electromagn. Waves Appl.* **29**, 2444–2453 (2015).
19. Elwi, T. A. Novel UWB printed metamaterial microstrip antenna based organic substrates for RF-energy harvesting applications. *AEU - Int. J. Electron. Commun.* **101**, 44–53 (2019).
20. Nowak, M. Metamaterial-based sub-microwave electromagnetic field energy harvesting system. *Energies* **14**, (2021).
21. Tan, T. et al. Renewable energy harvesting and absorbing via multi-scale metamaterial systems for Internet of things. *Appl. Energy* **254**, 113717 (2019).
22. Costanzo, S. & Venneri, F. Polarization-insensitive fractal metamaterial surface for energy harvesting in iot applications. *Electron.* **9**, 1–12 (2020)
23. Akarsu, G. et al. A Novel 5G Wideband Metamaterial Based Absorber for Microwave Energy Harvesting Applications. *2021 8th Int. Conf. Electr. Electron. Eng. ICEEE 2021* 309–312 (2021) doi:10.1109/ICEEE52452.2021.9415941.

11
Smart Robotic Arm

Rangit Ray[1], Koustav Das[1], Akash Adhikary[1], Akash Pandey[1], Ananthakrishnan V.[1] and O.V. Gnana Swathika[2]*

[1]*School of Electrical Engineering, Vellore Institute of Technology, Chennai, India*
[2]*Centre for Smart Grid Technologies, School of Electrical Engineering, Vellore Institute of Technology, Chennai, India*

Abstract

In response to the booming world of industrial automation, robotic arms have gained popularity in recent times. The primary focus of the project is to present a design and develop a robotic arm which could be used in industries for lifting and welding purposes. The designed robotic arm has four degrees of freedom which allows it to access objects in front of it in the polar co-ordinates. The developed robotic arm can be used to lift small-scale objects with its claws or may be fitted with a welding tip for welding purposes accordingly. A bot is also made to make the robotic arm movable. The four degrees of freedom has been achieved through the use of servo motors and Arduino, an open-sourced computer hardware and software that is being used to drive these motors. Additional sensors have been provided to detect the object in 3D space and also to detect human intervention for safety purposes.

Keywords: Robotic arm, degree of freedom, torque

Abbreviations and Nomenclature

DOF   Degree of Freedom
W1   weight of link 1
W2   weight of motor
W4   weight of motor

---

*Corresponding author: gnanaswathika.ov@vit.ac.in

W3   weight of link 2
W5   weight of gripper
W6   weight of load
L1   span of link 1
L2   span of link 2
L3   span of link 3
M   Torque
F   Force

11.1 Introduction

In the past few decades, global industrial automation has seen a positive growth. Global industrial giants are favoring automation because it helps to reduce human labor and thereby cost of production. Also, automation helps deliver quality product with enhanced consistency. For these reasons robots have become an integral part of the automation industry. A robot is an electro-mechanical device which is controlled with the help of software programs, using motors actuators and sensors. They are used to perform various operations like welding, picking up objects, colorings and serving various purposes in different work environments. Our main focus at Vellore Institute of Technology is to develop a prototype of joint arm configuration type with four degrees of freedom. These robots can be developed at minimal cost with high accuracy and superior control. Robotic arms that have four degrees of freedom can be used in industry to carry light materials. Moreover, to make the robot mobile we use a bot at its base.

11.1.1 Motivation

The usage of robotic arms rapidly increased in the 20th century and they are extensively used in several applications for the manufacturing sector. The arms are applied for various purposes where the process requires more accuracy, precision and perfection than a human being. The arms can be used not only for welding and painting auto parts but also for other utility-based processes. The arm design resembles the arm of a human but their application may depend on the mechanism defined by the designer of the robot. As an example, the designer may define the joint which connects the parts of a robotic arm such that it can move like a hinge or rotate. The arms are programmed to carry out a specific task or as an end effector as per the requirement of the field of its application.

11.1.2 Objectives

The aim of this project is to develop a robot which can be used for pick and place application. The robotic arm and the bot will be communicated with the help of Bluetooth protocol. There are several points to be considered for this particular project.

1. The robotic arm will have the feature of four Degrees of Freedom.
2. The arm should have the capability of pick up objects which have weight of 50 g.
3. The length of the arm will be approximately 33 cm.
4. The robotic arm shall be used in the applications for picking up and placing the object in the required place.
5. To place the object at the desired location a bot will be used which will carry the arm.

11.1.3 Scope of the Work

Robotic Arms have a very promising future in the years ahead. In the near future the arms will be able to perform every task that humans do with greater accuracy and efficiency. The number of accidents is also going to decrease as the moving of heavier objects will be carried out by the robots.

11.1.4 Organization

The robotic arm has 4 servo motors, each motor for on 1 Degrees of Freedom (DOF) and the motors operate at 5 volts. A robotic arm system quite often consists of actuators, joints, links, sensors, controllers, etc. In sensors we are using ultrasonic sensors with a range of 21 cm and a Bluetooth module to control the bot.

The bot consists of a Bluetooth module, DC motor, Arduino mega and L293D Motor Driver. Table 11.1 discusses the design specifications.

Table 11.1 Design specifications.

Components	Specifications
1. Arduino ATmega2560	Operating voltage-5V Digital I/O Pins 54 (14 of them are for PWM output)
2. Servo Motor	Voltage-4.8V-6V
3. Ultrasonc Sensors	Supply voltage-5V Global current consumption-15mA Range-21cm
4. DC Motor	Operating voltage-3V RPM-2400
5. Bluetooth Module	Power supply: +3.3V DC 50mA Frequency: 2.4GHz ISM band

11.2 Design of Robotic Arm with a Bot

11.2.1 Design Approach

The proposed robotic arm has been designed based on functions very similar to that of a human hand as shown in Figure 11.1 through Figure 11.6. An open kinematic chain has been formed with several links which has been connected with the help of a number of links. The sequence starts from the base of the robot and ends at the gripper to hold the object. The arms are made movable with the help of servo motors. The whole robotic arm is sited on a Bot which will be controlled by a Bluetooth module.

11.2.1.1 Codes and Standards

Arduino Code for Servo motor and Ultrasonic Sensor (in the Robotic Arm).

11.2.1.2 Realistic Constraints

The Robotic Arm will consist of Servo Motors, Ultrasonic Sensors, DC Motors and Bluetooth Module to control the bot.

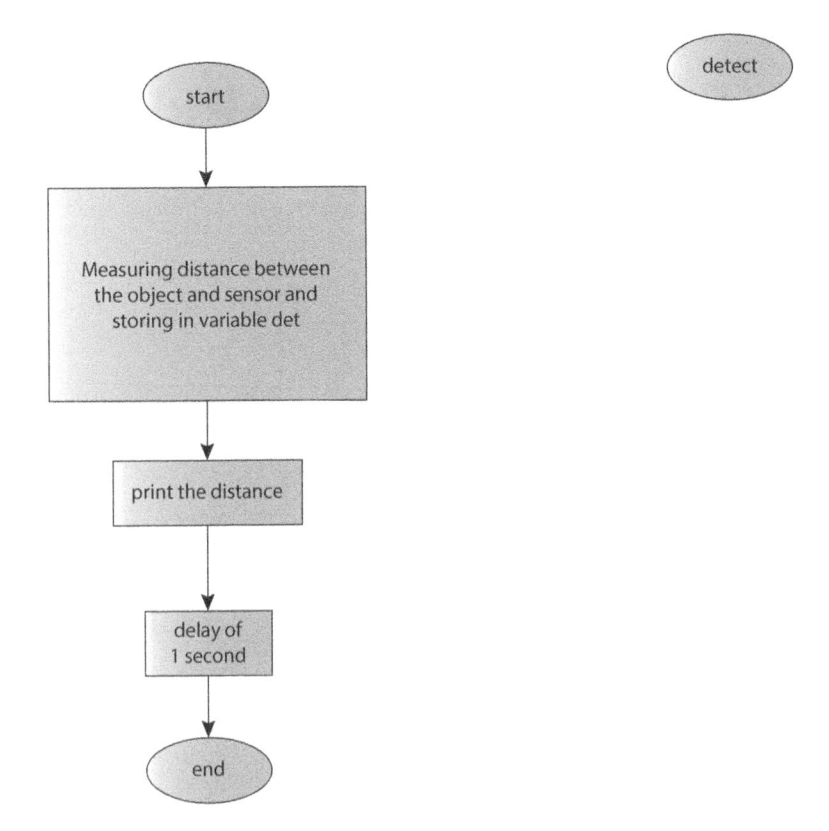

Figure 11.1 Flowchart to detect the object.

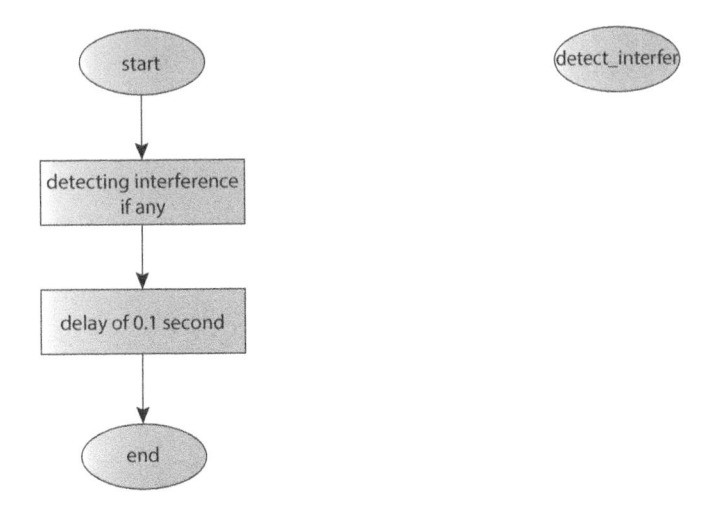

Figure 11.2 Flowchart to detect human interference.

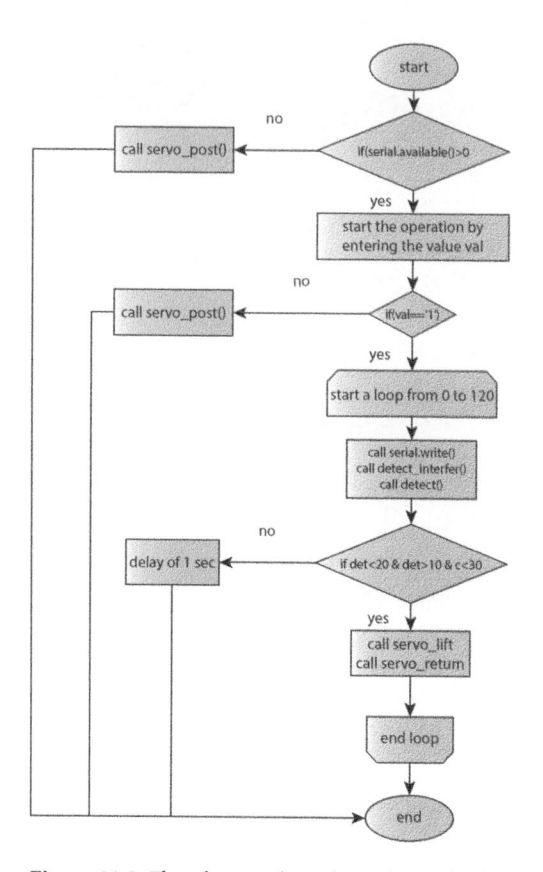

Figure 11.3 Flowchart to show the main method.

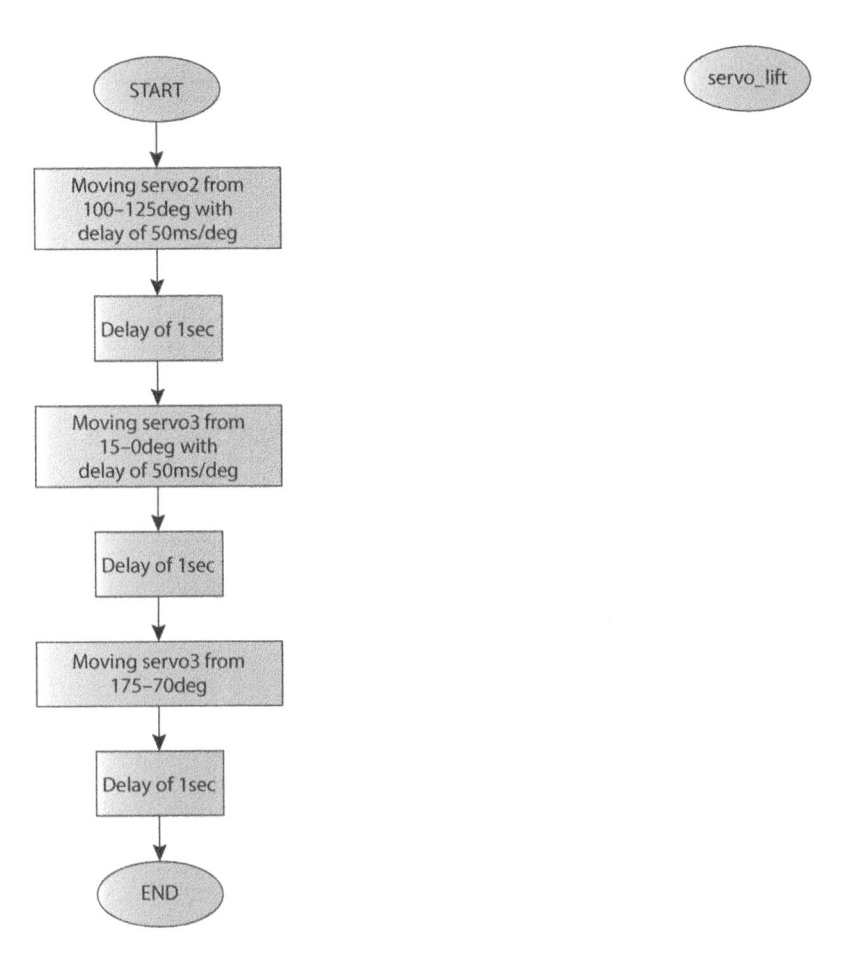

```
            servo_lift

   START
     │
     ▼
┌─────────────────┐
│ Moving servo2 from│
│ 100–125deg with  │
│ delay of 50ms/deg│
└─────────────────┘
     │
     ▼
┌─────────────────┐
│  Delay of 1sec  │
└─────────────────┘
     │
     ▼
┌─────────────────┐
│ Moving servo3 from│
│  15–0deg with    │
│ delay of 50ms/deg│
└─────────────────┘
     │
     ▼
┌─────────────────┐
│  Delay of 1sec  │
└─────────────────┘
     │
     ▼
┌─────────────────┐
│ Moving servo3 from│
│   175–70deg      │
└─────────────────┘
     │
     ▼
┌─────────────────┐
│  Delay of 1sec  │
└─────────────────┘
     │
     ▼
    END
```

Figure 11.4 Flowchart to make the servo lift the object.

```
         start                              servo_setpost()
           │
           ▼
  ┌─────────────────┐
  │ Bringing servo1 │
  │ to initial position │
  └─────────────────┘
           │
           ▼
  ┌─────────────────┐
  │  Delay of 1sec  │
  └─────────────────┘
           │
           ▼
  ┌─────────────────┐
  │ Bringing servo2 │
  │ to initial position │
  └─────────────────┘
           │
           ▼
  ┌─────────────────┐
  │  Delay of 1sec  │
  └─────────────────┘
           │
           ▼
  ┌─────────────────┐
  │ Bringing servo3 │
  │ to initial position │
  └─────────────────┘
           │
           ▼
  ┌─────────────────┐
  │  Delay of 1sec  │
  └─────────────────┘
           │
           ▼
  ┌─────────────────┐
  │ Bringing servo4 │
  │ to initial position │
  └─────────────────┘
           │
           ▼
  ┌─────────────────┐
  │  Delay of 1sec  │
  └─────────────────┘
           │
           ▼
          end
```

Figure 11.5 Initialise the parameters.

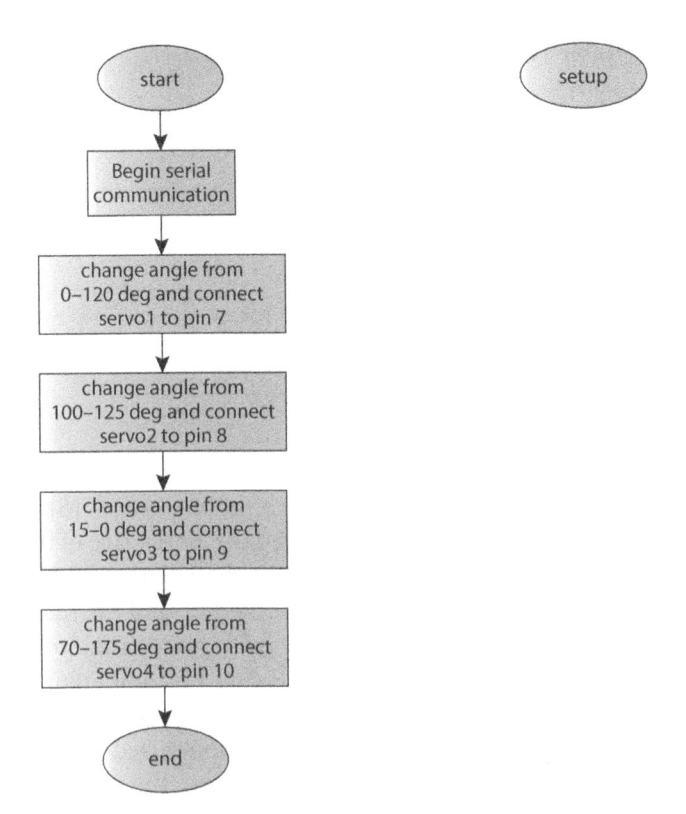

Figure 11.6 Initialise the serial.

11.2.2 Design Specifications

The suggested robotic arm has been designed such that it has functions very similar to a human hand [1, 2-4]. An open kinematic chain has been formed with several links which has been connected with the help of a number of links. The sequence starts from the base of the robot and ends at the gripper to hold the object. This design has a very similar structure to that of a human hand, which performs tasks for interaction with the real world [5, 6]. Generally there are two types of joint, prismatic and rotary joints, and it connects neighboring link. In Table 11.2 the degrees of freedom of the robotic arm has been shown [7]. A Robotic arm with four degrees of freedom as shown in Table 11.2 has been designed as it is adequate for all the movements in the required axis, and moreover it is cost effective, which is also a main factor. With the increase in the number of degrees of freedom, the number of actuators will also increase, which will

Table 11.2 Degrees of freedom of motors with angles.

Degrees of freedom (DOF)	Degrees
1. Base	(0-120)
2. Shoulder	(100-125)
3. Elbow	(0-15
4. Gripper	(70-175)

again increase the cost. The number of independent joint variables determines the number of degrees of freedom in a robotic system [8].

The area that the gripper can cover is called the workspace of the robotic arm. It depends on several specifications like the length of the links, the DOF, the angle at which the object will be picked up and the main configuration of the robot. Figure 11.8 shows the work place of the robotic arm with four degree of freedom (4 DOF) as in Figure 11.7 and Figure 11.8 [9].

The arm is fixed with the base with a support. Popsicle sticks are used as the main frame of the arm. A servo motor is connected at each joint so that the arm can move itself to pick up the object. The claw is designed such that it can be extended maximum so that the object can be picked up easily. A thread is connected with the claw to control it properly. The other end of the thread is connected with a servo motor. By rotating the servo, the claw is operated.

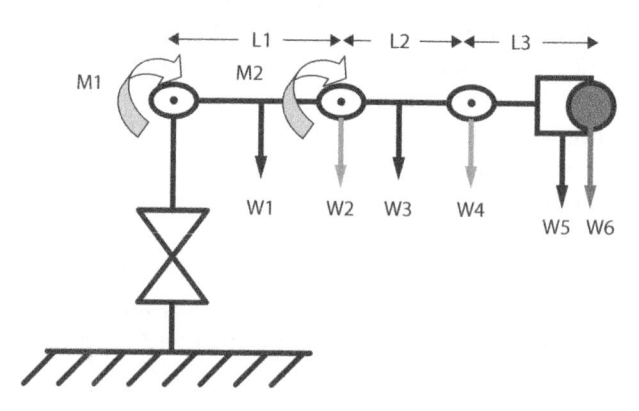

Figure 11.7 Free body diagram of the robotic arm [9].

```
         B                    C
    ┌─────────────────────────┐
    │         Link 1          │
    └─────────────────────────┘
         ↑         ↓         ↓
        BY        W1        CY
```

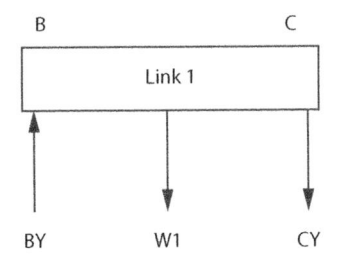

Figure 11.8 Work region of the robotic arm [9].

It is very important to choose the suitable motor. Based on torque and speed calculations of the motor, the design has been prepared. The exact force calculation is very important as it is a deciding factor for the chosen motor and to support the weight of the arm and the load [6]. Servo motors are generally used when the application requires a high amount of torque and a very precise speed control.

Torque is the tendency of a twisting force that causes rotation of an object about a particular axis. Mathematically it is given as the cross product of lever arm distance and force, which produces rotation.

$$T = F \times L \text{ Nm}$$

Where, F = Force applied on the motor
L = length of the shaft
Force, F is defined as,

$$F = m \times g \text{ N}$$

Where, m = mass of the load g = gravitational constant = 9.8 m/s
View of loads/moments on joints (right) and force diagram of link 1(left [1]).

The specifications used for the calculations of torque:

$W1 = 2.98$ gm
$W2 = W4 = 9$ gm
$W3 = 2.98$ gm
$W5 = 12.5$ gm
$W6 = 50$g
$L1 = 11$ cm

$L2 = 11$ cm
$L3 = 18.5$ cm

Calculating the sum of forces in Y-axis,

$$\Sigma Fy = (W6 + W5 + W4 + W3 + W2)\,g - CY = 0 \quad (11.1)$$

$$CY = 0.818\text{ Nm} \quad (11.2)$$

$$\Sigma Fy = (W6 + W5 + W4 + W3 + W2 + W1)\,g - BY = 0 \quad (11.3)$$

$$BY = 0.847\text{ Nm} \quad (11.4)$$

$$\Sigma Mc = -(W3L2/2) - WW5(L2 + L3/2) - W6\,(L2 + L3) - W4\,(L2) + Mc = 0 \quad (11.5)$$

$$\Sigma Mb = -W6\,(L1 + L2 + L3) - W5(L1 + L2 + L3/2) - W4(L1 + L2) - W3(L1 + L2/2) - W2(L1) - W1(L1/2) + Mb = 0 \quad (11.6)$$

$$Mc = 0.0138\text{ Nm} \quad (11.7)$$

$$Mb = 0.02778\text{ Nm} \quad (11.8)$$

As per the moment calculated proper servo motor has been selected, is, which has a torque of 2.5 kg/cm, i.e., 0.245 Nm to comply with the torque requirement of the robotic arm.

They can turn 0-180 o with a speed of 0.1s/60 o and weight of 9 gms each.

11.3 Project Demonstration

11.3.1 Introduction

The proposed robotic arm has been designed with functional properties very similar to as of a human hand which performs tasks to interact with the real world. A Robotic arm with four degrees of freedom has been designed as it satisfies the purpose of all required movements and also it is cost effective, which is also an important aspect. With the increase in the number of degrees of freedom, the number of actuators will also increase, which will again increase the cost.

11.3.2 Analytical Results

Under the reference of many survey papers of many other researchers on robotic arm we have been able to understand the parameters of performance and how it can be improved for better efficiency. Whatever the type of robotic arm it may be, it is essential to make sure its efficiency is high, otherwise the practical applications of the specified robotic arm cannot be industrially implemented. The type of robotic arm to be implemented and its modeling are mainly application specific; that is, a similar type of robotic arm can have slight modeling differences based on application of the arm. A robotic arm is used for wielding, gripping and spinning. The robotic arm's most basic operation is pick and place. While a human worker does the pick and place operation, they are reliable but they do not have mechanical timing. But in case of robotic arm we do not have to worry about that. Robotic arms provide consistent and measured quality. The robotic arm completes the same units of tasks in the same period of time uniformly without any significant time differences. Some activists say that with the emergence of robotic arm, human labor will decrease and unemployment will increase. It can be said that this is partially true since the robotic arm does a particular task with more durability, which in the case of a human might be a dangerous job. Thus human labor is pushed to a higher level of control, where humans will control the robotic arms for better success rate and the technological upgrading of the industry.

11.3.3 Simulation Results

The object that is to be picked is kept at a fixed distance from the bot anywhere on the periphery of the region of operation of the arms. The base servo motor is then programmed to rotate from its initial position that is 0 degree to 180 degrees slowly enabling the ultrasonic sensor to detect the presence of any object on its way. On detecting the object the bot will check for any human interference by checking the input from a second ultrasonic sensor. If it finds any interference the bot will stop working automatically or it will proceed for the next set of instructions. The bot will slowly bend down to a defined level by two mounted servo motors on the Arm and the fourth one will be used to open or close the grip of the Arm on the object. After picking up the object the Arm will come back to its initial position,

Release the object gently, Return back to the position where it found the object and resume its scanning process.

11.3.4 Hardware Results

In our project, the prototype model that we made as in Figure 11.9 has the following notable hardware results:

- The base ultrasonic can detect any object within a range of 21 cm from its mouth.
- The used can rest the position of the arm at any time.

In our project we modelled a simple pick and place robotic arm which will be controlled via Arduino and then we discussed and calculated the specifications of the arm so that the efficiency of the robotic arm increases. Lastly, we are trying to improve on the arm by working on effective algorithms and simulations and making the necessary modifications in the robotic arm.

Figure 11.9 Prototype of the robotic arm with the bot.

11.4 Conclusion

11.4.1 Cost Analysis

- The Arduino Mega which has a model number ATmega2560 is a microcontroller-based board.
- A servomotor is a rotating device that provides control for angular motion with great accuracy and precision. In the proposed model the servomotor has an Operating Voltage is +5V and Rotation ranges from 0° to 180°. The motor is designed such that it is coupled to a sensor for position feedback. MATLAB controlled Robotics Arm includes four such Servo motors at four different hinge points which allow the Arm to rotate and pick Objects.
- Ultrasonic sensor is an electronic device enabling the distance of a target object to be measured.
- The sensor emits an ultrasonic sound wave and with the help of a transducer it converts the reflected sound from the target object into electrical signal. The Sensor measures the distance to the target object by calculating the time between the emission and reception. The Robotic Arm include two ultrasonic sensors, one for detecting the position of the object and the other for detecting any interference during working of the model.

11.4.2 Scope of Work

The product developed can be further equipped with a camera which can effectively help in distinguishing between the objects to be detected and others that needs to be avoided. An image processing tool can be used on this occasion.

11.4.3 Summary

A robotic arm is used for wielding, gripping and spinning. The robotic arm's most basic operation is pick and place. While a human worker does the pick and place operation, they are reliable but they do not have mechanical timing. But in the case of a robotic arm we do not have to worry about that. Robotic arms provide consistent and measured quality. The robotic arm completes the same units of tasks in the same period of time uniformly without any significant time differences.

References

1. C. Carignan, G. Gefke, and B. Roberts, "Intro to Space Mission Design: Space Robotics," in *Seminar of Space Robotics*, University of Maryland, Baltimore, 2002.
2. R. Boyce and J. Mull, *Complying with the Occupational Safety and Health Administration: Guidelines for the Dental Office*, Dental Clinics of North America, vol. 52, pp. 653-668, 2008.
3. B. Siciliano, L. Sciavicco, L. Villani, and G. Oriolo, *Robotics: Modelling, Planning and Control*, Springer Science & Business Media, 2009.
4. M. A. Rahman, A. H. Khan, T. Ahmed, and M. M. Sajjad, "Design, Analysis and Implementation of a Robotic Arm-The Animator."
5. S. Yu, X. Yu, B. Shirinzadeh, and Z. Man, "Continuous finite-time control for robotic manipulators with terminal sliding mode," *Automatica*, vol. 41, pp. 1957-1964, 2005.
6. P. Singh, A. Kumar, and M. Vashisth, "Design of a Robotic Arm with Gripper & End Effector for Spot Welding," *Universal Journal of Mechanical Engineering*, vol. 1, pp. 92-97, 2013.
7. A.N.W.QI, K.L. Voon, M.A. Ismail, N. Mustafa, M.H. Ismail, "Design and Development of a Mechanism of Robotic Arm for Lifting Part1", *2nd Integrated Design Project Conference (IDPC) 2015*.
8. M. W. Spong, Vidyasagar, *Robot Dynamics and Control*, 1989.
9. A. Elfasakhany, E. Yanez, K. Baylon, and R. Salgado, "Design and development of a competitive low-cost robot arm with four degrees of freedom," *Modern Mechanical Engineering*, vol. 1, p. 47, 2011.

12

Energy Technologies and Pricing Policies: Case Study

Shanmugha S. and Milind Shrinivas Dangate

Chemistry Division, School of Advance Sciences, Vellore Institute of Technology, Chennai, India

Abstract

In this context we explain the irrigation technologies, energy inputs and conservation policies to irrigated agricultural production. We utilized empirical and simulation modelling to understand the impact of non-linear energy pricing on groundwater use decisions in the Republican River Basin in Colorado, USA. My hope is that this context will contribute to the overall understanding of the relationship between energy and economic growth. The importance of energy-related economic concerns is not likely to diminish in the near future. During the research process I have come to the realization that this work marks the first step in a career-long journey. Rather than ending the inquiry, each question answered has led to multiple others.

Keywords: Energy inputs, irrigation technology, energy pricing, energy management, economic growth

12.1 Introduction

Inputs from common pool nature of groundwater precipitates with inefficient rates of extraction that may exceed recharge, leading to posing the growing scarcity of aquifer resources in many areas. The objective of this chapter is to understand how non-linear input price schedules potentially

---

*Corresponding author: mili_ncl@yahoo.com;
Shanmugha S.: https://orcid.org/0000-0001-8555-7542
Milind Shrinivas Dangate: https://orcid.org/0000-0002-0210-5768

exacerbate common pool resource problems using groundwater extraction for irrigation as a theoretical and empirical illustration.

Groundwater utilization requires significant amounts of energy and prices often constitute the only pricing mechanism guiding groundwater use. Many energy providers employ non-linear pricing strategies to meet cost recovery and revenue smoothing objectives. However, past literature examining the relationship between energy price and water use considers only constant marginal pricing regimes and relatively little is known about the impact of non-linear pricing. This is a crucial difference as characterizing and estimating demand differs from cases at constant marginal pricing. We fill the gap in literature by understanding in which way non-linear energy pricing affects groundwater extraction decisions.

Specifically, we empirically estimate demand responsiveness for the case in which energy pricing follows a decreasing block rate (DBR) schedule. Our empirical analysis exploits a unique dataset which pairs spatially explicit groundwater demand with evolving energy price structures. We use empirical results to simulate the implications of depletion. Simulation results demonstrate that 4-7.5% of groundwater use can be attributed to the incentives created by DBR energy pricing. These are economically significant results as they demonstrate how energy pricing potentially exacerbates the common pool resource challenges of shared aquifers while also elucidating a more general relationship between priced inputs and unpriced environmental goods and resources. The relationship between energy and groundwater extraction is an example of a broader economic problem wherein priced inputs are complementary [1] to unprized resources in the production of a good or service. The characteristics of many environmental goods and resources preclude or complicate trade within a market, implying that these goods. Yet many environmental resources, from air quality to biodiversity, serve as vital inputs to Production. Missing pricing mechanisms imply salient the price of complementary and substitute inputs, and influence how firms and households use environmental and resource inputs. In this context, input pricing decisions generate effects that reverberate throughout vulnerable ecosystems and scarce natural resource stocks. We analysed these effects in River Basin, Colorado, USA, a sub-basin of the High Plains Aquifer (hereafter referred to as HPA), which made us understand how energy pricing regimes potentially exacerbate the challenges associated with common pool resource management. This research builds on three distinct strands of literature. First, we contribute to the broader non-linear pricing literature by examining the distributional impacts of heterogeneous rate structures. Second, we advance the water demand research by exploring the common pool resource implications of agricultural water use subject to DBR pricing. Finally, this paper enriches the literature analysing priced inputs and unpriced environmental goods and resources by questioning how complementarity in production affects resource outcomes. The paper proceeds as follows: we provide background on HPA and Republican River Basin of Colorado a cursory survey of the literature [2]. We describe a simple theoretical model of water demand under DBR energy pricing given profit maximization motives while the exploring the distributional impacts of a transition to constant energy pricing. We describe an empirical model which estimates farming water demand. We describe data sources, present empirical modelling results, and develop and present results from a counterfactual simulation model examining the impacts of a shift to constant energy pricing. Finally, we provide a conclusion summarizing the paper's findings and explore implications for water conservation policymaking.

12.2 Literature Review

In the United States of America, the largest aquifer at present is HPA, which provides Figure 12.1a, the Republican River Basin (dark blue), and the study area whose data are employed in later empirical modelling (indicated by red frame). The future of the HPA, and the rural agricultural economies which it supports, is uncertain as rates of extraction exceed recharge in many regions of the aquifer [1]. This context is about the prediction of HPA which reaches inducing a shift to less productive dryland agriculture and diminishing land values across the region. Groundwater extraction in the HPA depends on access to energy. Electricity is the most important source of energy for irrigated agricultural production in the HPA states. Over 70 Rural Electric Cooperatives (RECs) provide electricity to residential, commercial, and irrigation customers throughout the HPA region. Groundwater pumping comprises an important part of REC operations, representing distributed. While this energy access has facilitated rural and agricultural development, it has also contributed to the depletion of the region's groundwater resources [2].

The utilization of DBR energy pricing structures, which over half the RECS of the HPA utilize for irrigation customers, relates to the economies of scale inherent in electricity distribution as well as cost recovery objectives. The cooperative nature of REC ownership customers intermittently, as such, cost-recovery is a primary objective guiding REC pricing. Figure 12.1b presents the spatial distribution of REC pricing regimes in the HPA and demonstrates the prevailing use of DBR energy pricing in the region.

(a) High Plains Aquifer **(b)** REC pricing

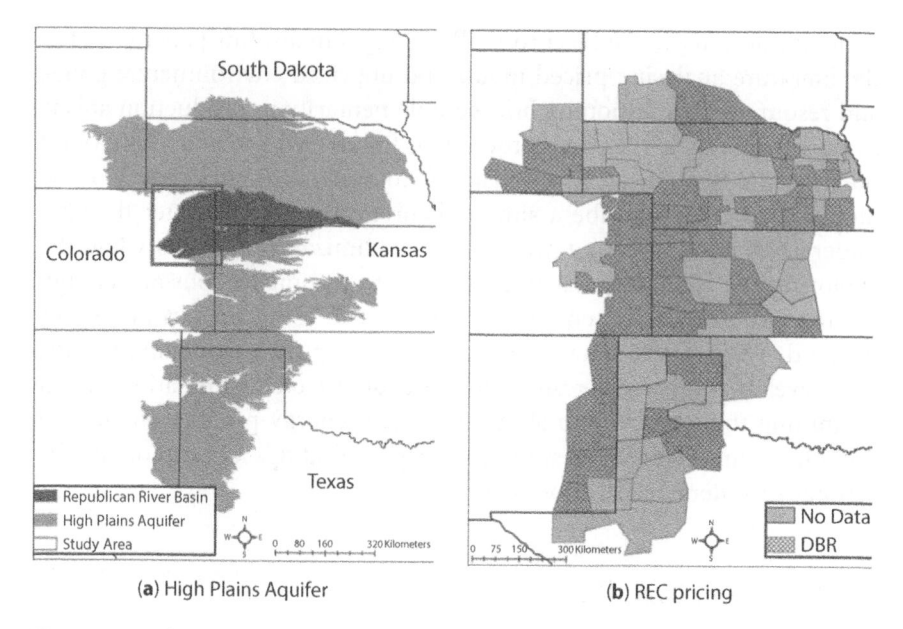

Figure 12.1 Electricity in the HPA 2.

The Basin embodies many of the characteristics which define agriculture throughout the HPA. As in many regions of the HPA, the rural economy of the Basin relies on irrigated agriculture supported almost exclusively by groundwater resources [3] powered with electricity provided by RECs [4]. Similar to the rest of the HPA, irrigation customers in the Basin are an important part of REC operations [5].

Electricity pricing in the Basin follows regional trends in that RECs utilize DBR pricing for irrigation customers wherein the thresholds of the price schedule are a function of well pump characteristics, namely horsepower (HP). Specifically, RECs in the Basin utilize thresholds defined in terms of HP [6]. As such, well pumps with more HP must utilize more energy for moving to lower the price of consumption. Figure 12.2 depicts one of the Basin's REC's (Y-W) rate structure in 2016. Well pumps constitute a significant long-term capital investment and once installed, well pump HP remains fixed. Therefore, while HP is a choice that producers make, it is a long-run decision that remains fixed over shorter time horizons. In the data used in our analysis, less than 6% of wells altered their HP between 2009 and 2017. As such, we treat well pump HP as predetermined and thus exogenous to annual water demand variation. We evaluate the robustness of our results to this assumption where we limit our analysis

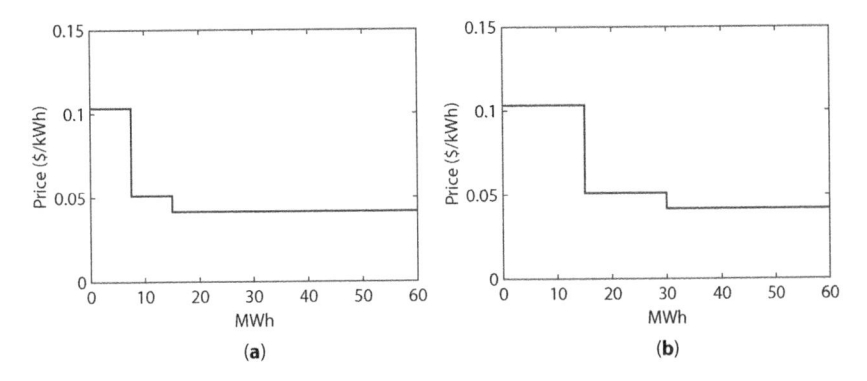

Figure 12.2 Irrigation rate system for 50 HP (a) and 100 HP (b) well pumps.

those their HP. RECs but demand. Therefore, communicate namely where along the rate structure the previous month's demand was located. This in real-time at electricity tallies demand. The RECs structure to beginning research builds upon the methodological contributions and results of the non-linear pricing literature, research on agricultural water demand, and a broader literature on the relationship between priced inputs and unpriced resources.

12.3 Non-Linear Pricing

There exists an extensive literature exploring the effect of non-linear pricing in a diverse array of applications from labour supply to household electricity demand. The methodological crux of this literature lies in appropriately addressing the endogeneity that exists between price and consumer or firm decision-making. Past literature employs structural and reduced-form empirical approaches to address potential endogeneity. We contribute to this literature by examining the distributional implications of price structures.

Structural approaches in the literature employ likelihood-based methods to model demand under non-linear pricing structures. The genesis of these structural econometric methods lies at the intersection of labour supply and consumer demand literatures. Leverages the modelling insights employed in the lab or supply literature to create a framework for consumer demand, commonly referred to as the discrete/continuous choice (DCC) model, which assumes that consumers jointly respond to marginal prices and thresholds when choosing demand.

The reduced-form literature instruments for price with rate structure parameters (e.g., thresholds, price levels) relying on the identifying assumption that rate structure parameters are exogenous to individual demand. Present an alternative, three stages least squares approach which uses predicted demand quantities as an instrument for price. More recently, investigates household electricity demand under varying non-linear price structures and tests whether households respond to marginal or average price. Results reveal that households respond to average rather than marginal price, demonstrating how information costs determine the salient price signal determining demand. This result is important as it calls into question the DCC model's assumption that consumers are fully informed of their position within the price schedule.

We contribute to this non-linear pricing literature by analysing the impact of rate structures on the distribution of resource use and welfare. Past literature examines heterogeneous rate structures and potential endogeneity in utility rate structure choice. We build on these insights by leveraging within REC variation in rate structure based on well HP to evaluate the impact rate structure heterogeneity.

12.4 Agricultural Water Demand

This paper also advances the literature that examines how agricultural water use responds to price signals. A large swath of this literature aims to measure the price elasticity of demand for irrigation water. Broadly, this literature finds that demand for irrigation water is relatively inelastic, suggesting the high value of on-farm water use and lack of available substitutes. We build on this literature by exploring the agricultural water demand under DBR pricing. Given that many scarce water resources remain unpriced, researchers analyse the impact of other price signals (e.g., energy inputs) to understand how producers respond to changing water prices. Model water demand as a function of extensive margin choices and estimate the price responsiveness of agricultural water use via heterogeneous pumping costs. Similarly evaluate how groundwater users respond to varying energy prices but utilize a sample selection model to incorporate cropping decisions. However, both limit their analysis to constant marginal pricing regimes despite the pervasive use of non-linear energy pricing in the HPA (see Figure 12.1b). Other literature examines the impact of energy subsidies on groundwater use in developing country settings and further demonstrates the relationship between energy price and groundwater depletion.

Explore how measurement error in imputed irrigation costs may bias empirical estimates of the price elasticity of agricultural groundwater demand. Their results are important as much of the groundwater demand literature relies on assumptions of uniformity with respect to irrigation cost parameters and the spatial interpolation of key hydrologic characteristics (e.g., depth to water). This analysis surmounts the measurement error issues presented by utilizing novel data from well tests measuring the water, rather than imputed irrigation costs.

A relatively smaller literature evaluates the impact of non-linear energy pricing on agricultural water use. Notably, D6TG en55velop a linear programming model to analyse the use of DBR electricity pricing for irrigation customers in the same region studied in this paper. Their results highlight the tension between REC revenue smoothing objectives and groundwater conservation. In a related paper, they analyse the use of increasing block rate (IBR) water pricing on conservation outcomes using the DCC modelling framework. Simulation results demonstrate that IBR water pricing generates groundwater conservation compared to the counterfactual scenario of constant water pricing. However, their analysis relies on the assumption that water users have perfect information regarding their position within the price schedule, an assumption questioned by results presented. This paper builds on the agricultural water demand literature by empirically evaluating price responsiveness under a DBR structure, which to our knowledge has not been addressed in the literature.

12.5 Priced Inputs and Unpriced Resources

Finally, this paper builds on a literature exploring the connection between priced inputs and unpriced environmental goods and resources. This paper contributes to this literature by analysing the case where priced inputs are complementary to unpriced resources. Previous literature examinees how inputs can serve as substitutes for resources as stocks become depleted. Explore implications of substitutability or complementarity between natural resources and capital/labour within industries which rely on renewable and non-renewable resources. Empirically treat the substitutability between natural and human capital in fisheries and conclude that increased levels of labour productivity counteract the effects of shrinking resource stocks. In the context of water and agriculture, explore how other inputs can serve as substitutes for water and ease the constraints of water scarcity.

A related literature focuses on the case of complementarity between priced inputs and resources and demonstrates how the price of complementary goods can increase adoption rates for resource conserving technologies and promote the provision of ecosystem services. Within the context of water and energy, it recognizes how energy pricing influences groundwater use and calls for pricing regimes to confront resource depletion and aid adaptations to climate change. It uses modelling and experimental methods to examine how changing energy subsidy policies impact groundwater depletion. A similar paper further explores this connection by analysing how a shift to volumetric, or marginal, electricity pricing can address India's pervasive groundwater depletion problems.

This literature begins to outline how the connection between priced inputs and unpriced re-sources applies within the context of water and energy. This paper furthers that understanding by exploring the role of input price structure in groundwater use decisions while cultivating greater knowledge of the factors which aid or hinder resource and environmental sustainability.

12.6 Proposed Set Up on Paper

This set up is of water demand under a non-linear input pricing regime. This model is then utilized to generate hypotheses regarding the impact of a shift to a constant input pricing regime on the distribution of water use and welfare across agricultural producers.

Suppose an agricultural produce a water extract which inputs required for extraction. Assume that output is sold at a constant price ρ, inputs (Q); Q = Aw production as it relates to water conditional on exogenous characteristics (e.g., resource constraints, soil type, weather, etc.) which are represented by the time variant parameter Z. Let the demand [7] price of water and w is water quantity. Note that extensive margin choices are not contained in Z, thus neither D(p; Z) or Γ(w; Z) are conditional on crop choices. Rather, cropping choices are implicit in demand and potentially vary with water price changes. The profit maximizing agricultural producer chooses optimal water demand such that the marginal benefit of water equals marginal price [8] decisions. Figure 12.3 is of water users whose demands differ according to exogenous characteristics defined by Z_1 and Z_2. Since time-variant Z captures the effect of weather, demand also varies across time and a given agricultural producer's optimal demand may be located on the highest or lowest marginal price block across time [8]. Now suppose that the energy provider shifts their pricing regime from P (w) to a constant marginal price, pc. The exact price defining the new constant marginal pricing regime depends on both the cost structure and

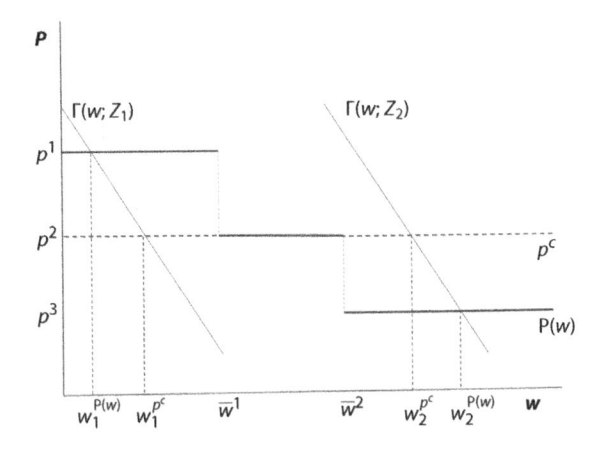

Figure 12.3 Water requirement theoretical calculation.

objectives of the energy provider as well as the price elasticity of water demand. If the energy provider aims to generate the same revenue with the new constant price and initial demand is distributed across all price blocks, then the constant price must fall between p1 and p3. For clarity we abstract away from the energy provider's choice of a revenue-neutral constant price and assume this price equals p2.

This new changes in for distribution demand across. Constant the lower water-using agricultural producers, Γ(w;Z1), increase water demand from w P(w) 1 to w p c 1 price determining demand. Similarly, higher water-using agricultural producers, Γ(w; Z2), decrease water demand from w2P(w) to w2pc as the price signal influencing demand on the margin increases from p3 to pc.

A transition to constant energy pricing also generates changes in the short- and long-run welfare of agricultural, constant [9]. While constant pricing in time remains dynamics through. Namely, costs and benefits accrue to producers in future time periods when constant energy and water pricing induces changes in resource availability. The theoretical and empirical chapter constant Figure 12.4 characterizes the short-run welfare implications of constant pricing for higher water using agricultural producers with inverse demand, Γ(w; Z2). For example, in Figure 12.4 the welfare losses demand $\int\_w2pc^{\wedge}(w2p(w))[\tau(w;Z2)-P3]dw$. However, consider as the infra-marginal price.

Increases on the interval [w2,wp2] while decreasing on the interval [0,w1] when pricing is constant. As such, short-run changes in welfare are a function of both the change in demand, or demand effects, and the difference between infra-marginal prices and the constant price, or our

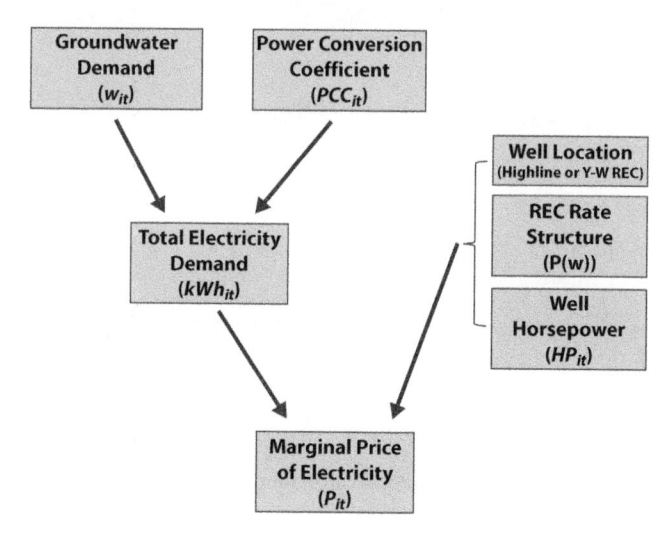

```
                ┌──────────────┐   ┌──────────────────┐
                │ Groundwater  │   │ Power Conversion │
                │   Demand     │   │   Coefficient    │
                │   (w_it)     │   │     (PCC_it)     │
                └──────┬───────┘   └────────┬─────────┘
                       │                    │
                       │                    │        ┌─────────────────────┐
                       │                    │        │   Well Location     │
                       ▼                    ▼        │(Highline or Y-W REC)│
                ┌──────────────────┐                 ├─────────────────────┤
                │ Total Electricity│                 │     REC Rate        │
                │     Demand       │ ◄────           │    Structure        │
                │     (kWh_it)     │                 │      (P(w))         │
                └────────┬─────────┘                 ├─────────────────────┤
                         │                           │       Well          │
                         │                           │    Horsepower       │
                         │                           │      (HP_it)        │
                         │                           └─────────────────────┘
                         ▼
                ┌──────────────────┐
                │  Marginal Price  │
                │  of Electricity  │
                │     (P_it)       │
                └──────────────────┘
```

Figure 12.4 Operating mechanism.

characterization schedule P (w). Welfare changes arising from a shift to constant pricing are then given by

$$\Delta \text{Welfare} = \int_{\max(Wp^c pc, Wp^{(w)})}^{\min(Wp^c, Wp^{(w)})} \left[\Gamma(w;z) - \min\left(P\left(Wp^{(w)}\right), P^C\right) \right] dw + \int_0^{\min(WP^C, Wp^{(w)})} [P(w) - P^C] \, dw$$

(12.1)

(w) gives at an arbitrary w when water pricing is DBR.

The first term in the equation depicts welfare gains and losses effects. The interval of the definite integral sign's welfare changes as the agricultural producer's indirect utility function is increasing in water implying negative demand effects when wpc < wp^((w)) and positive demand effects otherwise [10]. The second term of the equation characterizes gains and losses in.

Assuming that pc falls somewhere both terms in the equation are positive. Water demand effects infra-marginal price changes in Figure 12.4, the constant water effects arising marginal when pricing followed a structure. Theoretical prediction the likelihood of effects higher using producers, we hypothesize that negative welfare impacts concentrate among those producers that demand the most groundwater. Short-run welfare gains are guaranteed for producers of the rate schedule P (w) as they face a lower marginal price and increase their demand. However, the welfare implications remain unclear; it depends on the producer's price elasticity of demand for water and the parameters of P (w) relative to pc. We address this theoretical uncertainty in our empirical and simulation modelling to measure the magnitude of these effects and their implications for short-run welfare changes under constant pricing. Our theoretical model illustrates the distributional impacts of a shift from DBR to constant pricing. We hypothesize that decreases in water demand concentrate among high water-using producers as they face higher marginal prices under constant pricing. Similarly, low water-using agricultural producers increase their water use constant marginal price. As such, the impact of a transition to constant pricing on aggregate energy and water demand depends on the initial distribution of demand among the rate structure's blocks. For example, if the majority of producers demand within the rate structure's first block, then a transition to constant pricing potentially increases aggregate water demand as the majority of users experience a decrease in price. If most producers demand in later blocks of the price structure, then it is more likely that the constant pricing regime will result in diminished energy and water demand. The precise impact of constant pricing depends crucially on producer's responsiveness to changes in price, which we address empirically in our later modelling to uncover the impact of constant pricing on aggregate demand.

12.7 Empirical Model

In this section, we develop a water energy prices (P) and exogenous factors (Z) aiming to test the hypotheses generated. The model estimates how input prices and other exogenous factors determine water demand for agricultural producers and allows us to resolve theoretical ambiguity about impacts of a change from DBR pricing to a constant price regime. Let water demand by the ith well [11] in time t be given by

$$\log(w_{it}) = \alpha_i + \delta_t + \gamma \log(P_{it}) + \beta Z_{it} + \varepsilon_{it} \qquad (12.2)$$

α_i, time fixed effects, δ_t, and the vector of covariates [12], Z_{it}, with associated parameter vector β. Pit is the marginal energy price and, as the model is estimated in logs, the parameter idiosyncratic error term, employ the extensive.

We follow standard economic theory and estimate the effect of marginal price, rather than average price, on water demand. Recent literature suggests that consumers may respond to average price rather than marginal price signals when information costs associated with knowing where demand falls on the rate structure are significant and total expenditures on the good are a small proportion of their budget. However, in the context of agricultural water demand in the Basin, energy costs constitute a significant proportion [13] of farm expenses and most producers' well pump technology readily supplies information on cumulative water use.

12.8 Identification Strategy

There exist several sources of potential endogeneity concerning our estimation of groundwater demand. First, the utilization of non-linear price schedules for electricity in the study area introduces reverse causality between price and demand as a producer's pumping decision influences their marginal price. Second, given that electricity rate structures vary as a function of well pump characteristics, specifically well HP producers potentially have the ability to affect their rate structure by endogenously choosing their well's HP. Finally, well location decisions and the structure of REC governance present additional sources of endogeneity as these avenues potentially allow producers to influence their energy price schedule. To address this potential for endogeneity, we employ identification strategy as our preferred model specification [25]. Our FE approach controls for time-invariant unobservable like management capacity and soil attributes which potentially affect groundwater demand. We utilize DBR rate energy. Instrumenting for marginal price addresses the first source of endogeneity attributable to non-linear energy pricing by breaking the reverse causality between price and demand using exogenous rate structure parameters.

Endogenous energy ($p1 - p3$) required wells with the same HP as the observation. For example, we determine the value a given well with 100 HP by grouping all water threshold observations within the same REC and year that and finding water threshold. Our water threshold instrument addresses the second potential source of endogeneity introduced by energy price structures which vary according to a well's HP. By instrumenting with average water threshold values that exclude the cohort of wells with the same HP our instrument removes bias related to producers endogenously determining their well pump's HP. We further test the robustness of our results to this potential source of bias by restricting our sample to those wells which do not report a change in HP during our study period. Endogenous well location decisions and REC governance structures also present potential concerns for obtaining unbiased coefficient estimates.

Well location decisions that depend on differences between REC rate structures threaten the validity of our instrumental variable approach as inter-REC variation in rate structure parameters is no longer exogenous to individual groundwater demand but instead depends on the cohort of wells choosing each REC as their energy supplier. However, this potential source of endogeneity is unlikely given that well locations are fixed once installed and well location decisions were taken long before any recent differences in REC pricing or management were revealed to producers [14]. This suggests that other factors like land quality, distance to markets, and groundwater availability, for which there is significant heterogeneity across the Basin, were more relevant in the initial well location decisions rather than differences between RECs in their energy pricing.

The governance structure of RECs also presents an additional source of endogeneity that potentially threatens our identification strategy. RECs in the Basin are governed by elected boards comprised of REC constituents [15]. As such, there is the possibility that producers in the Basin could use REC boards to influence REC rate structure decisions to benefit agricultural operations. This is unlikely for several reasons. First, the composition of REC boards reflects the breadth of customer classes (e.g., residential, commercial, irrigation, municipal) served by the REC [16]. As such, irrigation customer representation does not constitute a majority in REC governing boards and any changes to irrigation rate structures would require the support of representatives from other customer classes. Second, REC cost recovery objectives constrain board members from significantly changing rate structures for their benefit as the REC must still generate sufficient revenue to cover the fixed and variable costs of distribution [17]. Third, there exists significant heterogeneity in resource availability and well pump technology across agricultural producers in the Basin. As such, changes in the rate structure that would benefit a particular producer is unlikely to benefit all producers which undermines the probability that producers act collectively to influence electricity rate structures [18]. Finally, we observe only one change in block threshold values among the two RECs in Basin between 2011-2017 providing further evidence that political capture by irrigation customer representatives does not significantly influence variation in rate structures.

We also scrutinize the stability of our results to our preferred model specification, FE-IV, by estimating a pooled OLS, instrumental variable (POLS-IV) model which is also common in the water demand literature. These robustness checks reveal that our results remain qualitatively similar across differing price instruments and restricted samples.

12.9 Data

Necessary data transformations required for our empirical modelling faces annually. However, electricity demand data is not directly available given privacy concerns among RECs in the Basin, rather we impute total leveraging characteristics and efficiency of imputing marginal prices extraction collected by the State of Colorado, which is largely 2017 (Figure 12.5), demonstrating the extent of pumping cost heterogeneity. The pairing of well-level groundwater demand and PCC measurements yields imputed annual electricity demand (Figure 12.6). Next, we utilize 35° Celsius. SSURGO irrigation capability class data is matched to irrigated parcels to account for cross-sectional variation in soil quality which we utilize in our

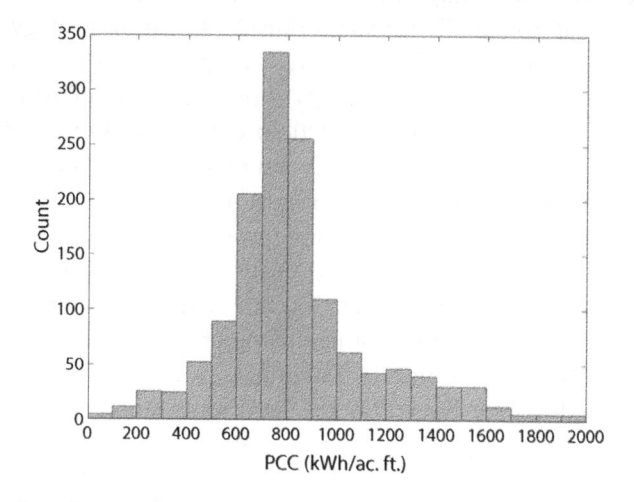

Figure 12.5 Distribution of PCC, 2017.

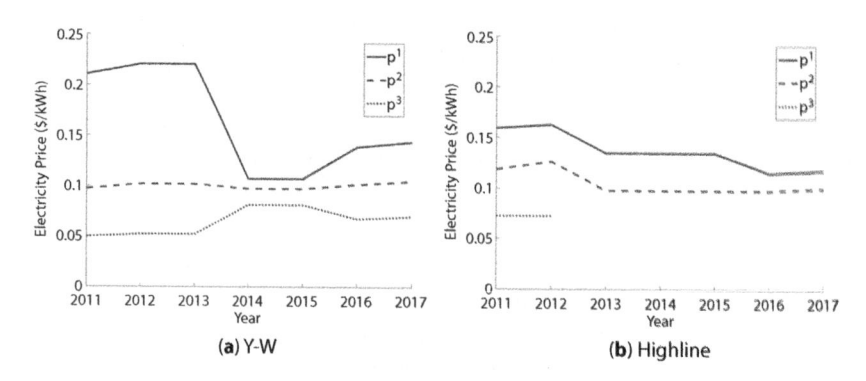

(a) Y-W

(b) Highline

Figure 12.6 REC electricity prices, 2011-2017 [19].

non-FE model specifications. Irrigation capability class data classifies soil types according to their suitability for irrigation (1 = most suitable, 8 = least suitable). Groundwater and well pump technology exhibited across wells within the Basin [20].

12.10 Empirical Results

This result is qualitatively robust across differing modelling specifications presented in columns (1)-(3). Model results presented in columns (1) and (3),

Table 12.1 Summary statistics.

Parameters	Average	Probable error	From	To
Available results				
Demand of water (acre feet/year)	229.32	103.18	0	1046.83
Electricity Demand (mWh/year)	116.16	52.11	0	630.40
Electricity Cost ($)	1488.24	6281.95	0	65556.87
Well Capacity (gallons/minute)	793.10	314.89	7.76	2852.28
Precipitation (inches/growing season)	11.75	3.78	3.73	21.72
Temperature (# days w/temp > 35°C)	18.99	14.16	0	55
Irrigation Capability Class factor variable)	2.026	0.8451	1	6
Well Pump Characteristics 1863.94				
Power Conversion Coefficient (PCC) (kWh/acre foot)	511.23700	127.34	126.60	1863.94
Horsepower (HP) (work/time)	112.36	42.67	10	700
Discharge Pressure (kPa)	44.11	25.57	5.00	567.00

which do not instrument for price, reveal the upward bias of price elasticity estimates when price endogeneity is not addressed. Finally, we assume our panel dataset of 1,392 yearly observations across 7 years is sufficiently long in the time dimension to accurately identify well-level fixed effects used for estimation of the inverse demand curve parameters. These results are presented and qualitatively align with modelling results generated using the full sample (see Tables 12.1 and 12.2).

Coefficient estimates of other covariates included in the model follow intuition in their sign and significance. Across all model specifications well capacity coefficient estimates are consistently positive and statistically significant pointing to the importance of resource constraints in groundwater demand. We also find that growing season precipitation negatively affects ground water demand while the number of days with a maximum temperature above 35° Celsius increases groundwater demand. This follows

Table 12.2 Empirical modelling as per requirements.

| | Various parameters | | | |
| | (Pumping) data | | | |
	POLS	POLS-IV (2)	FE (3)	FE-IV (4)
Log (Price)	-0.8345***	-0.4531***	-07953***	-0.2519***
	(0.0654)	(0.0676)	(0.0661)	(0.0634)
Availability	0.0009***	0.0009***	0.0003***	0.0003***
	(0.00003)	(0.00003)	(0.00003)	(0.0001)
Saturation	-0.0008***	-0.0007***	-0.0008***	-0.0007***
	(0.0001)	(0.0001)	(0.0001)	(0.0001)
Thermal state	-0.0020*	0.0026*	-0.0013	0.0051***
	(0.0010)	(0.0010)	(0.0010)	(0.0010)
Irrigation Class	0.0277**	0.0331***		
	(0.0087)	(0.0088)		
Constant	2.7307***	3.5367***		
	(0.1590)	(0.1588)		
Observations	9400	9400	9400	9400
R^2	0.4570	0.4360	0.2886	0.2361
Adjusted R^2	0.4567	0.4357	0.1646	0.1030
F Statistic	15810860***	14410520***	811.8179***	605.6612***

*$p<0.05$; **$p<0.01$; ***$p<0.001$ standard error

Energy Technologies and Pricing Policies: Case Study   173

intuition Coefficient estimates for the factor variable irrigation class points to increasing demand for groundwater as soil quality diminishes (e.g., higher sand content). Finally, potential endogeneity rate structures by restricting which in either well or irrigation system discharge pressure [21]. These results are presented and qualitatively align with modelling results generated using the full sample.

12.11 Counterfactual Simulation A

In this section, we leverage the results from our preferred empirical modelling specification, FE-IV, to simulate the counterfactual scenario of constant electricity pricing and test the theoretical predictions delineated. Specifically, we simulate the counterfactual scenario wherein RECs in the Basin transition to a constant electricity pricing regime in 2017. Constant utilize methods constant across Basin.

We assume that allow recovery their distribution procurement costs [22]. Therefore, we calculate the REC-specific, 2017 constant price that achieves the expected generated between 2011 and 2016. More specifically, for the kth REC, whose wells are indexed by j (j = 1 ... J), we determine the 2017 constant pricing regime, pc, which minimizes the difference between predicted average REC revenue under DBR pricing, $\bar{T}RP(w)$, and predicted average REC revenue under constant pricing, $\bar{T}Rpc$, between 2011 and 2016. We multiply preferred model specification demand predictions [23], $\hat{w}P(w)$, with PCC_{it} to yield predicted annual well-level electricity demand. To calculate REC revenue under DBR pricing, we match each well's predicted annual electricity demand to their REC's rate structure in time t, $P_{kt}(w)$, to produce total annual well-level electricity expenditures. Finally, predicted expenditures are aggregated across J wells and T years (T = 6), and averaged across T to give $\bar{T}RP(w)$. More formally, we calculate average REC revenue using the following equation

$$\bar{T}R_K^{P(w)} = \frac{[\sum_{t=1}^{T} \sum_{j=1}^{J} \hat{w}_{it}^{P(w)} * PCC_{it} * P_{kt}(w)]}{T} \quad (12.3)$$

We follow a similar approach to calculate $\bar{T}R\_k\^{(p\^c)}$ but utilize model parameter estimates to simulate well-level groundwater demand as a function of constant price, pc. Specifically, well-level simulations of groundwater demand under constant electricity pricing are given by the following equation

$$\bar{T}R_K{}^{p^C} = \frac{[\sum_{t=1}^{T}\sum_{j=1}^{J} w^{\wedge p^c}_{it} * PCC_{it} * P_C^K]}{T} \qquad (12.4)$$

Finally, for the kth REC we determine the 2017 constant price which minimizes the difference between average annual revenue between 2011 and 2016 under DBR and constant pricing according to the following optimization problem

$$\text{Minimize } p_k^{c \wedge} = [TR_k^{p(w)} - TR_k^{p(c)}]^2 \qquad (12.5)$$

which, for the kth REC, finds the value of pc that minimizes the difference between average total REC revenue under DBR and constant pricing regimes between 2011 and 2016. Predicted constant [24].

12.12 Counterfactual Simulation B

We utilize the 2017 constant prices derived by the optimization problem in equation 12.6 to simulate well-level demand using equation 12.1. We then sum these simulated demand quantities across wells to determine aggregate demand under constant pricing and compare to aggregate demand under DBR pricing. Constant in 2017 decreases aggregate groundwater pumping by approximately 5% compared to the alternative scenario of DBR pricing in 2017.

Figure 12.7 plots 2017 pumping differentiating between wells based on which price block defined their observed 2017 pumping. Only wells which demand water on the first, highest marginal price block in 2017 increased groundwater demand as a result of the transition to constant pricing. Under constant pricing these wells experience a decrease in the price signal determining their demand as the derived revenue-neutral constant prices are less than both REC's first block's marginal price. Wells that demand water on the third marginal price block [25] experience the largest decreases in groundwater demand under constant pricing in 2017.

These large reductions in demand derive from a relatively large increase in marginal price when electricity pricing is constant. Furthermore, these reductions in demand increase with well capacity suggesting that REC

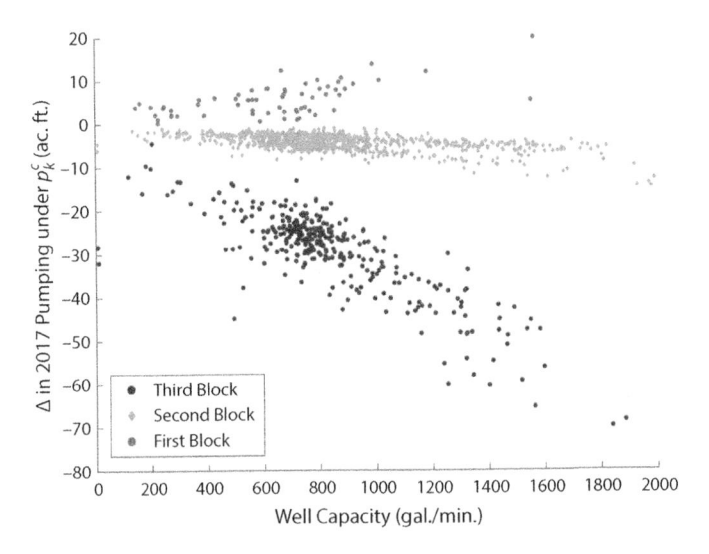

Figure 12.7 Change in 2017 pumping vs. well capacity.

rate structures and resource availability jointly determine the distribution of water use impacts across wells. Finally, wells which demand on the second price block [26] experience relatively minimal reductions in demand that also increase with higher well capacity. These minimal decreases in demand are related to the relatively small increase in marginal price experienced by these wells under constant pricing as both REC's second block's marginal prices are only slightly larger than the derived constant prices.

We also explore how the implementation of a constant pricing regime influences aggregate demand in years previous to 2017. To generate the appropriate comparison, we determine what each well's demand under DBR pricing would be between 2011 and 2016 if their REC priced electricity according to the rate structure used in 2017 and aggregate demand across wells within a given year. We then compare these annual, aggregate pumping decisions under the 2017 DBR rate structure to annual simulated aggregate pumping when electricity pricing is constant throughout the 2011 to 2017 time period and equal to the constant prices derived previously. This approach allows us to identify the impact of constant pricing separate from variation in REC rate structures across time.

Our constant reduces annual aggregate groundwater demand by between 4% and 7.5% depending on the year. Figure 12.8 plots annual

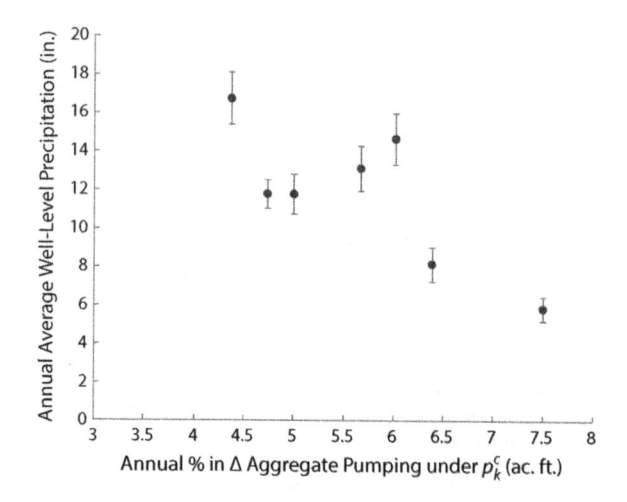

Figure 12.8 Change in annual pumping vs. average precipitation.

percent decreases in aggregate pumping against annual averages [27] of well-level precipitation demonstrating how exogenous growing season weather affects the conservation potential of constant electricity pricing. Generally, years with less than average well-level precipitation experience greater reductions in aggregate pumping when electricity pricing is constant. This result is related to the fact that wells demand more groundwater in years with less precipitation, thus increasing the impact of the constant pricing regime as more wells the DBR and experience a larger increase in price under the constant pricing regime. Results on the impact of constant pricing presented should be interpreted as the sum of both intensive and extensive margin adjustments. However, past research finds that extensive margin adjustments are relatively small compared to the total effect of energy prices on groundwater demand.

12.13 Counterfactual Simulation: Costs of Reduced Groundwater Demand

A transition from DBR to constant electricity pricing in 2017 also yields short-run changes in welfare provides a theoretical treatment of welfare changes, positing lower using agricultural producers [28]. Welfare impacts of constant pricing are based on demand and infra-marginal effects.

ENERGY TECHNOLOGIES AND PRICING POLICIES: CASE STUDY   177

The change in short-run welfare for the ith well in 2017 is given by

$$Welfare_{i,2017} = \int_{\min(\hat{w}_{i,2017}^{p(w)},\hat{w}_{i,2017}^{p_k^c})}^{\max(\hat{w}_{i,2017}^{p(w)},\hat{w}_{i,2017}^{p_k^c})} [\tau(w;\hat{\alpha}_i,\hat{\delta}_t,\hat{\beta},Z_{i,2017},\hat{y}) - \min(P_{I,2017},P_{ik}^c)]dw$$

$$+ \text{ Demand Effects}_{i,2017}$$

$$\underbrace{\int_0^{\min\; \hat{w}_{i,2017}^{p_k^c},\hat{w}_{i,2017}^{p(w)}} [P_{K,2017}(w) - P_{ik}^c]dw}_{Infra-marginal\; Effects_{i,2017}} \quad (12.6)$$

where pc is the constant price faced by the ith well served by the kth REC, Pi,2017 is marginal price under DBR in 2017, and Pk,2017(w) is a function that outputs 2017 marginal price of the kth REC for an arbitrary w. As outlined the first term in equation 6 depicts the welfare changes associated with altered water demand, or demand effects, while the second term accounts for changes in welfare arising from differences between the infra-marginal prices of Pk,2017(w) and pc, or inframarginal effects.

We utilize equation 12.6 to calculate short-run welfare changes for each well in our sample. wells demonstrates that some wells benefit from constant electricity pricing. Figure 12.9 depicts the distribution of Δ Welfare$_i$, 2017, revealing that while average welfare impacts are relatively small and negative some wells experience large losses and gains in welfare in the counterfactual scenario.

Equation 12.8 describes how well-level changes in welfare are a function of demand and infra-marginal effects. Section 12.4 explores these effects and posits that their relative magnitudes jointly determine welfare costs and benefits. Specifically, demand effects increase welfare only for those wells that increase their demand under constant pricing which is relatively uncommon given the changes in well-level demand predicted (see Figure 12.7). The magnitude and sign of infra-marginal effects are less clear as they depend on Pi,2017 and the difference between inframarginal prices and the constant price. Differentiating between demand and inframarginal effects aids an understanding of who accrues the welfare benefits and costs displayed in Figure 12.9. Namely, welfare benefits accrue to wells when positive inframarginal effects [29] outweigh negative demand effects [30] while welfare costs occur when negative demand and infra-marginal effects outweigh positive infra-marginal effects.

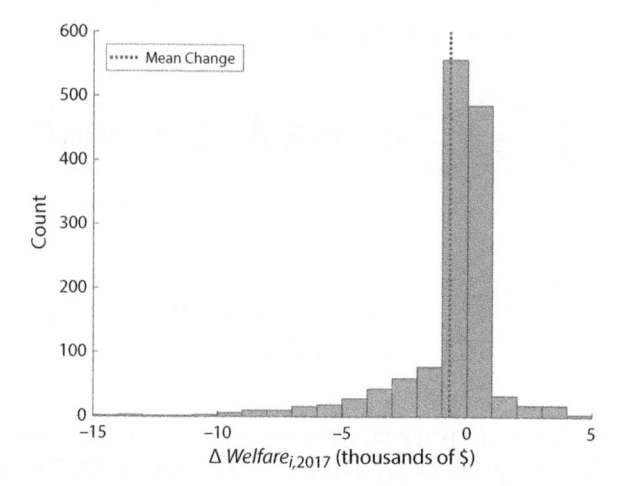

Figure 12.9 Distribution of $\Delta\ Welfare_{it}$.

We explore how these disparate effects determine welfare outcomes in Figures 12.10a and 12.10b which plot the distribution of demand and infra-marginal effects, respectively. Demand effects are largely negative given the paucity of wells which increase their water use in the counterfactual and the abundance of wells which experience a higher marginal price under constant pricing. Infra-marginal effects presented in Figure 12.10b demonstrate how differences between infra-marginal prices and

(a) Demand effects

(b) Infra-marginal effects

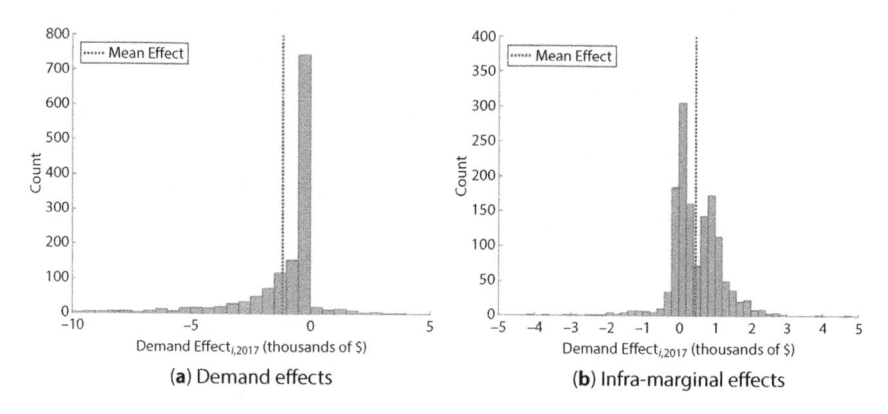

Figure 12.10 Demand and infra-marginal welfare effects.

the constant price generate average benefits for producers. Finally, comparing Figures 12.10a and 12.10b with producers using less water under the constant pricing regime experience increases in their total welfare in 2017.

The abundance of wells which experience a higher marginal price under constant pricing. Infra-marginal effects presented in Figure 12.10b demonstrate how differences between infra-marginal prices and the constant price generate average benefits for producers. Finally, comparing Figures 12.10a and 12.10b with producers using less water under the constant pricing regime experience increases in their total welfare in 2017.

We analyse the spatial distribution of welfare impacts of constant electricity pricing in Figure 12.11 which maps well-level results, saturated thickness [31], and REC boundaries in the Basin. For visual simplicity, we classify each well as either experiencing a welfare gain or loss as a result of the transition to constant pricing in 2017. The distribution of welfare impacts presented in Figure 12.11 provides some visual evidence of the importance of spatially variable resource stocks, measured by saturated thickness, in determining welfare outcomes. Specifically, wells that experience short-run welfare losses have, on average, 6 feet more saturated

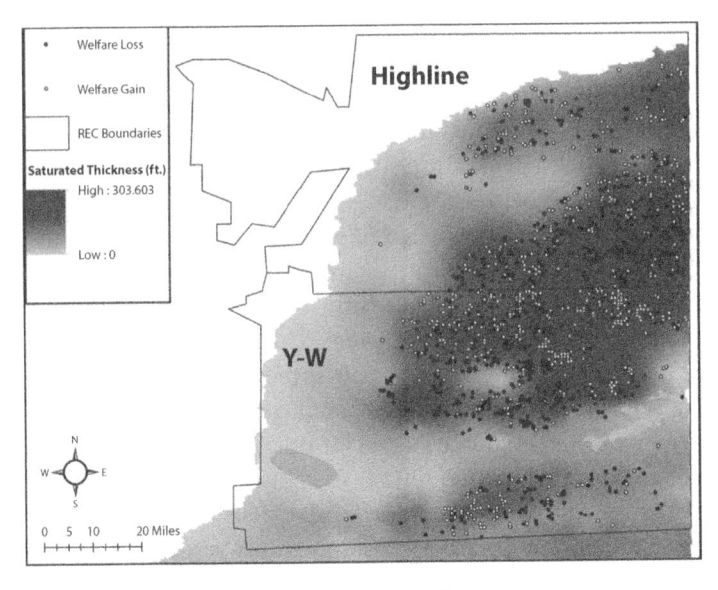

Figure 12.11 Spatial distribution of average welfare effects.

thickness and 50 gal./min. more well capacity [32] than wells that gain under constant pricing.

The relationship between resource availability and changes in welfare and groundwater demand (see Figure 12.7) point to a potential inefficiency introduced by constant pricing. Namely, the constant pricing regime does not consider heterogeneity in the social costs of pumping across space [33]. Previous research finds that the long-run gains from reduced groundwater pumping depend on initial aquifer conditions wherein producers with minimal initial groundwater stocks accrue more gains from diminished pumping. This result demonstrates that in a dynamic setting, the social costs of pumping costs are likely higher for wells with relatively lower capacity. Constant electricity can introduce inefficiencies by increasing water use in regions where external pumping costs are highest (i.e., low water users). While decreased water use concentrates in areas with higher well capacity and abundant groundwater availability where external costs are likely lower.

12.14 Conclusion

Paper non-linear (DBR) utilize groundwater the counterfactual constant how demonstrating through stocks. context agriculture Basin given concerns groundwater. electricity pricing However, on average these welfare costs are minimal, only constituting approximately 6% of average annual well-level energy expenditures. Welfare costs are minimal because infra-marginal price effects diminish the welfare effects associated with reduced groundwater withdrawals. Differentiating between demand and infra-marginal effects uncovers that in some cases producers who demand less groundwater under constant pricing experience an increase in their short-run welfare as less expensive infra-marginal water prices compensate producers induced to conserve water. This result has significant policy relevance as it demonstrates how lump-sum transfers, potentially generated using revenues from price-based management policies, can mitigate the welfare impacts of resource conservation efforts.

The welfare effects of conservation. Also, our model does not explicitly differentiate between intensive and extensive margin adjustments induced by the constant pricing regime. While previous research finds extensive margin impacts are relatively small compared to the total effect of energy prices, the literature has not evaluated how non-linear energy pricing influences extensive margin choices. Future research in this area should disentangle the intensive and extensive margin to analyse how non-linear energy pricing impacts cropping patterns.

References

1. [A. Colin Cameron, 2005] A. Colin Cameron, P. K. T. (2005). *Microeconometrics*. Cambridge University Press.
2. [Acemoglu et al., 2007] Acemoglu, D., Antràs, P., and Helpman, E. (2007). Contracts and technology adoption. *American Economic Review*, 97(3):916–943.
3. [Agthe et al., 1986] Agthe, D. E., Billings, R. B., Dobra, J. L., and Raffiee, K. (1986). A simultaneous equation demand model for block rates. *Water Resources Research*, 22(1):1–4.
4. [Arellano, 2003] Arellano, M. (2003). *Panel data econometrics*. Oxford University Press.
5. [Arsanjani et al., 2015] Arsanjani, J. J., Mooney, P., Zipf, A., and Schauss, A. (2015). Quality assessment of the contributed land use information from OpenStreetMap versus authoritative datasets. In *Lecture Notes in Geoinformation and Cartography*, pp. 37–58. Springer International Publishing.
6. [Badraoui and Dahan, 2011] Badraoui, M. and Dahan, R. (2011). The green morocco plan in relation to food security and climate change. Food Security and Climate Change in Dry Areas, p. 61.
7. [Bar-Shira et al., 2006] Bar-Shira, Z., Finkelshtain, I., and Simhon, A. (2006). Block-rate versus uniform water pricing in agriculture: An empirical analysis. *American Journal of Agricultural Economics*, 88(4):986–999.
8. [Barron et al., 2013] Barron, C., Neis, P., and Zipf, A. (2013). A comprehensive framework for intrinsic OpenStreetMap quality analysis. *Transactions in GIS*, 18(6):877–895.
9. [Besley and Case, 1993] Besley, T. and Case, A. (1993). Modeling technology adoption in developing countries. *American Economic Review*, 83(2):396–402.
10. [Bollinger et al., 2018] Bollinger, B., Burkhardt, J., and Gillingham, K. (2018). Peer effects in water conservation: Evidence from consumer migration. Technical report, National Bureau of Economic Research.
11. [Bollinger and Gillingham, 2012a] Bollinger, B. and Gillingham, K. (2012a). Peer effects in the diffusion of solar photovoltaic panels. *Marketing Science*, 31(6):900–912.
12. [Bollinger and Gillingham, 2012b] Bollinger, B. and Gillingham, K. (2012b). Peer effects in the diffusion of solar photovoltaic panels. *Marketing Science*, 31(6):900–912.

13. [Brill and Burness, 1994] Brill, T. C. and Burness, H. S. (1994). Planning versus competitive rates of groundwater pumping. *Water Resources Research*, 30(6):1873–1880.
14. [Brown, 1980] Brown, D. C. D. C. (1980). *Electricity for rural America: the fight for the REA. Contributions in economics and economic history*, no. 29. Greenwood Press, Westport, Conn.
15. [Brozovic et al., 2010] Brozovic, N., Sunding, D. L., and Zilberman, D. (2010). On the spatial nature of the groundwater pumping externality. *Resource and Energy Economics*, 32(2):154–164.
16. [Burke and Sass, 2013] Burke, M. A. and Sass, T. R. (2013). Classroom peer effects and student achievement. *Journal of Labor Economics*, 31(1):51–82.
17. [Burtless and Hausman, 1978] Burtless, G. and Hausman, J. A. (1978). The effect of taxation on labor supply: Evaluating the gary negative income tax experiment. *Journal of Political Economy*, 86(6):1103–1130.
18. [Cai et al., 2008] Cai, X., Ringler, C., and You, J.-Y. (2008). Substitution between water and other agricultural inputs: Implications for water conservation in a river basin context. *Ecological Economics*, 66(1):38–50.
19. [Camp, 1998] Camp, C. R. (1998). Subsurface Drip Irrigation: A Review. *Transactions of the ASAE*, 41(5):1353–1367.
20. [Carey and Zilberman, 2002] Carey, J. M. and Zilberman, D. (2002). A model of investment under uncertainty: modern irrigation technology and emerging markets in water. *American Journal of Agricultural Economics*, 84(1):171–183.
21. [Caswell and Zilberman, 1983] Caswell, M. and Zilberman, D. (1983). The economics of land augmenting irrigation technologies. Working Paper, Giannini Foundation of Agricultural Economics, California Agricultural Experiment Station (USA).
22. [Caswell and Zilberman, 1985] Caswell, M. and Zilberman, D. (1985). The choices of irrigation technologies in California. *American Journal of Agricultural Economics*, 67(2):224.
23. [CDNR, 2015] CDNR (2015). Final permit - designated basins. availabile: http://cdss.state.co.us/GIS/Pages/AllGISData.aspx.
24. [CDNR, 2017] CDNR (2017). Structures – with water rights. availabile: http://cdss.state.co.us/GIS/Pages/AllGISData.aspx.
25. [CDNR, 2018] CDNR (2018). Final permit - designated basins. availabile: https://dnrweb.state.co.us/cdss/WellPermits.
26. [Chen, 2009] Chen, Y. (2009). Does a regional greenhouse gas policy make sense? a case study of carbon leakage and emissions spillover. *Energy Economics*, 31(5):667–675.
27. [Chicoine et al., 1986] Chicoine, D. L., Deller, S. C., and Ramamurthy, G. (1986). Water demand estimation under block rate pricing: A simultaneous equation approach. *Water Resources Research*, 22(6):859–863.

28. [Cohen-Cole and Fletcher, 2008] Cohen-Cole, E. and Fletcher, J. M. (2008). Is obesity contagious? social networks vs. environmental factors in the obesity epidemic. Journal of Health Economics, 27(5):1382–1387.
29. [Commission, 2008] Commission, B. C. U. (2008). 2008 residential inclining block rate application.
30. [Conley and Udry, 2010] Conley, T. G. and Udry, C. R. (2010). Learning about a new technology: Pineapple in ghana. *American Economic Review*, 100(1):35–69.
31. [Cowan, 2010] Cowan, T. (2010). An overview of usda rural development programs. Washington, DC.
32. [CSU, 2013] CSU, E. (2013). Agricultural and business management; crop enterprise budgets.
33. [Cunningham *et al.*, 2016] Cunningham, S., Bennear, L. S., and Smith, M. D. (2016). Spillovers in regional fisheries management: Do catch shares cause leakage? *Land Economics*, 92(2):344–362.

13
Energy Availability and Resource Management: Case Study

Shanmugha S. and Milind Shrinivas Dangate*

Chemistry Division, School of Advance Sciences, Vellore Institute of Technology, Chennai, India

---

Abstract

This chapter empirically investigates how peer effects and resource availability influence a producer's choice to adopt a resource-conserving irrigation technology using data from the Trifa Plain of Morocco. These results demonstrate the potential impact of conservation policies that incentivize individuals to account for resource scarcity via pricing or other resource management policies on the adoption of efficiency-enhancing technologies. More specifically, our results suggest that drip irrigation subsidy rates that vary according to resource availability could potentially increase drip irrigation adoption rates among the least resource constrained individuals. Future research should investigate how the source of water available to producers influences the likelihood of engaging in pro-conservation behaviour and adopting an efficiency enhancing irrigation technology.

Keywords: Energy policies, impact on energy policies, drip irrigation, energy management, pro-conservation of energy sources, efficient irrigation technologies

13.1 Introduction

A growing world population and changing climate place increasing pressure on agricultural production and scarce water resources. Promoting the adoption of efficient irrigation technologies is a favoured policy option to

---

Corresponding author: mili_ncl@yahoo.com
Shanmugha S.: https://orcid.org/0000-0001-8555-7542
Milind Shrinivas Dangate: https://orcid.org/0000-0002-0210-5768

conserve water resources and sustainably intensify agricultural production to confront food security concerns. Despite these efforts, global adoption of efficient irrigation technology remains low while interest in investigating the determinants of adoption have surged in both policy and research communities. This paper contributes to this literature by exploring how peer effects and resource availability jointly influence the adoption of an irrigation technology that conserves natural resources.

A growing literature recognizes how social interactions are important in determining technology diffusion patterns. This literature posits that individual technology adoption decisions depend on peer group adoption rates which allow individuals to learn about the potential returns of the technology. A similar literature argues that economic, demographic, and environmental or resource characteristics determine technology adoption choices within the context of irrigation. In many scenarios, individual adoption decisions generate outside impacts, particularly when adoption influences conservation behaviour among individuals utilizing a common pool resource (CPR). This paper contributes to the technology adoption literature by recognizing these outside impacts and investigating how peer effects and resource availability affect adoption and conservation behaviours. Specifically, we investigate the adoption of drip irrigation systems which potentially alter how individuals utilize common pool water resources.

Drip irrigation increases the application efficiency of irrigated agricultural production by directly applying water to the plant's root zone thereby minimizing application losses. A large body of agronomic literature finds that drip irrigation adoption increases crop yields and potentially reduces variable input costs. Furthermore, the adoption of drip irrigation generates benefits for other resource users if efficiency gains translate into conservation [1] when water resources are scarce and common pool. As such, drip irrigation adoption potentially constitutes a change in individual conservation behaviour. This paper evaluates how peer effects and resource availability influence drip irrigation adoption decisions and alter conservation behaviour using data from the Trifa Plain of northeastern Morocco.

The characteristics of agriculture in the Trifa Plain provide an ideal setting to explore the relationship between resource availability (e.g., groundwater) and peer effects. First, the region has a long history of furrow, flood-irrigated agriculture dating back to French colonization. Recently, the Trifa Plain has seen an increase in drip irrigation system adoption which research suggests increases water use efficiency and agricultural productivity. Second, Moroccan agricultural policy extends generous subsidies to cover the cost of drip irrigation systems, suggesting that capital constraints are less binding in adoption. Third, while surface water availability is ubiquitous throughout the Trifa Plain, the distribution of groundwater is heterogeneous. Finally, agricultural production in the Trifa Plain's climate requires irrigation for most high-value crops. We exploit these characteristics to measure how groundwater availability and peer effects jointly determine the rate of drip irrigation adoption.

This paper utilizes a novel panel dataset of parcel-level drip irrigation system adoption decisions to estimate the effect of peer group adoption and groundwater availability on the probability of adoption. Empirical results provide modest evidence regarding the importance of social learning and peer effects in irrigation technology adoption. We also find evidence that resource availability decreases the likelihood of adoption which aligns with past research results and demonstrates the role of resource constraints in determining conservation behaviour and adoption decisions.

This paper proceeds as follows: in the next section, we survey relevant peer effect and conservation technology literature and situate the paper's contribution within that literature. We provide an overview of irrigated agriculture within the Trifa Plain. We develop an empirical framework to model the adoption of a resource conserving irrigation technology. We describe the data utilized to estimate our empirical model of irrigation technology adoption. Finally, in this context we present results detailing the relationship between drip irrigation adoption, peer effects and resource availability and conclude with a discussion of the policy implications of our results.

13.2 Literature Review

This paper builds on several veins of literature exploring the determinants of technology adoption. In this section, we survey this literature beginning with more general treatments of technology adoption and ending with applied research efforts exploring the adoption of irrigation technology. Finally, we provide an overview of the peer effects literature and discuss how this literature addresses identification challenges.

Economists and social scientists have long been concerned with individual technology adoption decisions, which is often cited as the seminal treatment of technology adoption within the context of agriculture. Recent literature builds on this by empirically and theoretically modelling technology adoption under differing institutional settings. A separate but related literature investigates aggregate technology adoption decisions aiming to understand why developing countries exhibit low adoption rates for productivity and profit enhancing technologies.

A related literature focuses on technology adoption among agricultural producers, generating results that reveal how environmental, economic, and demographic characteristics determine adoption decisions. Of particular importance for our paper are the applied research efforts examining the determinants of irrigation technology adoption in a similar paper in this vein of research which develops a stylized theoretical framework to understand an agricultural producer's irrigation technology adoption decision extends this framework to the case of irrigation technology adoption with non-renewable resource extraction (e.g., groundwater). Provide an exhaustive review of the drip irrigation technology adoption literature. This literature generates several hypotheses pertinent to our analysis. Specifically, it is found that well-depth and its associated pumping costs are a significant determinant of irrigation technology adoption decisions. As water resources become scarcer, the benefits of adopting an efficient irrigation technology, like drip irrigation, increase. Our paper empirically tests this hypothesis by evaluating how the availability of groundwater affects drip irrigation system adoptions. Similarly, it posits that the learning costs associated with drip irrigation technology influence adoption decisions. We test this hypothesis by incorporating the effect of peer group adoption on individual adoption decisions, leveraging the methodological advances of recent empirical literature investigating the role peer effects play in individual decision-making.

A related literature recognizes the importance of peer effects in individual adoption decisions. Generally, this literature posits that social learning is the primary mechanism through which peer effects influence adoption decisions. This literature identifies that the spatial clustering of outcomes arises from both contextual and endogenous effects. Exogenous contextual effects generate clustering in outcomes as characteristics shared among groups or spatial units generate similar outcomes. To control for these contextual effects, we follow those who investigate the role of peer effects in the adoption of groundwater-fed irrigated agriculture in Kansas using a rich set of spatial fixed- effects and trends to control for the possibility of peer self-selection. In particular, we utilize common correlated effects (CCE) developed by to account for region-specific trends that influence groundwater adoption but are unrelated to peer effects.

Endogenous effects include those interactions wherein the decision of an individual is causally affected by the behaviour of other individuals in their peer group. For example, an agricultural producer may learn about the benefits of drip irrigation from adopting members of their peer group. This relates to what is identified as the "reflection problem" wherein an individual's decision influences group outcomes and vice versa. However, in our context it is unlikely that individual choices affect group behaviour within a given time period as an individual's choice to adopt drip irrigation likely only affects group behaviour through a lag, given the time needed to install a drip irrigation system. We incorporate these endogenous effects by following and controlling for peer group adoptions, or the installed base, in our empirical modelling. In our context, installed base refers to the lagged number of adoptions within an individual's peer group.

13.3 Study Area

The Trifa Plain is the most productive agricultural region of northeastern Morocco with over 39,000 ha of cultivated land irrigated in a semi-arid climate adjoining the Mediterranean Sea. Figure 13.1 situates the Trifa Plain and its principal source of water, the Moulouya River, within North Africa. The economy of the region is built around the cultivation of perennial fruit, particularly citrus, and annual crops, such as potatoes, sugar beet, loquat and vegetables. Over 60% of the region's cultivated land is planted in citrus. The region's climate is characterized by cool, wet winters and hot, dry summers making irrigation a necessity for most crops with the exception of some cereals and forage. The Trifa Plain traditionally relied on imported water [2] and groundwater wells to support the region's

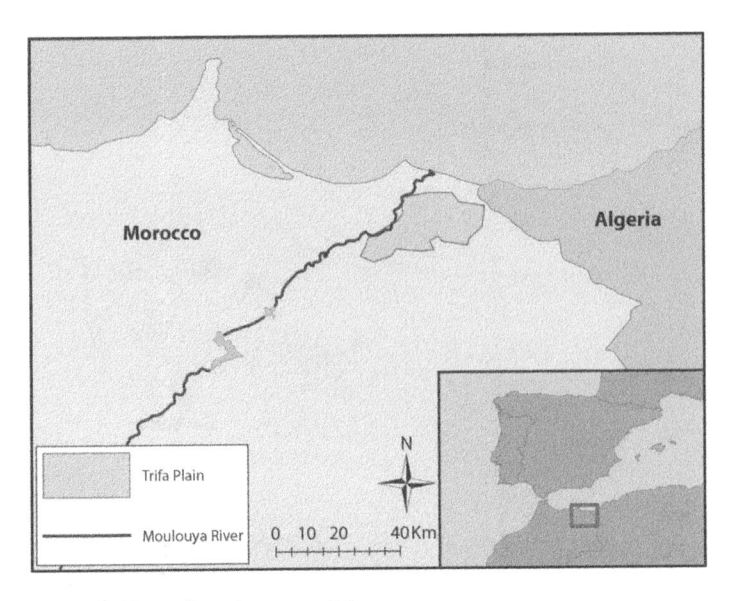

Figure 13.1 Trifa Plain of north-eastern Morocco.

agricultural economy. Figure 13.2 maps the primary [3] irrigation canal that imports water from the Moulouya river into the Trifa Plain. Figure 13.2 also maps the location of the 834 active groundwater wells in the Trifa, and gives an approximation of the aquifer extents. Aquifer locations are an approximation as they are based upon water table data collected from existing groundwater wells; as such, we cannot preclude the existence of groundwater in other locations in the Trifa. Therefore, in our later empirical analysis, we treat the existence of an active groundwater well within a parcel as the indicator of groundwater availability.

Growing irrigation demand and climatic variability catalysed governmental efforts to promote water conservation and agricultural productivity through drip irrigation and water storage basin adoption. These policymaking efforts resulted in the implementation of a generous subsidy program administered by the Ministry of Agriculture to support the adoption of drip irrigation systems among Moroccan farmers which include producers in the Trifa Plain. The subsidy program covers between 60% and 100% of the costs of drip irrigation system installation, depending on the timing of adoption and farm size [4]. The subsidy program also requires and covers the installation costs of water storage basins. The necessity of water storage basins is related to water quality issues that require water to settle in a basin before application through the drip irrigation system.

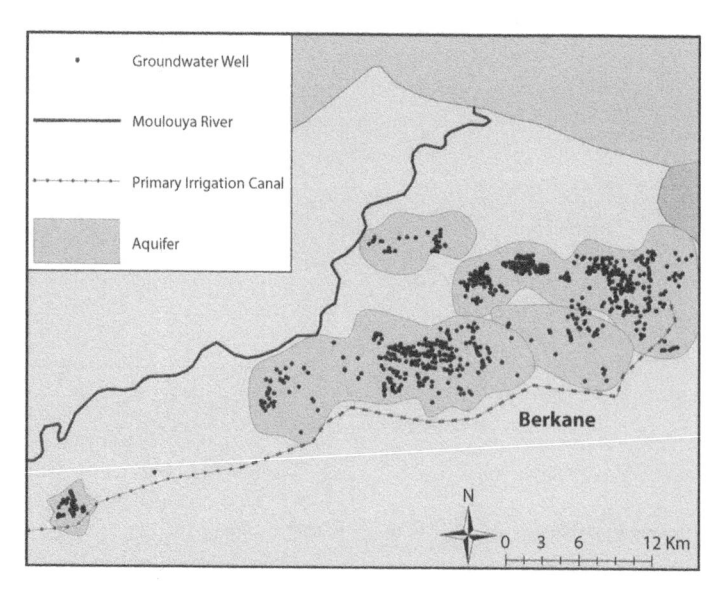

Figure 13.2 Aquifer, wells, and irrigation canals of the Trifa Plain.

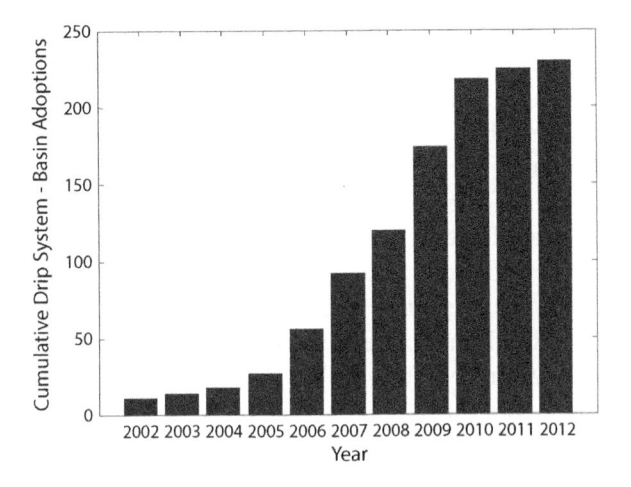

Energy Availability and Resource Management: Case Study   191

Figure 13.3 Cumulative drip irrigation system adoptions, 2002-2012.

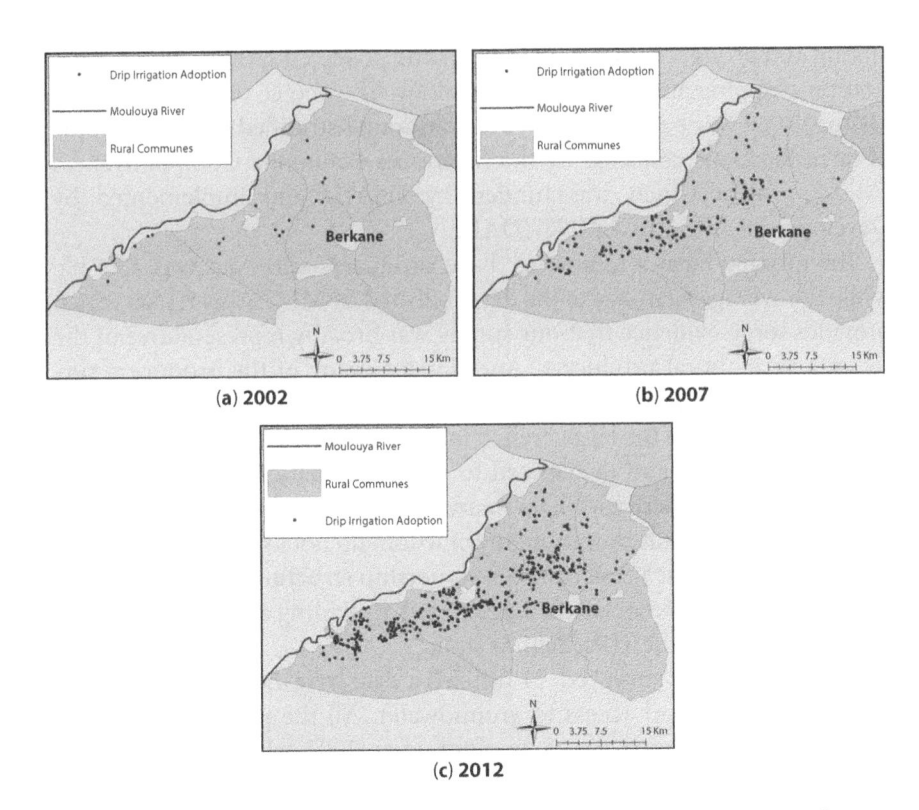

Figure 13.4 Spatial distribution of drip irrigation system adoptions, 2002, 2007 and 2012.

As such, water storage basins which increase the productivity of irrigation are synonymous with drip irrigation systems in the study area.

Figure 13.3 presents the cumulative adoption of drip irrigation systems within the study area between 2002 and 2012 while Figure 13.4 depicts the spatial distribution of adoptions in 2002, 2007, and 2012 as well presenting the boundaries of the Trifa Plain's rural communes [5]. These figures demonstrate the rapid uptake of drip irrigation systems and the spatial distribution of these adoptions within the Trifa Plain. Despite these recent increases, research suggests that aggregate drip irrigation adoption rates in the Trifa Plain and Morocco remain low. This paper aims to understand how peer effects and resource availability determine patterns of drip irrigation adoption across time.

13.3.1 Producer Survey

To better understand agricultural production and the determinants of irrigation technology adoption decisions in the study area, an in-person survey was conducted during the spring of 2018. The survey was implemented among 100 producers in the Trifa Plain and collected farm-level information on cropping and irrigation technology. The choice of which producers to survey was based upon a rural commune stratified random sample of farm locations collected by the Moroccan Economic Competitiveness (MEC) project which was funded by USAID and implemented by Development Alternatives INC (DAI).

The average farm size among those producers surveyed was 13.2 ha. while the average farm size of the data collected by MEC was 11.3 ha. which provides some evidence that our survey was broadly representative of the region's agricultural producers. Anecdotally, many of the producers surveyed farmed on land their family received after the end of French colonization in 1956 when the large French farms which once existed in the Trifa Plain were split up and distributed to Moroccan nationals. Furthermore, given Moroccan inheritance laws many farms are owned by multiple individuals within the same family, many of whom do not live in the Trifa Plain or work on the farm. These complex ownership structures complicate land transactions as all owners must agree to sell, providing some evidence that land tenure within the Trifa Plain is static.

Of the producers surveyed, 51 utilized a drip irrigation system on their operation and 45 had access to groundwater. All the producers utilizing drip irrigation adopted the technology after 2002 when the Moroccan government's drip irrigation subsidy program began, and many producers noted the importance of subsidies in their choice to adopt drip irrigation.

68% of producers surveyed planted the majority of their cultivated land in perennial crops, primarily differing varieties of citrus (e.g., mandarins, navel oranges, tangerines, etc.) which aligns with regional trends regarding perennial crop cultivation. Among the 45 producers surveyed whose operation has a groundwater irrigation well, only 6 (13%) had adopted a drip irrigation system by 2018.

Many surveyed producers that adopted drip irrigation mentioned the increased management effort needed to operate their system. Specifically, producers recounted that transitioning their operation to drip irrigation demanded additional management of water quality given the potential for nutrient loading and system blockage. A few of the producers surveyed also lived on the plots they farmed, opting instead to live in nearby towns and commute to their fields. As such, the neighbours which constitute their peer group are potentially more spatially dispersed than those producers which farm parcels near their own.

13.4 Empirical Model of Adoption

In this section, we develop an empirical model of irrigation technology adoption using notation and discuss the suite of spatial and time controls we utilize to account for common contextual effects. Suppose the ith individual faces the decision of whether to adopt a drip irrigation system in each period t. Let $d_{it} = 0$ denote the decision to continue farming without drip irrigation and $d_{it} = 1$ denote the decision to adopt drip irrigation. The perceived profit associated with each decision is given by π_{it}^{dt}.

The returns to adopting a drip irrigation system (π_{it}^1) consist of the perceived present and future value of increased irrigation application efficiency less installation costs. The returns of not adopting (π_{it}^0) consist of the expected profit of current irrigated farming (surface or groundwater) plus the value of the future option to adopt. The net profit of drip irrigation system adoption is then given by $\pi_{it} = \pi_{it}^1 - \pi_{it}^0$ and the agricultural producer adopts when $\pi_{it} > 0$. We conceptualize an optimal stopping model wherein an individual chooses when, if ever, to adopt a drip irrigation system. Previous literature has dealt with such models of adoption, or more broadly, when to start or end an activity through two approaches. One strand directly estimates the parameters of the individual's dynamic decision-making process. However, this approach can be computationally intensive particularly with a large sample. Another approach involves approximating the individual's dynamic decision-making process with a reduced-form, limited dependent variable model. Past research shows that reduced-form models perform as well as structural models in terms of prediction. As such, we employ a reduced-form, random effect approach in estimating drip irrigation system adoptions. We utilize a random effects model to account for the likely case that unobserved heterogeneity exists within our sample.

Let the latent return function, π_it, expressing the ith individual's adoption decision in time t, be a given by the following function:

$$\pi_{it} = \underbrace{y_{i(t-1)} + \beta'x_{it} + \theta_{it} + \mu_i + \varepsilon_{it}}_{V_{it}} \tag{13.1}$$

Where $y_{i(t-1)}$ represents the installed base of adoptions in the ith individual's peer group in the previous time period, x_{it} is vector of observable covariates, θ_{it} is a vector of regionally-specific time trends and common correlated effects, and μ_i is an unobserved, random individual effect where $\mu_i \rightarrow N(0, \sigma_\mu^2)$ and ε_{it} it is the model error term. Given our formulation of the latent function, π_{it}, the probability that $d_{it} = 1$ is given by the following logit expression.

$$P_{it} = \frac{e_{it}}{1+e_{it}^v} \tag{13.2}$$

Where p^{it} represents the probability that the ith individual adopts in time t. Given these probabilities, model parameters are estimated using maximum likelihood. The installed based characterizing the endogenous effect of peer adoption on individual adoption decisions are defined as $y_{i(t-1)} = \Sigma h \in g[i] \, F_{j(t-1)}$ where $F_{j(t-1)} = 1$ if the jth peer within the ith individual's peer group, g[i], adopted drip irrigation in a time period before t. The vector of observable covariates, x_{it}, consists of variables that account for distance to surface water canals and wholesale markets as well as size of their operation and the subsidy program their operation qualifies for. We account for the possibility of other unobserved factors by including a rich set of fixed effects and regional time trends captured in the parameter θit. Specifically, we include rural commune dummy variables and interactions between those fixed effects and quadratic time trends. We also address potentially unobserved spatially-temporally varying effects by specifying common correlated effects (CCE) for each rural commune as well as the entire Trifa Plain. We follow and define CCE as $\frac{\Sigma_i d_{it}}{I_t}$ where I_t represents the number of individuals that have yet to adopt drip irrigation in time t. We also account for unobserved heterogeneity amongst individuals in our preferred random effect model specification. Estimation of the random effect model rests on the following assumptions: 1) the random effect (RE), µi, and model covariates, x_{it}, are independent; 2) model covariates are strictly exogenous; 3) the random effect is normally distributed with variance, $\sigma 2$; and 4) there exist no serial correlation in the dependent variable conditional on model covariates and the random effect [6]. While these assumptions are stringent, the random effects model allows our empirical framework to account for further unobserved differences between individuals not captured in θ_{it}. Given the strict assumptions required for the RE specification, we also estimate pooled OLS (POLS) and linear probability specifications which rely on fewer assumptions but do not explicitly model unobserved heterogeneity among individuals.

There is potential concern regarding the assumed exogeneity between our model covariates and unobservable captured by the estimated random effect and the model's error term. If our covariates are not independent of these unobservable, then our model's ability to generate un- biased parameter estimates is suspect. Including a rich set of spatial and time controls partially address this selection on unobservable concerns in so far as these unobservable correlates at the regional level. However, the concern remains for unobservable at the individual level, for example if producers with more management capacity choose to farm larger operations with groundwater available that are also closer to markets and the primary surface water canal. To that extent, we rely on anecdotal evidence regarding the fixed nature of land tenure in Trifa Plain to address these exogeneity concerns. Namely, the historical and institutional setting of land tenure in the Trifa Plains and Morocco as a whole wherein many agricultural land parcels are owned by many individuals within the same family who received the land after the end of French colonization suggest that what parcels producers farm is mostly a function of what family they were born into rather than attributes of the parcel of agricultural land.

Finally, there are several important features in our data that warrant discussion. Firstly, once we observe an individual adopt a drip irrigation system we do not observe if those individual stops using their system. This presents a challenge in our modelling regarding the appropriate manner to code the binary dependent variable representing adoption after the adoption decision is taken. We could potentially set d_{it} = 1 for all subsequent periods after adoption. However, this approach assumes the drip irrigation system is utilized in each subsequent year, which may not be the case and could potentially bias model estimates. Similarly, we could follow and code post-adoption decisions as zero but this approach assumes the individual utilizes their drip irrigation system for at most one year which seems unreasonable given the effort required to adopt. We follow and drop observations from the sample in subsequent time periods after their initial adoption while keeping these observations of adoption in our calculation of the installed based use Monte Carlo simulations to test how their method of coding adoption decisions affects parameter estimates and conclude that dropping post initial adoption decisions generates peer effect parameter estimates below the true estimate. As such, this approach to coding adoption decisions produces conservative peer effect parameter estimates.

13.5 Material and Methods

In this section, we describe the parcel-level drip irrigation system adoption data we utilize to estimate the econometric model of technology adoption presented. We integrate open-source parcel data with spatially referenced drip irrigation adoption decisions observed in the Trifa Plain to generate a panel dataset of 1,364 parcel-level observations across the 2002-2012 time period. We utilize geospatial data collected in the Trifa Plain by the Morocco Economic Competitiveness (MEC) Project implemented by Development Alternatives Inc. and funded by USAID. These primary data were collected between 2011 and 2012 and include a geospatially referenced inventory of all drip irrigation system adoptions in the Trifa Plain up to 2012. These data also include information on the timing of adoption as well as a full inventory of groundwater wells.

As in many developing country contexts, the Trifa Plain lacks reliable, georeferenced land tenure data which complicates matching drip irrigation adoption to individuals and their land. To surmount this challenge, we utilize publicly available OpenStreetMap [7] (OSM) data to define agricultural parcels within the Trifa Plain. A growing literature examines the validity and quality of open source, user-generated geospatial data, concluding that such data sources are quite accurate where metrics to assess their quality exist. A smaller literature analyses the use of OSM road network data to define land parcels in urban and rural settings and finds OSM derived parcel data a useful alternative when other parcel data sources are not available.

We utilize road networks and geospatial data on urban boundaries to define agricultural land parcels. Specifically, we utilize OSM road network data within land classified as agricultural in the Trifa Plain to generate 1,364 land parcels. This characterization of land parcels does not account for potential patterns of land ownership that differ from parcel boundaries. For example, it is likely that some agricultural producers own or manage multiple parcels nearby each other. However, we do not observe these patterns of land tenure. In our later empirical modelling, we assume that the effect of an individual adopting drip irrigation on one of their nearby parcels is the same as the peer effect of another producer adopting on a nearby parcel.

We match individual land parcels to the 236 drip irrigation adoption decisions observed in our data. We define the installed base for a given parcel in time t as the number of parcels within a 0.5,1,2, or 3 km buffer [8] of the parcel's centroid that adopted drip irrigation before and also control for the number of parcels within a given buffer to account for variation in the number of neighbouring parcels or peer group size. Summary statistics regarding the average number of adopters or the installed based are provided in Table 13.1, which demonstrates the gradual increase in adoptions from 2002 to 2012.

We also utilize additional geospatial data to control for groundwater availability, parcel size, drip irrigation adoption subsidies, and market

Table 13.1 Summary statistics on peer group drip irrigation adoptions.

Year	Neighbouring adopter inside			
	½ Km	1 Km	2 Km	3 Km
2002	0.05	0.16	0.37	0.60
2003	0.07	0.20	0.47	0.77
2004	0.09	0.26	0.62	1.01
2005	0.14	0.38	0.92	1.48
2006	0.26	0.72	1.86	2.72
2007	0.43	1.15	3.11	4.56
2008	0.56	1.50	4.08	6.07
2009	0.79	2.15	5.92	8.83
2010	1.00	2.75	7.46	11.15
2011	1.03	2.78	7.65	11.51
2012	1.04	2.79	7.63	11.52

Table 13.2 Summary statistics on parcel characteristics.

Variable	Mean	Std. Dev.	Min.	Max.
GW Available	0.18	0.13	0	1
Parcel Size (Ha)	20.10	30.38	0.604	194.19
Parcels < 5Ha	0.14	0.34	0	1
Distance to Berkane (Km)	11.99	5.40	1.26	23.44
Distance to Irrigation Canal (Km)	5.13	4.06	0	19.07

and surface water access. We define parcel-level resource/groundwater availability as a dummy variable which equals one if a groundwater well is observed within a given parcel at any point in time between 2002 and 2012, and zero otherwise. In total, we observe 437 parcels with groundwater availability. We assume that groundwater availability is exogenous given anecdotal evidence that producers farm on land initially distributed to their families after the end of French colonization, over 70 years ago and before the advent of large-scale groundwater-fed irrigation in Trifa Plain.

We control for variation in drip irrigation adoption subsidies by including a dummy variable which indicates whether a parcel is less than 5 hectares, which is the threshold to qualify for the 100% drip irrigation adoption subsidy. Parcels larger than 5 ha. also qualify for subsidies but these cover a smaller proportion of adoption expenses. Finally, we control for parcel-level surface water and market access by calculating distance to the largest wholesale market in the Trifa Plain, which is located in Berkane, and the distance to the nearest large surface water canal for each parcel in the sample. These variables capture differences in transportation costs and conveyance losses in surface water delivery. Table 13.2 presents summary statistics for these time-invariant covariates.

13.6 Results

Tables 13.3 and 13.4 present random effect (RE) model specification results for the econometric framework developed in section 13.4. Specifically, Tables 13.3 and 13.4 show model results when the peer group is defined by the number of neighbours adopting within 1 km and 3 km, respectively, of an individual's parcel. The columns in Tables 13.3 and 13.4 present model results with increasing levels of spatial and time controls wherein column 1

Table 13.3 Drip irrigation adoption model with peer group defined as parcels within 1 km.

	(1)	(2)	(3)	(4)
# of Peers Adopting W/I 1 km	0.295*** (0.0484)	0.180*** (0.0347)	0.166*** (0.0438)	0.130* (0.0548)
GW Available	-0.219 (0.382)	-0.435 (0.278)	-0.460 (0.302)	-0.705+ (0.428)
# of Peers Adopting X GW	-0.0910 (0.104)	-0.0689 (0.0836)	-0.0750 (0.0897)	-0.0878 (0.118)
Parcel Size	-0.0346*** (0.0101)	0.0173* (0.00832)	0.0189* (0.00908)	0.0286* (0.0128)
Parcel Size^2	0.000209** (0.0000689)	-0.000123+ (0.0000626)	-0.000132+ (0.0000678)	-0.000194* (0.0000950)
Less than 5 Ha.	-0.996** (0.0366)	0.104 (0.271)	0.119 (0.294)	0.262 (0.404)
Distance to Canal	-0.261*** (0.0369)	-0.0527 (0.0345)	-0.0667+ (0.0382)	-0.129* (0.0527)
Distance to Market	-0.128*** (0.0208)	-0.0447+ (0.0255)	-0.0477+ (0.0280)	-0.0701+ (0.0389)
# of Parcels W/I 1 km	-0..962*** (0.00957)	-0.00649 (0.00854)	-0.00545 (0.00933)	-0.000638 (0.0129)
σ_μ^2	1.785*** (0.232)	0.560* (0.270)	0.949** (0.323)	2.330*** (0.140)
Commune Dummies	X	C	C	C
Commune Dummies X Trend^2	X	X	C	C
CCE	X	X	X	C
Observations	14059	14059	14059	14059

Standard errors in parentheses, Parcel Size2 is parcel size squared, Trend2 is a quadratic time trend + $p < 0.1$, * $p < 0.05$, ** $p < 0.01$, *** $p < 0.001$.

Table 13.4 Drip irrigation adoption model with peer group defined as parcels within 3 km.

	(1)	(2)	(3)	(4)
# of Peers Adopting W/I 1 km	0.148*** (0.0203)	0.0960*** (0.0187)	0.150*** (0.0292)	0.591* (0.0320)
GW Available	-0.464 (0.550)	-0.776* (0.373)	-0.751* (0.383)	-1.052* (0.504)
# of Peers Adopting X GW	-0.0160 (0.0402)	-0.0190 (0.0321)	-0.0135 (0.0332)	-0.0269 (0.0429)
Parcel Size	-0.0476*** (0.0112)	0.0201* (0.00922)	0.0206* (0.00951)	0.0277* (0.0121)
Parcel Size^2	0.000276** (0.0000756)	-0.000136+ (0.0000683)	-0.000138+ (0.0000702)	-0.000184* (0.0000892)
Less than 5 Ha.	-1.268** (0406)	0.0937 (0.300)	0.104 (0.308)	0.213 (0.386)
Distance to Canal	-0.291*** (0.0397)	-0.0768* (0.0385)	-0.0787+ (0.0404)	-0.128* (0.0502)
Distance to Market	-0.140*** (0.0216)	-0.0503+ (0.0285)	-0.0513+ (0.0294)	-0.0631+ (0.0370)
# of Parcels W/I 1 km	-0.0324*** (0.00329)	-0.00207 (0.00283)	-0.00337 (0.00297)	-0.000213 (0.00381)
σ_μ^2	2.443*** (0.150)	1.045* (0.320)	1.135** (0.364)	2.059*** (0.177)
Commune Dummies	X	C	C	C
Commune Dummies X Trend^2	X	X	C	C
CCE	X	X	X	C
Observations	14059	14059	14059	14059

Standard errors in parentheses, Parcel Size2 is parcel size squared, Trend2 is a quadratic time trend + $p < 0.1$, * $p < 0.05$, ** $p < 0.01$, *** $p < 0.001$.

uses no additional spatial or time controls while column 4 uses the full suite of controls. Our preferred model specification results are contained in column 4 of Tables 13.3 and 13.4 which control for commune-specific effects, commune-specific quadratic trends, and study area and commune-specific common correlated effects (CCE). Our empirical model finds modest evidence regarding the positive impact of peer effects on drip irrigation adoption. Namely, in both model specifications as increasing levels of spatial and time controls are included the impact of neighbouring adoptions, or peer effects, remains positive and statistically significant, at least the 10% level. The significance of peer effects diminishes as additional spatial and time controls are added to the model specification. Specifically, the inclusion of CCEs which control for time variant regional trends significantly reduces the statistical significance of peer group adoption rates on individual adoption decisions. This result demonstrates how the inclusion of a rich set of spatial and time controls reduces residual variation necessary to identify peer effects, particularly when dealing with smaller sample sizes.

We also evaluate the average marginal effect of peer group adoption and find that a producer is 0.23% more likely to adopt when one additional peer within 1 km adopts while a producer is 0.10% more likely to adopt when one additional peer with 3 km adopts, these average marginal effects are statistically significant at the 5% and 10% levels, respectively. These results follow intuition regarding the diminished marginal effect of peer adoption as the peer group increases in size. Furthermore, these estimated marginal effects align with marginal peer effect estimates in past literature, in both developed and developing country contexts, which finds that an additional peer adopting increases the likelihood of individual adoption between 0.1% and 0.76%.

Modelling results also reveal a consistent negative and statistically significant, at least the 10% level, relationship between the availability of groundwater and the likelihood of adoption. This result speaks to the importance of resource constraints and availability in determining adoption decisions and aligns with past empirical and theoretical research investigating the relationship between technology adoption and resource constraints and price. In our context, agricultural producers with groundwater are less resource constrained than their counterparts who rely solely on stochastically available surface water supplies. As such, the expected parcel-level returns of drip irrigation adoption are less when groundwater is available on the parcel than when groundwater is not available.

We also estimate marginal peer effects for both producers with and without groundwater available on their operation. These results reveal that average marginal peer effects are consistently greater for producers without access to groundwater on their operation. Specifically, for the model with peer groups defined by a 1 km buffer, the average marginal peer effects are 0.17% and 0.24% for producers with and without groundwater, respectively, which are both significant at the 5% level. For the model with peer groups defined by a 3 km buffer, average marginal peer effects are 0.06% and 0.11% for producers with and without groundwater, respectively. However, only the marginal peer effect for producers without groundwater is statistically significant at the 10% level. These results suggest that peer effects are potentially more salient among producers without access to groundwater.

Results also reveal a positive and statistically significant relationship between the probability of adoption and parcel size when the full set of spatial and time controls are included. Furthermore, the positive coefficient on parcel size squared implies that the magnitude of this positive effect is decreasing as parcel size increases. We also find that the 5 ha indicator variable is not statistically significant in any of the specifications which include spatial or time controls. This result suggests that differences in drip irrigation subsidies are not a significant factor driving adoption decisions.

Finally, our results indicate a negative relationship between the probability of drip irrigation adoption and distance to market, implying that access to markets and the government services in Berkane are a significant determinant of adoption decisions. Our results also reveal a relatively consistent and statistically significant relationship between distance to surface water canal and the probability of adoption which is somewhat counterintuitive given we would expect that parcels further from a canal would be more resource constrained given water conveyance losses and thus experience increased returns from adoption. Rather, it is likely that this variable is capturing a separate effect, particularly if the location of surface water canals was determined by land quality attributes which are otherwise not controlled for in our model.

We also estimate our empirical model of drip irrigation adoption using the percentage of peers adopting within the peer group. These results demonstrate that the impact of peer adoption is less robust when the peer effect variable enters the model as percentage rather than a level. This result provides evidence that there may be differing mechanisms based on percentage rather than level of peer group adoption through which peer effects influence adoption decisions.

We test the robustness of our empirical results presents model results for differing spatial buffers defining peer groups, model specifications, and formulations of the peer effect variable. wherein peer groups are defined by 1/2 and 2 km buffers are similar to those presented here, providing evidence that our results are not particularly sensitive to the distance defining a parcel's peer group. Displays empirical model results using a pooled OLS model specification which qualitatively align with the random effects specification results presented above.

13.7 Conclusion

This paper investigates agricultural producers' decision to adopt an efficient irrigation technology and how peer effects and resource constraints determine this choice. We utilize parcel-level data on drip irrigation adoption decisions from the Trifa Plain of northeastern Morocco to empirically estimate this relationship. Our results reveal that peer effects, based on spatial proximity, have a limited impact on drip irrigation adoption decisions when spatial and time controls are included in our empirical modelling. We also find that groundwater availability negatively influences drip irrigation adoption.

Our results regarding the impact of peer effects on drip irrigation adoption provide modest evidence supporting the notion that peer group adoption rates influence individual adoption decisions. Specifically, we find that an additional peer adopting increases the likelihood of individual adoption between 0.10% and 0.23%, depending on the spatial buffer defining the peer group. This result aligns with past technology adoption literature which finds peer group adoption a significant determinant of adoption decisions. We describe our results as modest given the diminished level of statistical significance of peer effects as increasing levels of spatial and time controls are included. However, this result may be related to our relatively small sample size and the minimal residual variation left to identify peer effects after controlling for unobservable with the full suite of controls common in the literature (e.g. regional time trends, CCEs). Future research should explore peer effect identification issues within the context of small sample sizes common in data poor regions.

Our results also reveal that resource availability decreases the probability of drip irrigation adoption which aligns with past conclusions in the literature. The availability of alternate water supplies measured by access to groundwater diminishes resource constraints and reduces the returns to drip irrigation adoption. We also find that marginal peer effects are diminished for producers with access to groundwater compared to those without access, implying that peer effects are potentially more salient for more resource constrained producers. These results demonstrate the potential impact of conservation policies that incentivize individuals to account for resource scarcity via pricing or other resource management policies on the adoption of efficiency-enhancing technologies. More specifically, our results suggest that drip irrigation subsidy rates that vary according to resource availability could potentially increase drip irrigation adoption rates among the least resource-constrained individuals. Future research should investigate how the source of water available to producers influences the likelihood of engaging in pro-conservation behaviour and adopting an efficiency enhancing irrigation technology.

A weakness of our analysis lies in our characterization of land parcels based upon road boundaries. It is likely that actual patterns of land tenure and management do not strictly align with road-based land parcels. This potentially introduces some biases in our modelling if other avenues (e.g., land ownership) outside peer effects influence how adoption decisions in a given parcel affect neighbouring parcels. Future research should evaluate the impact of using road-based parcels rather than observed land ownership patterns in models of technology adoption and land-use.

References

1. [Cunningham et al., 2016] Cunningham, S., Bennear, L. S., and Smith, M. D. (2016). Spillovers in regional fisheries management: Do catch shares cause leakage? *Land Economics*, 92(2):344–362.
2. [Daoud and Engler, 1981] Daoud, Z. and Engler, I. (1981). Agrarian capitalism and the moroccan crisis. *MERIP Reports*, 100(99):27.
3. [den Broeck and Dercon, 2011] den Broeck, K. V. and Dercon, S. (2011). Information flows and social externalities in a Tanzanian banana growing village. *Journal of Development Studies*, 47(2):231–252.
4. [Dinar and Yaron, 1990] Dinar, A. and Yaron, D. (1990). Influence of quality and scarcity of inputs on the adoption of modern irrigation technologies. *Western Journal of Agricultural Economics*, 15(2):224–233.
5. [Dridi and Khanna, 2005] Dridi, C. and Khanna, M. (2005). Irrigation technology adoption and gains from water trading under asymmetric information. American *Journal of Agricultural Economics*, 87(2):289–301.
6. [Drysdale and Hendricks, 2018] Drysdale, K. M. and Hendricks, N. P. (2018). Adaptation to an irrigation water restriction imposed through local governance. *Journal of Environmental Economics and Management*, 91:150–165. 100
7. [Duflo et al., 2011] Duflo, E., Kremer, M., and Robinson, J. (2011). Nudging farmers to use fertilizer: Theory and experimental evidence from Kenya. *American Economic Review*, 101(6):2350–2390.

8. [Dumler *et al.*, 2007] Dumler, T. J., Rogers, D. H., and O'Brien, D. M. (2007). Irrigation capital requirements and energy costs. Agricultural Experiment Station and Cooperative Extension Service, Kansas State University.

14

Energy-Efficient Dough Rolling Machine

Nerella Venkata Sai Charan[1], Abhishek Antony Mathew[1], Adnan Ahamad Syed[1], Nallavelli Preetham Reddy[1], Anantha Krishnan V.[1] and O.V. Gnana Swathika[2]*

[1]*School of Electrical Engineering, Vellore Institute of Technology, Chennai, India*
[2]*Centre for Smart Grid Technologies, School of Electrical Engineering, Vellore Institute of Technology, Chennai, India*

Abstract

We all know that Flat breads are very popular, especially in those parts of the world where bread constitutes a major source of dietary protein and calories. In India, wheat is a daily staple, consumed in different forms. With urbanization and industrialization, there is a rise in the demand for ready-to-eat and easy-to-make products, but making the equipment available at a reasonable cost, energy efficient and making it work under optimal conditions was a big task.

In this paper we are designing a dough rolling machine which is energy efficient, does not consume much power and comprises a simple mechanical process. It is lightweight and is available for an affordable cost. Sheeting is usually carried out by a pair of sheeting rolls with rotation motion. A portable and manually operated, energy-efficient dough rolling machine was designed and fabricated.

The analysis of the experimental results revealed that the machine consistently provided the necessary products by consuming a low amount of power. The proposed equipment not only allows to efficiently obtain a product but also allows us to define the optimal shape. Thus, the main aim of this paper is to reduce the power consumed by the equipment and also the speed at which the rolling process is done.

Power efficiency helps to save on electrical costs and is therefore a necessary consideration to take. We use an efficient SMPC supply and a simple gear setup in order to minimise losses and hence raise efficiency. While there are many dough

---

*Corresponding author: gnanaswathika.ov@vit.ac.in

sheeters in the market, ours is targeted towards usage in ordinary homes rather than in mess halls or cafeterias, hence the simple but efficient design.

Keywords: Switch mode power supply, gears, speed reduction, rotating platform and rollers, main base structure

14.1 Introduction

Energy-Efficient Dough Rolling machine is a solution that comes to the minds of those who understand the difficulty of Indian women living in rural areas. Not only Indian women but also those in the country that live below the poverty line need time to make food. This project aims to help them make it faster and with more ease. Roti is one of the most common types of food in India.

In making a dough rolling machine, it is very important to manage all the parameters like energy efficiency, design, appearance, cost, strength, power consumption and user friendliness, thus making it optimum. We can add many more features to the machine like further reducing its weight, integrating electrical protection for the system, adding control circuitry, etc.

14.2 Methodology

The following block diagram shows the power flow of the process from alternating power source to the rotating platform (see Figure 14.1).

The entire setup consists of an SMPC, a DC motor, a pair of gears, a rotating platform, a roller, and a structure to hold all of the above. The SMPC converts the AC supply to a DC supply for the DC motor. The two gears are in a size ratio of 1:3.3 [2], with the smaller gear (driven gear) connected to the DC motor and the larger gear (driving gear) connected to the rotating platform by means of shaft, and the shaft is connected to the rotating platform. The roller is attached to the structure in such a way

SMPS (switch mode power supply) → Electric motor → Gears coupling → Rotating platform

Figure 14.1 Flow chart.

that the dough can be placed on the rotating platform and removed from it with ease.

The roller is placed on the rotating platform in such a way that the friction due to the rotational motion would prevent the platform from rotating properly and even stop if left idle. In order to increase torque, attach the motor to the smaller of the two gears and then couple the two gears. The power remains unchanged, but the speed of rotation and in the larger gear is slower than in the smaller gear while the torque is increased. This is because torque and rotational speed are inversely proportional at constant power.

$$P = \omega * \tau \qquad (14.1)$$

Where P is power, ω is angular speed, τ is torque.

In modifying the torque and speed delivered by the motor as per the requirement, conventional gear system is followed. On considering the fact that any mechanical driven system has losses but the energy efficiency of the system matters too, single-stage spur gear system is used to develop the necessary torque. Spur gears in the machine reduces the load on the motor which helps in reduction of current drawn which proportionally decreases the losses in the system. Energy would still be lost in the form of heat and sound; in order to deal with that, cast iron is a material used for gears considering efficiency and properties of wear and tear.

By means of gear coupling, we have transferred the power while minimising the loss in the overall system and thus raised the torque and decreased the speed of the rotational platform. With the decreased speed, the dough will not be thrown off due to high-speed spinning. With the increased torque, the roller will not be able to stop the rotation of the rotating platform.

The roller is actually two separate cylindrical rods – each capable of rotation – which are connected together. When the dough is placed on the rotating platform and pressed down on by the roller, there will be much more resistance to the rotation of the platform if the rod was a single solid piece that did not rotate, and the dough would not be flattened properly if the rod was a single rotating piece which might not even rotate at all and hence would be not much different than a non-rotating rod. The direction of the rollers is opposite to each other and the directions are clockwise and anti-clockwise with respect to the axis of the rollers, respectively.

14.3 Specifications

14.3.1 Motor

A simple DC motor with a gear box is attached to the shaft of a mechanically coupled electric motor powered by direct current. It provides a massive torque of 15kgcm. The shaft has metal-bushings for wear resistance. A motor of constant speed is very helpful in modifying the speed as per the requirements. The motor is fitted to the main base structure. Specifications of the motor are shown in Table 14.1.

14.3.2 Switch Mode Power Supply (SMPS)

The motor which drives the platform runs on direct current (DC), so to provide it we use SMPS, which converts AC supply to direct current (DC). The motor is connected to SMPS to power it [4] (Table 14.2).

Table 14.1 Specifications of motor.

Speed	1000RPM
Voltage	12 V
Shaft diameter	6 mm
Shaft length	15 mm
Stall torque	15 kg cm
No load current	800 mA
Stall current	Up to 9.5A (maximum)

Table 14.2 Specifications of SMPS.

Input voltage	230 V(AC)
Output voltage	12 V(DC)
Maximum current	5A

14.3.3 Speed Reduction

Through some assumptions and tests we estimated that the motor requires three times more torque. It is obvious that the dough will not always be at the same consistency, especially when considering the different type of flours that could be used, so there are no appropriate values of speed and torque. Speed of the motor has less relevance than its torque, because when the platform rotates, its rotation will be opposed by the friction. Based on the dough, the friction will vary, so we change the applied force accordingly. In order to increase the torque by compromising the speed of the rotating platform, we used the conventional speed reduction technique using gears. In other words, we can say that power is transferred through the gears. Using the gears, we are reducing the shaft speed by keeping the power transferred constant, which helps in amplifying the shaft torque.

$$P = \omega * \tau \qquad (14.1)$$

where, ω = speed, τ = torque Power of the gear coupled to motor = power of the gear coupled to shaft (14.2)

$$\omega 1 * \tau 1 = \omega 2 * \tau 2 \qquad (14.2)$$

From equation (14.2) we can find the angular velocity of the gear which is coupled to shaft that is

$$\omega 2 = \frac{\omega 1 * \tau 1}{\tau 2} \qquad (14.3)$$

As per the requirement, we modified the speed-torque characteristics of the shaft using gears. Now the speed and torque of the motor are 1000rpm and 15kgcm and the speed and torque of the shaft are 333.33 rpm and 45kgcm. Now, for the analytical part we designed reduction gear. As we need to decrease the speed of the shaft the number of teeth of the gear connected to the shaft should be more.

$$\text{Gear reduction ratio} = x/y \qquad (14.4)$$

where, x = number of teeth on motor side y = number of teeth on shaft side. The gear ratio of the reduction box is 1:3.

14.3.4 Coupler

The process of coupling plays a major role in power transfer by means of mechanical parts. One gear is connected to motor and another gear is attached to the shaft. The steel shaft of 6mm diameter is used in the process of coupling. As the steel shaft has more flexibility of usage and shows good strength, toughness, and wear resistance, nylon gears were used. The shaft is fitted to the chassis by means of ball bearings at the ends. The gears are coupled and then the shaft is attached to the platform by means of flanged aluminium coupler with bore diameter of 6mm. Couplers of this type are generally used to couple 6mm shaft with anything like link or crank or another 6mm shaft.

14.3.5 Main Base Structure

The main base structure provides strength, balance, and makes up the outer appearance. In this project, we are developing strength, balance, and appearance of the chassis using conventional box (cuboid) structure. A box structure provides a larger area which makes it simple to attain balance. The walls of the box structure face less shear stress compared to any other structure. Dimensions of the main base structure are shown in Table 14.3.

14.3.6 Rotating Platform and Rollers

The rotating platform and rollers are what is used to roll the dough. The rotating platform is nothing much; it is just a disc which rotates by means of some rotating drives. Roller is a cylinder that rotates about its central axis to spread or flatten the dough. There will be two rollers with common axis joined together by means of shaft and placed directly above the rotating platform. The center of the two rollers is to be matched with the centre of the rotating platform. When we place dough between rotating platform

Table 14.3 Dimensions of the cuboidal (box) structure.

Length	13 inch
Breadth	5 inch
Height	6 inch
Thickness of the wall	0.2 inch

and rollers and activate the machine, the rollers will begin to rotate, then the platform will rotate due to friction caused by rotation of the rollers. This causes the dough to be flattened.

Chassis: The word chassis refers to the stationary parts of the structure which give strength to the whole structure. The design aspects of the chassis play an important role in building any structure. We opt for 3D printing method in constructing the dough rolling machine. The 3D structural design of dough rolling machine is designed in solid works software. 3D printing provides very accurate structure which is very important. If the structure is not well made, the movement of mechanical components will be stiffer, acoustics and wear and tear will be present, which will lead to energy loss. In building any structures, the structure is only as good as the material used to build it. Polyastic Acid (PLA) is used in building the structure as it is eco-friendly and a rather strong plastic among the available plastics. The structure of dough rolling machine is composed of a main base structure, rotating platform, and rollers [7].

14.3.7 Rotating Platform

The rotating platform is a disc which is coupled to the shaft. This is where the dough is placed. It is very important to select the proper material for the rotating platform to provide friction to the cover which is placed on the dough. We can use PLA or well-sanded wood. Dimensions of the rotating platform are shown in Table 14.4 [7–9].

14.3.8 Rollers

Rollers are the structures which flatten the dough by coordinating with rolling platform [1, 3, 6]. Two cylindrical structures are passed through a steel rod and the cylinders are coupled with ball bearings to rotate smoothly [5]. PVC pipes can be used as the rollers as they are very easily available in the market. The rollers are attached to a hinged handle which helps in pressing

Table 14.4 Dimensions of rotating platform.

Dimeter	10 inch
Thickness of rotating platform	0.2 inch

Table 14.5 Dimensions of roller.

Diameter of roller	1 inch
Length of each roller	6.5 inch

the roller as well as removing them from the dough [10]. Dimensions of the rollers are shown in Table 14.5.

Assembly: Motor is fitted to the bottom of the main base structure. Then bearings are fitted on both sides of the main base structure, then the shaft is passed through both bearings with a gear on the shaft and then both gears are coupled by taking appropriate measures. Now we fix the flanged aluminium coupler to the shaft and then we couple both flanged coupler and rotating platform. A steel rod is passed through supporting elevations on the back of the main base structure. Rollers are fitted to the shaft such that the rollers are diametrically parallel to the rotating platform. The gears are lubricated and the SMPS is connected to the motor. The combination of the three structures: main base structure, rotating platform, rollers with necessary support will be the final structure of the dough rolling machine [7–9].

14.4 Result and Discussion

As per loading test of the machine the maximum electrical power consumed by the machine, i.e., at stalling condition, is 26.91 W. The sufficient applied manual force is in the range of 1500g to 2200g weight equivalent of force. On considering 2200g as constant load on the machine for

Table 14.6 Loading specifications of the motor.

Load	Voltage (V)	Current (A)	Power (W)
No load	11.6	0.23	2.668
750	11.7	0.3	3.51
1500	11.8	0.7	8.26
2200	11.8	0.6	7.08
stall	11.7	2.3	26.91

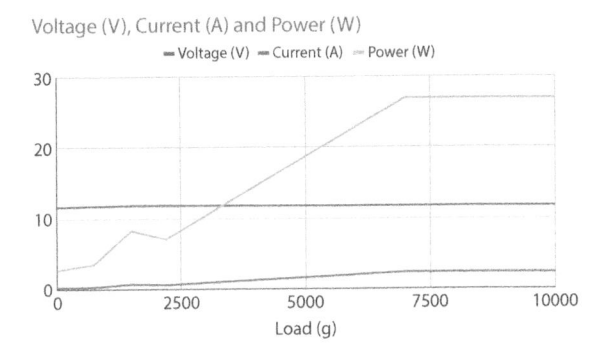

Figure 14.2 Power consumption at different loadings.

an hour per day and it is consuming only 7.08W of power. Scaling it to per year consumes 2,584.2W (2.5KW) equal to 2.5 units. As per Andhra Pradesh Electricity Regulatory Commission (APERC) the average cost of unit power is 6.37 rupees and tariff is 16.4613 rupees. Many commercial dough rolling machines are using 250W motor. If we consider this type of machine the tariff will be too high it will consume 91.25KW of power per year and the tariff is 581.2625 rupees. The spur type gear fabricated from nylon is used in the machine and it can reduce the energy losses and stress on the axis of the motor. The cost of manufacturing the machine is 1500 rupees. The weight of the machine is 2.5 Kg. The observations are tabulated (Table 14.6) and plotted (see Figure 14.2).

14.5 Conclusion

The dough rolling machine is simple and optimal because it satisfies the considered parameters power consumption, energy efficiency, cost, weight and user friendliness. This simple machine will help to roll dough in a simple manner and it can roll evenly, which is the major consideration in making sheets. Comparing these Dough Rolling Machines with commercial machines, they are very low tariff and with low and easy maintenance.

References

1. B. K. Yoo and D. S. Yoo, "A roll gap adjustment for uniform thickness of dough sheets of instant ramen noodles," 2011 *IEEE/SICE International Symposium on System Integration (SII)*, 2011, pp. 1003-1005, doi: 10.1109/SII.2011.6147586.

2. E. Park, C. Kim, S. Jung and Y. Kim, "Dual Magnetic Gear for Improved Power Density in High-Gear-Ratio Applications," *2018 21st International Conference on Electrical Machines and Systems (ICEMS)*, 2018, pp. 2529-2532, doi: 10.23919/ICEMS.2018.8549367.
3. F. -. Wang, O. Kwan and T. -. Yi, "Dynamic modeling of rotating flexible platforms," [1992] *Proceedings of the 31st IEEE Conference on Decision and Control, 1992*, pp. 1315-1316 vol.2, doi: 10.1109/CDC.1992.371500.
4. G. Ortenzi and J. Antenor, "Switch mode power supply applied to very low cost electronic board of home appliances," *2009 Brazilian Power Electronics Conference*, 2009, pp. 291-297, doi: 10.1109/COBEP.2009.5347640.
5. H. Triyono, I. Priadythama and F. Fahma, "Conceptual design for dough processing integrated machine to obtain uniformity of karak size in traditional karak industries," *Proceedings of the Joint International Conference on Electric Vehicular Technology and Industrial, Mechanical, Electrical and Chemical Engineering (ICEVT & IMECE)*, 2015, pp. 229-232, doi: 10.1109/ICEVTIMECE.2015.7496679.
6. Voicu, Gheorghe & Casandroiu, T & Constantin, Gabriel & Stefan, Elena-Madalina & Munteanu, Mariana. (2017). Considerations about the sheeting of dough pieces to obtain bread.
7. Kai Jun Chen, Joseph D. Wood, Idris K. Mohammed, Shirley Echendu, David Jones, Kate Northam, Maria N. Charalambides, Mechanical Characterisation and modelling of the rolling process of potato-based dough.
8. Engmann J., Peck M.C., Wilson D.I. (2005). An experimental and theoretical investigation of bread dough sheeting, *Food and Bioproducts Processing*, 83(C3): 175–184.
9. Levine L., Drew B.A. (1990). Rheological and engineering aspects of the sheeting and laminating of doughs. In: Faridi, H.M., Fanbion, J.M. (Eds.), *Dough Rheology and Baked Product Texture*, New York: van Nostrand Reinhold, pp. 513–555, Chapter 14.
10. Wang C., Dai S., Tanner R.I. (2006). On the compressibility of bread dough, *Korea-Australia Rheology Journal*, 18(3): 127-131.

15

Peak Load Management System Using Node-Red Software Considering Peak Load Analysis

Mohit Sharan, Prantika Das, Harsh Gupta, S. Angalaeswari*, T. Deepa, P. Balamurugan and D. Subbulekshmi

School of Electrical Engineering, Vellore Institute of Technology, Chennai, Tamil Nadu, India

Abstract

India is a very fast-developing country. Over 1.5 billion people reside in it and in the 21st century every activity associated with growth and development is complemented with very high energy demand. Hence, it is very challenging to ensure "Power for all" with the existing systems backed by non-renewable energy sources and a naïve system, which exists under the name of "Energy Management". With the realization of the above-stated fact, it is high time to look into a smarter approach to energy management and step towards the goal of ensuring "Power to all". By introducing a transparent operational system between the utility and the consumers we can implement a smarter technique in energy management, which benefits both the consumer and the utility in every possible way and also helps in improving customer-provider relations. In order to achieve the goal, we have come up with a solution by creating an interface between the utility and the consumer, enabling the customers to observe and map the peaks throughout the day. Based on that, a dynamic billing pattern can be set which will guide consumers towards a judicial usage of the existing power delivery and consumption system.

Keywords: Peak Load Management (PLM), load curve, load profile, Kilowatt hour (KWh), Kilowatt (KW), dynamic billing system, Internet of Things (IoT), Time of Day (TOD)

---

*Corresponding author: angalaeswari.s@vit.ac.in

Milind Shrinivas Dangate, W.S. Sampath, O.V. Gnana Swathika and P. Sanjeevikumar (eds.) Integrated Green Energy Solutions Volume 1, (217–228) © 2023 Scrivener Publishing LLC

15.1 Introduction

Energy Management is a procedure to track and adjust the use of energy consumption in a locality or a specified area. Planning and operational management of energy consumption and production is a part of energy management. The few approachable and implementable methods of energy management are metering energy consumption, data acquisition, and finding and quantifying opportunities to save energy. Hence, a Smart Energy Management System can help the reduction of monetary value while still meeting the electrical power demands of the people. Various advanced technologies are being implemented in the field of Internet of Things (IoT) and Data Analytics which might be utilized in better managing energy in the near future, leading to the emergence of Smart Homes where data collected from each electrical device is sent to a centralized server for further analysis to optimize energy consumption. The analysis of large chunks of information from various devices in a residential block of the area is where Big Data will come into play using Business Intelligence (BI) platforms.

Since recently, Home Energy Management (HEM) Systems and Smart Grid (SG) technology have been majorly used for efficiency control demand and supply of power. HEM Schemes are primarily used to reduce Peak Average Ratio, electrical consumption cost, peak demand, and wastage of energy. Various cost-efficient techniques such as Real-Time Pricing (RTP), Critical Peak Pricing (CPP) can be inculcated to meet the objectives of reduction of Energy consumption cost both on the utility end as well as the consumer end. Smart Grids contribute to smart features which are used to increase the efficiency of the electrical distribution system which makes the whole infrastructure of power consumption more sustainable and reliable.

IoT is among the most crucial technologies that have evolved in a short span of time due to its nature of connecting a large range of people with devices around them. The internet-based technology will help in connecting one grid station to another by rectifying any faults if any, thus removing interruptions that exist during data transmission. The IoT platform allows devices to be connected and controlled remotely using network infrastructure.

A system, which brings both the consumers and utilities together in implementing energy conservation in a smarter way without compromising on any basic need of the user in general and hence, contributing to an efficient energy management system is something that can majorly impact the energy sector. If we look at the global level, energy management is a global concern. The need of the hour is to reduce the damage already done on the planet. Conventional sources are limited hence with repeated exploitation of these sources, they are prone to get exhausted very quickly. So, it is a global need to save energy resources and target sustainable development. Dependency on fossil fuels can be reduced by implementing proper energy management techniques. For the same, a few general optimizations and consumption tweaking is needed to be done in order to achieve the desired output.

The main motivation behind this work is the importance of understanding why it is the need of the hour to stress managing the available energy resources. Power is a very important parameter in a developing country like India as it is an essential aspect which we should look into so that we are able to ensure "power to all" and contribute to pacing up the development process. A better saving rate does not only mean less consumption but in the long run it could also mean more investments resulting in greater economic growth. Hence, the rate of GDP growth is directly proportional to personal savings.

In this paper, an idea is proposed and elaborated which will eventually cut down the extensive use of energy in the long run. In this proposition, the data related to the power consumption on a daily basis is in the form of graphs serving as a visual representation, thus enabling when the peak in the power consumption load curve is attained. An IoT-based interface is implemented which will notify the consumers at what Time of Day (TOD) the power consumption should be reduced so as to attain a flat load curve. The flat load curve will optimize the energy consumption thus reducing the load on the power grid and also cut down the bill paid by the users. Hence, this is a bi-profitable system beneficial for both the utility and the consumer end. This interface also provides a "Dynamic billing system" giving pricing details based on the user's consumption for the rest of the day. This will enable the user to shift towards smarter energy consumption techniques throughout the period. The IoT interface is further reduced to a "Smart-Home" user-friendly app, which helps the user to control all home appliances from one place, by a single click reducing the delay in time to switch off the electrical equipment when the peak is attained.

15.2 Methodology

15.2.1 Peak Demand and Load Profile

Peak demand describes a period in which electrical power is expected to be provided for a sustained period at a significantly higher than usual supply level. This factor never remains constant throughout the day. They are subjected to changes and drastic fluctuation, thereby affecting the grid. Peak demand fluctuations may occur periodically such as on daily, monthly, seasonal and yearly cycles. The actual point of peak demand is a single half-hour or hourly period which represents the highest point of customer consumption of electricity, a variation in this only affects the grid. A load curve or load profile is a graphical representation of electrical load over a specified time period. The study or analysis of load curves helps the generation companies to decide how much electricity to generate and distribute. Hence, these two factors tell us about the current power demand as well as a trend in the load profile of a particular location. This, as a whole, dictates the required managerial steps for Peak Load Management.

15.2.2 Need of Peak Load Management (PLM)

By implementing Peak Load Management (PLM), we can clip the load peaks and significantly reduce electricity costs, both in the current period (delivery) and during the next capacity period (supply). The monthly payments are determined by both the actual energy you consume (kWh) and the amount of energy that needs to be available to serve your account based on your peak load KW demand. The mentioned factors not only contribute to consumer pocket management but also benefit the grids and the environment as well. When combined with capital improvements, peak load management can save hundreds of thousands of crores annually, while reducing GHG emissions by shifting load overnight.

15.2.3 Data Analysis

It can be clearly observed that Peak Load Management is a very important factor in Energy Management. It majorly impacts the conservation strategies and managerial approaches in the power sector. Dataset of power consumption from different distribution companies were taken and analyzed. Power consumption does not have a particular trend and it varies significantly throughout the day. This fluctuation affects the utility companies as the peak load curve obtained is not flat enough, which is the ideal requirement. We require a flat curve to maintain the requirement of the electricity demand from that particular utility at the same rate and also the utility must not be under pressure, otherwise the distribution companies will have to come up with new utility units and this will result in an increase in the cost of consumption, thereby hurting the pockets of the consumers.

15.2.4 Need to Flatten the Load Curve

Load management is a means to derive economic advantages in the operation of power plants. For this purpose, efforts are made to obtain as flat a load curve as possible. Flattening of load curve leads to avoidance of start-up and shutdown of power plants, less requirement of regulating capacity, reduction of transmission cost, and avoidance of overload of equipment in parts of the system.

15.2.5 Current Observations

When similar datasets as shown in Figure 15.1 were observed a common trend was observed in the power consumption pattern. It showcased drastic variations with Time of Day (ToD), i.e., at a particular hour of the day the KW value was low and at a particular hour of the day it was relatively very high hence, peaking. Such a variation tends to affect the utility company as the grid health gets affected drastically. In order to maintain the grid health, the utility companies tend to charge more from the consumer. Also, this can lead to power shedding thereby pilfering the rights of the consumer. Hence, from the above observation, we can infer how unpleasant the variation in power consumption can be for the utility as well as the consumers.

15.2.6 Equations

From the above set of observations in order to build a bi-profitable system, i.e., both the utility and the consumer are benefited in every aspect, it is important to average out the power consumption from the historically available dataset to flatten the load curve.

$$P_{avg} = \Sigma P_{obs}/n \qquad (15.1)$$

where:
P_{avg} = Averaged out power in KW
P_{obs} = Power Observed at particular time instant in KW
n = Total time instance of observation

The required analysis can be carried out after analyzing large sets of data over a long period of time to find the best-suited average value.

15.3 Model Specifications

In accordance with the current scenario, a peak load curve model will be developed using R- Studios and Node-RED and the user will get an idea

Hrs	Time	Central_Delhi	BRPL	BYPL	NDPL	NDMC	MES
1	0.0	00:00:00	4826.55	2110.31	1077.54	1419.07	147.67	26.09
2	0.5	00:30:00	4708.35	2059.85	1049.80	1381.80	146.37	25.88
3	1.0	01:00:00	4583.10	2006.26	1018.31	1343.42	145.84	25.55
4	1.5	01:30:00	4444.37	1938.01	994.88	1304.22	139.92	25.10
5	2.0	02:00:00	4335.78	1883.70	974.25	1274.16	142.46	24.62
6	2.5	02:30:00	4222.97	1828.79	944.26	1242.00	142.90	24.29
7	3.0	03:00:00	4123.68	1781.63	923.71	1212.61	142.61	23.87
8	3.5	03:30:00	4029.75	1739.83	895.42	1189.82	142.65	23.78
9	4.0	04:00:00	3957.36	1710.55	879.67	1160.65	142.75	23.63
10	4.5	04:30:00	3890.08	1687.56	857.06	1141.32	141.65	23.70
11	5.0	05:00:00	3819.34	1657.22	837.87	1122.71	141.14	24.02
12	5.5	05:30:00	3782.46	1639.23	827.61	1111.72	143.46	24.47
13	6.0	06:00:00	3725.69	1611.42	822.19	1087.25	144.39	24.28
14	6.5	06:30:00	3711.15	1586.31	838.92	1081.44	144.23	25.02
15	7.0	07:00:00	3671.42	1558.33	832.47	1069.91	148.90	26.24
16	7.5	07:30:00	3728.69	1560.20	851.07	1098.03	156.37	27.78
17	8.0	08:00:00	3731.48	1551.02	846.56	1109.89	159.29	29.42
18	8.5	08:30:00	3814.48	1591.73	845.33	1141.14	170.68	31.23
19	9.0	09:00:00	3903.71	1632.34	846.85	1167.88	187.24	32.93
20	9.5	09:30:00	4178.27	1727.03	887.35	1291.67	199.20	34.28
21	10.0	10:00:00	4417.28	1811.02	947.48	1372.82	209.45	34.54
22	10.5	10:30:00	4456.04	1831.79	947.73	1373.18	223.32	36.17
23	11.0	11:00:00	4530.49	1854.91	956.36	1408.15	232.31	35.85
24	11.5	11:30:00	4623.58	1888.48	971.24	1443.79	239.17	36.50
25	12.0	12:00:00	4834.03	1956.14	1030.52	1524.99	239.85	36.53
26	12.5	12:30:00	4829.77	1942.57	1044.02	1522.97	244.50	35.59
27	13.0	13:00:00	4717.93	1898.19	1052.96	1441.54	244.41	35.71
28	13.5	13:30:00	4483.71	1829.78	1006.13	1337.86	236.31	33.44
29	14.0	14:00:00	4556.20	1826.61	1015.30	1404.54	234.98	31.41
30	14.5	14:30:00	4729.42	1896.03	1036.78	1486.71	232.96	31.92
31	15.0	15:00:00	4877.13	1963.51	1063.21	1538.10	233.46	32.61

Figure 15.1 Snippet of the dataset, which gave the current 24-hour consumption of energy from different utilities in the Delhi Region.

(via a user interface developed using Node-RED) when peak load is about to reach; they can alter their usage pattern accordingly. In addition to this, a variable tariff pattern and prepaid billing system will be incorporated, thereby encouraging them for taking smart measures for energy consumption. In order to analyze the data and provide a control system clubbed

Peak Load Management System Using Load Analysis 223

with electricity consumption monitoring system was developed using R and Node-RED. With the induction of Node-RED into the development of the system, we look into much of an IoT-based approach of dealing with the problem of Energy Management hence making the system a lot smarter.

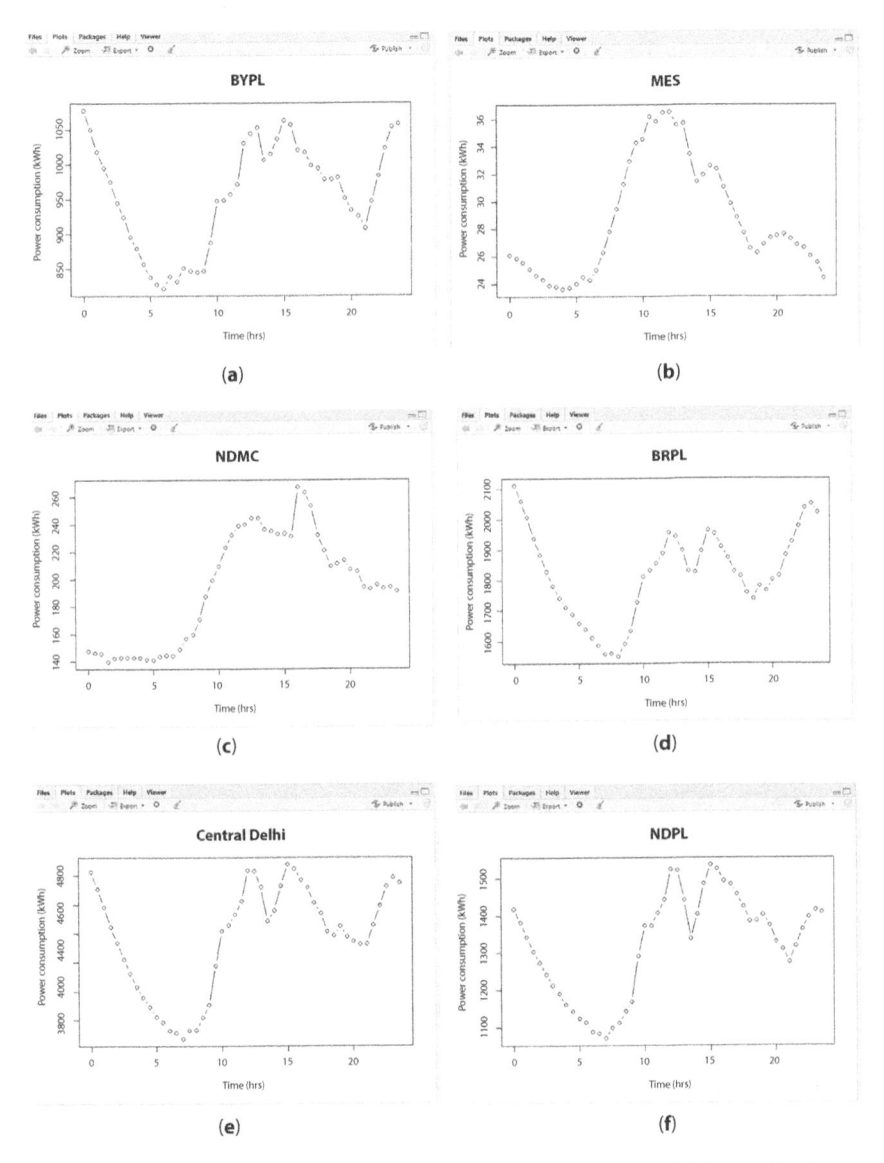

Figure 15.2 (a) Peak load curve of BYPL. (b) Peak load curve of MES. (c) Peak load curve of NDMC. (d) Peak load curve of BRPL. (e) Peak load curve of Central Delhi. (f) Peak load curve of NDPL.

A brief description of both the systems has been explained below: R (used for analyzing the problem), as mentioned earlier, we have the half-hourly energy consumption pattern of different distribution companies of Delhi state as shown from Figure 15.2(a) to Figure 15.2(f). The collected data was studied and results were obtained which gave us an idea of mean KW consumption which would help in setting up particulars like threshold limit of peak load warning, time of the alert, etc. Node-RED (for giving an IoT-based energy management set-up and the UI for the consumer), using Node-RED an interface has been developed for the customers where they could monitor their consumption, keep an eye on the load curve, track their optimum

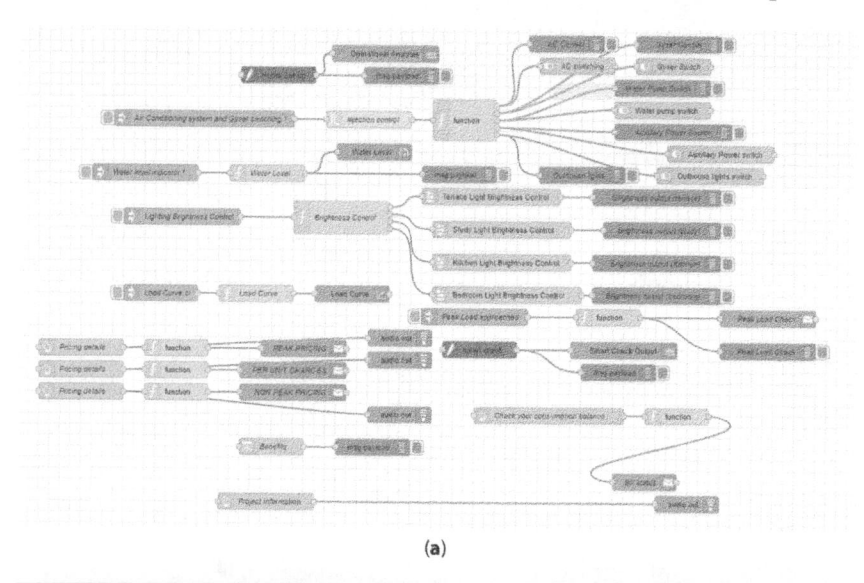

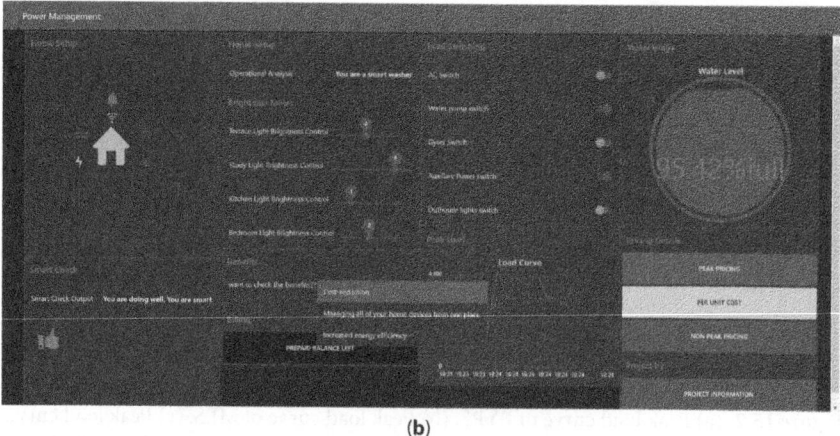

Figure 15.3 (a) Node control of the UI. (b) Overview of UI.

consumption value and also control the various appliances in their home and other basic smart house and energy management parameters incorporated.

From the peak load curve attained, at particular hours of the day the load curve is very high and during a few hours of the day it is on the lower side. This fluctuation needs to be compensated for the sake of energy conservation and management. Once the data analysis and problem assessment are done the load curve data (live data) would clown on the UI as displayed below.

The UI has been developed on a node-red platform by the means of "Visual Programming" where the backend codes are in Java Script language. As mentioned, visual programming has been exploited for making the UI. An interconnection of nodes as shown in Figure 15.3(a) depicts the back end for the development of the UI and the overview is given in Figure 15.3(b).

A certain threshold would be set for the peak KW data from the load curve and it would be used to send in the notification to the user and he could tweak their electricity consumption and benefit not only themselves but the utility as well.

The UI has been discussed below.

15.4 Features of UI Interface

The pricing detail tab has three buttons, which give the pricing details as per the conditions thereby informing about the "Dynamic billing system". The benefits tab highlights the benefits of the system hence, encouraging the user for shifting to smarter energy consumption techniques. The water usage tab comes under the monitoring aspect of the system. Here the user can check the water level and act accordingly if he wants to fill the water tank with respect to peak and non-peak conditions. The peak load curve clowns the load on the grid at the moment building a transparent relation between the grid and the consumers. The billing tab showcases the consumption balance left for the particular user and can notify him when to pay his due, thereby helping the grid in recovering from the financial losses due to late clearance of dues. The home set up and operational analysis tab showcase if the particular appliance is being used in an efficient and smart manner. Smart check tab is a tab, which highlights the overall performance of a particular household. Brightness meters and the load switching tab are being used for controlling the consumption depending upon the peak conditions.

15.4.1 App Prototype

An APP prototype has been developed based on the approach discussed previously named "ManagePrism". This APP consists of a Home Setup with various appliances connected to it. This home setup provides an operational analysis feature to check whether the usage of the electrical equipment is done in an optimized way or not. All the Monitoring, managing, and controlling of power consumption can be done from this Single Point

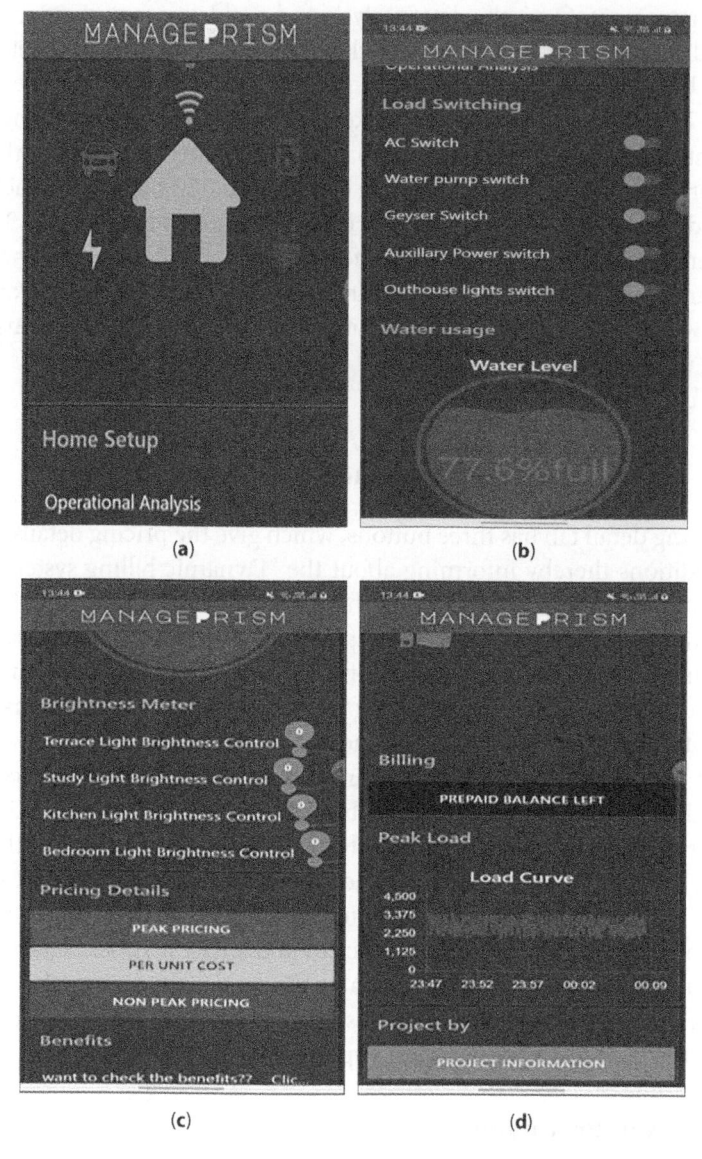

Figure 15.4 (a) Manage prism home setup. (b) Load switching, water usage level. (c) Brightness meter, pricing details, benefits of the proposed system. (d) Billing statistics, real-time peak load curve.

of contact. The functionalities of the App are depicted from Figure 15.4(a) to Figure 15.4(d).

The benefits of the APP are: It is a user-friendly APP, which will let the user control the various electrical equipment from one place assisting the user to maintain a check on his/her electricity consumption. It helps in managing the monthly savings of households. It would also be beneficial for the utility companies as this will indirectly decrease the load on the utilities during peak hours and increase the longevity of the grid. Such a smarter approach is the only way forward of achieving the goal of "POWER FOR ALL".

15.5 Conclusions

The aim is to implement an efficient Demand Side Management system for flattening the peak load curve. This would benefit both consumers and the utilities. Once this is implemented on a large scale this would drastically affect help in flattening the load curve thereby contributing to proper energy management and pacing up the process of ensuring power for all. After implementing the proposed idea, we can give a path for flattening the peak load curve and hence, increase the longevity of the generating stations and not only ensuring power for all but a sustainable infrastructure for the same. This is a gateway for smart energy consumption patterns for the users. As we are introducing a dynamic billing pattern this becomes a bi-profitable energy infrastructure. A variable tariff scheme can help consumers to save lakhs of rupees annually. Also, the utility companies could possibly ensure power for all and make money out of the scheme as the wearing out of the power plants would reduce drastically.

Bibliography

1. Mande Praveen, G.V. Sivakrishna Rao, "Ensuring the reduction in peak load based on DSM strategy for smart grid application", 1877-0509 © 2020 The Authors. Published by Elsevier B.V.
2. Rongxin Yin, M.A. Piette, Rish Ghatikar, "Big-Data Analytics for Electric Grid and Demand-Side Management", Technical Report, May 2019 DOI: 10.13140/RG.2.2.18087.70564.
3. Sidhant Chatterjee, Master of Science "Demand Side Management in Smart Grid using Big Data Analytics", All Graduate Plan B and other Reports. 1143.
4. Assoc. Prof. Dr. Sathish Kumar Selvaperumal, Waleed Al-Gumaei,Raed Abdulla, "Integrated Wireless Monitoring System using LoRa and

Node-Red for University Building", *Journal of Computational and Theoretical Nanoscience* Vol. 16, 3384–3394, 2019.
5. Alireza Ghasempour, "Internet of Things in Smart Grid: Architecture, Applications, Services, Key Technologies, and Challenges", *Inventions* 2019, 4, 22; doi:10.3390/inventions4010022.
6. Shahid Yousuf Wani, Shavet Sharma, "Internet of Things (IoT) Based Smart Grid", *IRJET* Vol. 05 Issue: 08, Aug 2018.
7. Venkata Ratnam Kolluru, Nerella Srimannarayana, "IOT based Smart Environment using Node-Red and MQTT", *Journal of Adv Research in Dynamical & Control Systems*, Vol. 12, No. 5, 2020.
8. KT M U Hemapala, "IOT Based Energy Management System for Standalone PV Systems", *Journal of Electrical Engineering and Technology*, June 2019.
9. M. Cirrincione, M. Cossentino, S. Gaglio, V. Hilaire, A. Koukam, M. Pucci, L. Sabatucci, G. Vitale, "Intelligent Energy Management System", Conference Paper, July 2009 DOI: 10.1109/INDIN.2009.5195809. Source: IEEE Xplore.
10. Marco L. Della Vedova, Tullio Facchinetti, "Real-Time Scheduling for industrial load management", Conference Paper, September 2012 DOI: 10.1109/EnergyCon.2012.6348243.
11. Nikola Lj. Rajakovic, Vladimir M. Shiljkuti, "Long-term forecasting of annual peak load considering effects of demand-side programs", *J. Mod. Power Syst. Clean Energy* (2018) 6(1):145–157.
12. Aiman Roslizar, Mohammad Alghoul, Ba Bakh, Nilofar Asim, "Annual Energy Usage Reduction and Cost Savings of a School: End-Use Energy Analysis", *Scientific World Journal*, November 2014 DOI: 10.1155/2014/310539 ·
13. Kunal Patel, Arun Khosla, "Home Energy Management Systems in Future Smart Grid Networks: A Systematic Review", *2015 1st International Conference on Next Generation Computing Technologies (NGCT-2015) Dehradun, India, 4-5 September 2015*.
14. Osama Majeed Butt, Muhammad Zulqarnain, Tallal Majeed Butt, "Recent advancement in smart grid technology: Future prospects in the electrical power network", *Ain Shams Engineering Journal*, Vol. 12, Issue 1, March 2021, pp. 687-695.
15. A. R. Al-Ali, Imran A. Zualkernan, Mohammed Rashid, Ragini Gupta, Mazin Alikarar, "A smart home energy management system using IoT and big data analytics approach", *IEEE Transactions on Consumer Electronics*, Vol. 63, No. 4, November 2017.
16. Behzad Lashkari, Yuxiang Chen, Petr Musilek, "Energy Management for Smart Homes- State of Art", *Applied Sciences*, 2019.

16

An Overview on the Energy Economics Associated with the Energy Industry

Adhithiya Venkatachalapati Thulasiraman* and M.J.A. Prince

Department of Petroleum Engineering, AMET Deemed to be University, Kanathur, Tamil Nadu, India

Abstract

This investigation will exemplify the auxiliary associated with the time worth of cash (Time value of Money), divergent classifications of cost, economic paradigm and analysis associated with the energy industry. Initially, the time utility of funds is nothing but the financial power to buy commodities that have the potential to vary in a time's revolving wheels. The prospective variation will be in the form of spending, reserving, acquiring and financing. To have an outlook of the commodity one should scrutinize the cost linked with that particular energy project. The total lifecycle of an industrial plant can be narrated with the use of the economic indicator. The economic specification comprises topics associated with return on costs, payback span, net present worth, discounted money flow and discounted money flow gain on an investment. This approach helps the energy industry to account for whether to opt for a project or to stop until any switch in the contemporary market transpires. Moreover, after cautious apprehension of the aforementioned conceptualization, it is important to employ analysis which includes incremental, sensitivity and replacement. Thus the overall perspective about the ins and outs of energy statistics can be studied in detail.

Keywords: Time worth of cash, net present worth, return on costs, discounted money flow, sensitivity analysis, incremental analysis and replacement analysis

---

*Corresponding author: vtadhithiya94@gmail.com

16.1 Time Value of Money

It is the financial power to buy a product or service that can vary in a time's revolving wheel. Theconcepts associated with the time value of money initially include the way generally people approach to make their money value to increase will be to spend, save, borrow and invest. Spending involves an equivalent amount of cash today and can help buy more products than in the future. Saving is generally followed by individuals to tackle emergencies. Borrowing is usually done to own a house and steadily repay it in the future with some interest. Investing is a method to increase your value to a greater extent.

People need to invest in multiple areas. So that even though they have a worst-case scenario they will still make a profit more than the interest paid by your bank monthly. Overall when we say the concept of time worth of cash we need to look at three important technical terms, namely the Present Value of an asset, Future Value of an Investment and Internal Rate of Return.

16.1.1 Present Value of an Asset

There are two generalised approaches towards present value. If it is invested in a bank, then people or an individual can make an interest of about 5.5% every year. But if you invest in the oil and gas industry for purchasing necessary equipment in that industry thenthe equipment depreciates each year at a constant rate and ends up in the curb. In the second scenario, you will have to see how much yield you are going to make during the entire tenure of that equipment to be precise.

16.1.2 Future Value of an Investment

Future value is mainly dependent upon the amount of cash invested initially and the time you are going to keep it in a financial institution. It is a sacrifice that you are making to earn interest from the bank. This value can be calculated by

$$FVI = PVA + IE$$

FVI – Future Value of an Investment
PVA – Present Value of an Asset
IE – Interest Earned

Energy Economics Associated with the Energy Industry 231

Table 16.1 Future value of an investment and their corresponding interest earned.

Year	FVI	IE
1	2110	110
2	2226.1	226.1
3	2348.5	348.5
4	2477.6	477.6
5	2613.9	613.9
6	2757.7	757.7
7	2909.4	909.4
8	3069.4	1069.4
9	3238.2	1238.2
10	3416.3	1416.3

In another form, the same FVI can be calculated as follows

$$FVI = PVA \,(1 + RI)$$

FVI – Future Value of an Investment
PVA – Present Value of an Asset
RI – Rate of Interest

For instance, if an individual invests 2000 inr at the rate of 5.5% over 10 years. The value can be calculated as mentioned in Table 16.1.

After 10 years the same individual will get a sum of 3416.30 inr.

16.1.3 Rule of 72

If an individual is eager to find when his or her invested money will get doubled, he can probably try to divide the 72 by RI. In this scenario we can take the same RI 5.5% then the answer would be 13.1 years. In the same way, we can tackle finding the number of years needed for an amount to get doubled as represented in Table 16.2.

Table 16.2 Rate of interest and their corresponding years taken to double the investment.

RI	5	5.5	6	6.5	7	7.5	8	8.5	9	9.5	10
Y	14.4	13.1	12.0	11.1	10.3	9.6	9.0	8.5	8.0	7.6	7.2

16.2 Classification of Cost

16.2.1 Fixed Cost of an Asset (FCA)

FCA is defined as the charge that is self-reliant on the end outcome. The fare is constant throughout the entire cycle of a project until the abandonment. There is some typical case which includes rental quotations of buildings, equipment that was hired for use, a wage of the working professionals, executive honorarium and more of the same.

16.2.2 Variable Cost of a Plant (VCP)

VCP is determined by the fare that is dependent on the net result. The toll is changing throughout the entire span until the project comes to a complete halt. Exemplars include feed-stock, power utilization in the form of electricity, solar, wind, hydel, oil/gas, wage associated with short-term employment, and so forth.

16.2.3 Total Cost of a Plant (TCP)

TCP is estimated/evaluated by just adding FCA and the VCP. In most cases, if the result or the final product is going to intensify there is a sheer possibility of the TCP to surge correspondingly. TCP is generally expressed as below:

$$TCP = FCA + VCP$$

TCP – Total cost of a plant
FCA – The fixed cost of an asset
VCP – Variable cost of a plant

16.2.4 Break-Even Location (BEL)

BEL is the place where the TCP is identical to the income generated by the end product. An example will be an oil and gas project that has managed to recover the total cost of initial capital at that particular stage; it is also known as BEL/Payback. Gain Location is a point where the end product A becomes considerably higher than the BEL. Loss Location is a place where product A becomes substantially lower than the BEL. A typical example of an oil and gas industry wherein the rates are divided into five stages namely the preparatory survey, procuring, exploring, developing, producing and abandoning. Each stage of the petroleum industry will have its own fixed and variable rates that come into play. But economic professionals need to prepare a flowchart and find out what might be the likely profit that might be acquired through this project or in case of worst scenarios how firms can save some amount for their future and so on.

16.3 Economic Specification

16.3.1 Return on Cost (ROC)

ROC aids in the measurement of profit or loss spawned from outlaying concerning the amount of cash that is been outlaid. The suitable approach is to find the ratio of the total net gain generated to that of the total money that is financed.

$$\text{ROC} = \frac{(\text{Net worth of investment} - \text{Primary worth of investment})}{(\text{Primary worth of investment})} *100$$

16.3.2 Payback Span

It is the period required to recuperate the primary worth of investment. This approach does not take into account the Time Value of Money as in other cases.

16.3.3 Net Present Worth

It is the gap between the current worth of money that is coming in to that of the current value of money that is going out over a specific period. Cash in the current perspective is valued higher than the equal amount in the future because of the alternative earning potential such as share market, real estate and whatnot could have been made possible during that leftover time.

NPW = Current worth of money that is entering inside
− Current value of money that is going outside

16.3.4 Discounted Money Flow (DMF)

It assists in evaluating the worth of an investment using the upcoming value of money flow. The current worth of expected upcoming money flows will materials by making use of discount rate to acquire the value of Discounted money flow. If the DMF is higher than the charge of the capital this might result in a gain on investment. Firms in most cases weighted average cost of capital for the discount rate as it takes into attention the anticipated price of remit from the stockholders. The only demerit of this approach is that it strongly relies on the forthcoming money flow which could manifest to be unreliable.

16.3.5 Internal Charge of Returns (ICR)

The discount rate at which the net current worth of all the money flows either it is coming in or out of a particular project tends to be a null character. This phenomenon does not account for the economic downturn or even the price of preliminary capital. This term is usually expressed as

$$ICR = DR @ \text{ which } NCW = 0$$

ICR – Internal charge of returns
NCW – Net Current Worth

16.4 Analysis

16.4.1 Incremental Analysis (IA)

Enterprises or industries wield this IA as a judging essential. Judgement on either to procure or trade a product or even to dump a project can be easily implemented using IA. This analysis is also known as pertinent cost. In some cases, this approach has also been referred to as differential or Marginal analysis.

16.4.1.1 Pertinent Cost (PC)

This will not only take the fixed charge but also take into account the variable levy that is associated with a particular project. This will also consider the opportunity cost.

16.4.1.2 Non-Pertinent Cost (NPC)

The other name of NPC is called a submerged charge. It aids enterprises to decide on whether to obtain the extra special order or to avoid it. The extra special order levy is usually less than the marketing price of that commodity. It also guarantees that an insufficient valuable is used to extreme merit.

An appropriate model for an IA in a Manufacturing firm
A firm sells an item for 500 inr. It pays 50 inr for variable overhead selling expenses, 200 inr for working professionals and 100 inr for feed-stock. The industry has also allocated around 100 inr per item as fixed overhead costs. The industry will not require to invest in equipment or overtimeto accept a special order it receives. Calculate the gain and loss % of the following scenarios to make a wise decision.

Scenario 1
The special order requests the purchase of 25 items for 370 inr each.
Scenario 2
The special order request is to purchase 30 items at 350 inr per item.
Scenario 3
The special order request is to purchase 50 items at 300 inr per item.

Scenario 1 will be selected as per the IA since the fixed overhead cost has to be paid even if you are not running the firm. So considering the other cost incurred we get a net gain of 500 inr. In scenario 2, the firm will not make any profit or loss. In scenario 3, the firm will make a loss of 2500 inr. The best feasible approach is to opt for scenario 1.

16.4.2 Sensitivity Analysis (SA)

In this model, the forecast is done using past historical records. It is also obligatory to know about the variables and the result. The other name of this kind of analysis is called the What-If technique. Key decisiveness about enterprises, economy, real estate and stock markets can be acquired. The preliminary variables that will have a consequence on the objective variable should be carefully studied.

A typical prototype of SA in a Business enterprise

Table 16.3 Sensitivity analysis problem statement.

SENSITIVITY ANALYSIS		
ASSUMPTIONS		
Tables Sold	1500	inr
Rate per Table	250	inr
Fee per Table (inr)	100	inr
Stockpile Rental Expenses (Fixed Rate) (inr)	5000	inr
Wages paid to workers (Fixed Rate) (inr)	60000	inr
GAIN AND LOSS DECLARATION		
Gross Sales = Tables Sold * Rate per Table	375000	inr
Worth of Sales = Tables sold * Fee per Table	150000	inr
Overall Gain = Gross Sales - Worth of Sales	225000	inr
SG and A = Stockpile Expenses + Wages paid	65000	inr
Utilising Gain	160000	inr

The aforementioned case was to analyse two sensitivity variables which include the number of tables sold (Horizontal Axis) and the rate of the table (Vertical Axis). For instance, if we assume the rate as 200 and sell about 3,000 tables this year that business model can yield 75,000 inr higher than the existing model in place. Due to some problem, the firm can sell only 1,000 chairs but if the rate is going to be 350 they can still make a gain of 25,000 more than the existing model (Table 16.3). The concept that we should have in mind is lower the price higher number of tables can be sold.

Table 16.4 Sensitivity analysis by varying both tables sold and utility gain to decide the marketing strategy.

		Tables sold						
Utilising Gain Rate	160000	500	1000	1500	2000	2500	3000
	150	-40000	-15000	10000	35000	60000	85000
	200	-15000	35000	85000	135000	185000	235000
	250	10000	85000	160000	235000	310000	385000
	300	35000	135000	235000	335000	435000	535000
	350	60000	185000	310000	435000	560000	685000
	400	85000	235000	385000	535000	685000	835000
	450	110000	285000	460000	635000	810000	985000
	500	135000	335000	535000	735000	935000	1135000

On the other hand, the higher the price (Table 16.4) you will probably sell a limited number of tables. This analysis can also be implemented for different applications.

16.4.3 Replacement Analysis (RA)

It is critical analysis in resource budgeting. Since a piece of equipment or a commodity's life span might be diminished due to corporal impairment, swap is cost-effective essential. The replacement of equipment offers enterprises a profit-making possibility. In this approach, we have two choices. One being the existing asset and the other the advanced contestant of the proposed model; it is more like a case held in a court.

A prototypical model of RA in an oil and gas environment

A piece of equipment cost 350,000 and it can be either swapped in 2 years or after 3 years. Hampering the exchange will considerably escalate the utilizing rates and simultaneous reduction in the dump value of that commodity.

The goal is to find whether the equipment needs replacement in 2 years or 3 years (Tables 16.5 to 16.9).

Table 16.5 Replacement analysis problem statement.

Year	Utilising price	Selling revenue
1	60000	290000
2	70000	270000
3	80000	250000

Table 16.6 Replacement analysis after 1 year.

The exchange did after 1st Year				
		Money flows	DMF @ 8.5 %	Current worth
0	Purchase	350000	1	350000
1	Utilising price	60000	0.922	55320
1	Selling Revenue	290000	0.922	267380
Current Worth				137940
Equivalent Yearly Rate (EYR)				149610

Table 16.7 Replacement analysis after 2 years.

The exchange did after 2nd Year				
		Money flows	DMF @ 8.5 %	Current worth
0	Purchase	350000	1	350000
1	Utilising price	60000	0.922	55320
2	Utilising price	70000	0.849	59430
2	Selling Revenue	270000	0.849	229230
Current Worth				235520
Equivalent Yearly Rate (EYR)				132987.01

Table 16.8 Replacement analysis after 3 years.

The exchange did after 3 Years				
		Money flows	DMF @ 8.5 %	Current worth
0	Purchase	350000	1	350000
1	Utilising price	60000	0.922	55320
2	Utilising price	70000	0.849	59430
3	Utilising price	80000	0.783	62640
3	Selling Revenue	250000	0.783	195750
Current Worth				331640
Equivalent Yearly Rate (EYR)				129851.21

Table 16.9 Comparison table for 1-3 years.

Equivalent Yearly Rate (EYR)	129851.21
Equivalent Yearly Rate (EYR)	132987.01
Equivalent Yearly Rate (EYR)	149609.54

As per the aforementioned evaluation, you can find the lowest Equivalent yearly rate in 3 years and the value is found to be 129851.21 inr. So the exchange of old equipment for new is recommended to be on the third year.

16.5 Conclusion

In the flourishing economic environment, countries like India, Japan and China need to concentrate more on the Time value of money. This will enable them to have enough time to quantify the divergent types of costs, which include Fixed-rate, Variable rate and Total price. Moreover, businesses, enterprises and industries need to know the economic technical jargon in a simpler way to tackle the current issues around the world. Last but not least, it is important to employ distinct approaches such as

Incremental, Sensitivity and Replacement techniques for different applications, so that it will aid the respective domain to choose a cost-effective approach.

Bibliography

1. John Hill, Time value of money: interest, bonds, money, market funds, Ch. 7 in *Fintech and the Remaking of Financial Institutions*, 2018, https://doi.org/10.1016/B978-0-12-813497- 9.00007-X, pp. 157-176.
2. Brian Kettel, The time value of money: The key to the evaluation of financial markets, *Economics for Financial Markets*, 2002, https://doi.org/10.1016/B978-075065384- 8.50002-5, pp. 33-46.
3. Saroj S. Shivagunde, Ashwani Nadapana, V. Vijaya Saradhi, Multi-view Incremental Discriminant Analysis, *Information Fusion*, 2021, https://doi.org/10.1016/j.inffus.2020.10.021, pp. 149-160.
4. Michael P. Brundage, Qing Chang, Jing Zhou, Yang Li, Jorge Arinez, Guoxian Xiao, Energy economics in the manufacturing industry: A return on investment strategy, *Energy*, 2015, http://dx.doi.org/10.1016/j.energy.2015.10.038, pp. 1426-1435.
5. Laveet Kumar, M.A.A. Mamun, M. Hassanuzzaman, Chapter 7, Energy Economics, *Energy for sustainable development: Demand, supply, conversion and management*, 2020, https://doi.org/10.1016/B978-0-12-814645-3.00007-9, pp. 167-178.
6. Rong Yuan, Tao Zhao, A combined input-output and sensitivity analysis of carbon-dioxide emission in the high consuming industries – A case study in China, *Atmospheric Pollution Research*, 2015, http://dx.doi.org/10.1016/j.apr.2015.10.003 1309-1042, pp. 1-11.

17

IoT-Based Unified Child Monitoring and Security System

A.R. Mirunalini, Shwetha. S., R. Priyanka and Berlin Hency V.*

School of Electronics Engineering (SENSE), VIT, Chennai, India

Abstract

In this fast-growing modern era, mothers are likely to work soon after childbirth, which makes it hard for them to render complete care to their child. Hiring a childminder is not just costly but also unsafe, especially during a global pandemic like Covid-19. Child abuse is also a major worry. This paper introduces an IoT-based Unified child monitoring and security system without any third-party involvement, thus addressing the parents' needs and concerns. The proposed system monitors the temperature and heart rate of the infant, humidity of the room, detects motion and sound produced by the baby and adopts suitable measures to notify the parent such as sending alert messages, live video streaming of the infant or turning on a motor to swing the cradle. This system also monitors the movement of toddlers using GPS- and GSM-enabled wrist-bands and continuously sends their live locations. A buzzer is also interfaced with the band to alert if any stranger is in close proximity with the toddler. This system enables parents to keep a watch on their children remotely and thus ensures the safety of the child from any type of abuse. An added feature of this band is that it also prompts to maintain social distancing from the toddlers. Overall, a reliable, continuous and real-time baby monitoring is ensured by the proposed system.

Keywords: IoT, child security, baby monitoring, GPS, unified system, wearable electronics, Arduino, wireless system

---

*Corresponding author: berlinhency.victor@vit.ac.in

Milind Shrinivas Dangate, W.S. Sampath, O.V. Gnana Swathika and P. Sanjeevikumar (eds.) Integrated Green Energy Solutions Volume 1, (241–262) © 2023 Scrivener Publishing LLC

17.1 Introduction

The days where the father of the family is the breadwinner and mothers stay at home looking after the family is no more prevalent. In the 21st century both the parents of the family need to work to ensure a sophisticated lifestyle. In almost 90 percent of the Indian households both the parents are working. This implicitly means that both parents remain busy all the time, which creates mental stress in striking a balance between work and family. Often, young mothers are concerned about the safety and healthy wellbeing of their toddlers when they are bound to leave them alone for a job. Especially in the pandemic times, leaving kids uncared or under the care of childminders is not entirely safe. Our proposed system intends to reduce this stress and to ease the job of working parents. This system monitors the baby 24x7 and notifiesparents in case of an emergency [1]. The system also provides live streaming of the baby througha camera module which can be accessed by parents anytime, even from their workplace. It also helps to track the movement of toddlers using wrist bands and enables parents to communicate with their kids in case of threats.Live location of the kid is sent to the parents once every 20 seconds. And if anybody gets within 30 centimeters of the toddler, an alert notification is sent and a buzzersound is made. This system empowers women to focus more on their career while not compromising their roles as mothers, thus enabling them to contribute more to satisfy the economic needs of the family.

The prime motive of the project is to help working parents by providing a single unified system satisfying all their needs. This objective is achieved by continuously measuring and monitoring the environment and the vitals of the baby using sound, humidity, temperature, heart rate and PIR sensors and storing them. The live location tracking is achieved by using GSM- and GPS-enabled wrist bands with a buzzer alert [20]. A Camera module is used for live video surveillance. All the data from sensors are monitored, stored in the cloud and sent in real time to parents. This is mainly to monitor the infant, avoid infant death syndrome, track the toddler if they go missing and also to protect them from abuse and from contacting disease. Apart from parents' usage this system has a good potential of usage in childcare hospitals and crèches where the alert messages can be sent to the caretakers.

The novelty of the proposed system is that it provides a single unified system that takes care of the baby right from the day of its birth, up to the toddler stage.

The system also focuses on making the proposed approach energy efficient. Eco-friendly rechargeable batteries are used instead of standard disposable batteries. Also, the DC motor of the infant's cradle system is only turned on when the baby's cries surpass a certain decibel. This way energy is conserved instead of running the motor continuously even in unnecessary situations. Also, as a future extension we plan to interface Bluetooth HC-05 module to allow the parents to choose between Bluetooth or Wi-Fi based on their vicinity. Switching over to Bluetooth whenever feasible will greatly minimize the energy usage.

The noteworthy features of this project in the infant stage includes real-time data collection, processing and storage in cloud, anomaly detection in temperature, humidity and heart rate. Drastic movements of the baby are also considered an anomaly and the mother is notified, as she also is if the baby's cries exceed certain decibel limits [10]. In the toddler stage the parents are given access to the live location of the toddler every 20 seconds, and Buzzer alert and notification if anyone comes in close contact with the toddler. The system employs Wi-Fi and Pushover app to notify in case of emergency, 24x7 surveillance by camera (ESP 32 module).

17.2 Literature Review

This health monitoring system [1] uses different sensors to measure temperature, heart rate, respiration rate and blood flow rate, which are sent in to the controller which stores it in a cloud. These data are then sent to the user's mobile phone, using Wi-Fi, notifying them about the status of their baby. This monitors the infant's body temperature, heart beat rate and diabetics check-up, etc. This system completely analyses the infant's health condition.

The system in [2] uses the obstacle detecting sensors to detect when the child is in a dangerous environment and alert the parents via a mobile phone. Ultrasonic Obstacle Sensor is waterproof, which is an added advantage over other similar sensors. This sensor is positioned on a simple locket-like structure that is given to the baby. This locket sends an alert message to the caretaker's mobile phone, in case of an emergency. The system detectsthe objects that surrounds the baby and alerts if the baby goes near any harmful objects.

This system [3] is similar to [7] with an added advantage of a Blynk server notifying the parent. Blynk app is customized by pre-setting the feeding time of the baby, and the caretaker is duly notified about the same. The feeding time can be set on the system on the basis of the age of the baby. Blynk Server also enables the parents to monitor their baby's cradle any time via their mobile phones.

This system [4] uses temperature sensor, heart rate sensor, motion sensor and voice sensor. The microcontroller continuously reads the sensor input data and sends the same to the cloud, to a specific URL/IP address. Parents can look into their child's health parameters simply by visiting a web site or URL. In this system, the major parameters such as body temperature, heart rate and the baby's movements are monitored, and this information is passed on to the parents.

This paper [5] focuses on the safety of women and girls, concentrating on the increase in abuse towards them. The paper proposes a safe, portable device that sends the child's location and activities to the parents via SMS. The project is implemented on an Arduino UNO or Nano, and uses GSM to facilitate SMS notification to the parents' mobile phones.

The system in [6] uses IoT, deep learning techniques. This device ensures that the baby's condition, such as the position of the body, temperature and the posture are correct by employing deep learning techniques, thus avoiding potentially hazardous events and possible injuries. This device also helps the parents to monitor them and know about the baby's condition, remotely using IoT.

A prototype for a baby's crib is designed and constructed along with automatic rotational webcam support and is implemented using the NodeMCU controller [7]. In addition, an automated mini fan, controlled by a built-in temperature sensor, is placed to avoid excessive heating. Parents will be able to observe the data such as the ambient temperature and control the external on-off switch, which is stored in the cloud, using the MQTT server. Gentle rocking of the cradle is also an additional feature in [7] which soothes the baby during distress.

The proposed system [8] uses a PIR sensor that processes the infrared light coming from objects to detect the movements of the child. A threshold noise level is set so as to differentiate the noise of a cry and the sound of screaming detected by the sensor. The built-in temperature sensor measures the ambient temperature and feeds it to the cloud. The main objective of project [8] is to minimize the physical interface and to reduce the burden for the working parents with more reliable performance, enhanced adaptive security and performance than the existing ones. The project employs GSM module, SBC (Arduino) as master control and Blynk cloud service to achieve this.

A special environment has been developed for devices that can monitor the child's room, aided with a two-way ZigBee wireless communication. The system [9] has temperature sensor, humidity sensor, wet alarm, microphone, camera based on TCP/IP protocol, infra-red based alarm and other security systems for the collection of information, as well as a control system for a music player. With the help of the data processing by smart drivers, parents can remotely monitor the parameters of the surrounding environment of the child in the home and in real time. [9] is an integrated system with bi-directional wireless communication, with fair security, relatively lower consumption of power and a decent scalability. The system not just monitors the parameters, but also allows the parent to control some of them, including the humidity and temperature of the room. The microphone and camera provide real-time and accurate data.

The project [10] makes use of GSM to send notifications to parents in real time. It detects the baby's crying using audio sensors, and the cry automatically powers the DC motor in the cradle to swing it. Also, a pre-recorded voice starts playing until the baby stops crying. An alert is sent as intimation to the parents. An FN-M16P module is used to recognize the voice of the mother and it plays it back every time the baby cries. This is one of the key innovations that distinguish this from other similar works.

The project in [11] uses SMS and GPS to build a platform that will help the caregivers to keep track of their child's geographical location. The parent sends text messages, using an obvious watchword such as "Area", "Temperature", etc. to which the system responds by sending back the child's area, the ambient temperature, etc. The project makes use of a micro-controller as the main control, which is comparatively slower than the SBC. One of the most important features is that the system is easy to use and will send real-time data back to the parent.

This advisory system [12] keeps constant track of the child and sends updates to the parents about the child's condition. These include the measurement of temperature, heart rate, and observation of physical activity. The system sends it to the server, where the data is processed. The server analyzes the received data and sends the processed information about the child to the parent and it also creates an early warning system if the child is found to be abnormal. In case of an emergency, this alarm message can be transmitted to the support centers, medical clinics in proximity. The notable feature of this project is that when an abnormality is detected, it not only alerts the parent but also sends an alarm to the nearby hospitals. It also provides advisory first-aid information to the parent. The hardware control is microcontroller based.

The system in [13] consists of a baby cot with a variety of sensors, such as the acoustic, the temperature and light intensity sensor. They are connected to the receiver end using ZigBee network. At the rear end of the cot an IP camera is fixed. The smartphone app has been designed to access and process of all the sensor data. In [13], the researchers have also developed a system that offers solutions for sudden infant death syndrome (SIDS), with the help of the smart cradle, a network of wireless sensors (WSN) and smartphones.

The method used in the device [14] is based on the PHVA cycle. It is broken down into four steps: Planning (P), Do (H), Verifying (V) and Act (A). The temperature and ambient noise level is measured in real time. If the device detects a rise in the body temperature or an increased noise level, it will send an alarm signal to the parents via mobile application.

In system [15], the prototype model of a smart room for the child with a PIR sensor, a gas sensor, and an acoustic detector is built. These sensors are connected to the Arduino. Detecting the movement in the room, the PIR sensor activates the Camera and turns on the light. A dangerous gas leak and a baby's cry is also sensed, and an alert message is sent to the parents.

The system in [16] is incorporated with acoustic sensor, humidity sensor and a buzzer. All the measured parameters are sent to the users' mobile through the cloud via mobile app. The baby's cradle oscillates automatically on detecting the infant's cry. If the infant cry does not stop after a specified time or if the sheet placed in the cradle is wet, a "cradlecannot handle" and similar alert messages are sent to parents. Video surveillance is enabled for continuous monitoring.

In [17], if the condenser microphone picks up a sound, or when the IR sensor detects the movement of a child, it sends a signal to the Raspberry Pi, which in turn switches on the camera, and the information is sent to the controller. The Raspberry Pi then processes the data and sends it to the LCD screen in video format. At the same time, an audible alarm will be activated and as a result, the buzzer will give an alarm and alert the parents.

In this paper [18], the hardware setup contains an RPI with a camera. RPI has wireless sensor that is connected to the wireless network in a smart home. Frames from the camera are acquired and two different motion detection algorithms are applied to compute motion soas to detect every slight motion of the baby. UCL refers to very intense motion and LCL refers to no motion at all in this scenario. Crossingany of these limits for a specific number of frames in a single interval yields an alert. The control chart constructed for a whole interval (currently 5 frames per interval) is checked for abnormal behavior. If the majority of the points crosses the upper or lower boundary in a single interval, it ismarked as abnormal and an alertis generated. The generated alert is transmitted to the guardian or parents for quick actions along with options to view the current state of the baby through live streaming.

In [19] as soon as the baby cries, the sound sensor senses the frequency and starts the motor of the cradle, resulting in oscillation of the cradle. If the baby's mattress is wet or if the baby's body temperature changes relative to the atmosphere, or if the baby is restless in a cradle, or a motion is detected by the PIR sensor, then an alarm message is sent as SMS to the parents via the GSM module.

Lots of crimes against kids are being reported nowadays. So, for tracking and monitoring children, a smart IoT device is developed in [20]. The system uses LinkIt ONE board programmed in embedded C and it is also interfaced with sensors such as temperature sensor, touch sensors, heartbeat sensor, GPS, GSM and camera modules. The system in [20] will automatically notify parents by sending them an SMS message when there is a threat to the child. The analysis of the parameters is done for the child's touch, temperature, and heart rate, and a graph of the results is plotted. This system will ensure safety of the children, even when parents are not in their vicinity.

17.3 Proposed System

17.3.1 Block Diagram

In the block diagram 1 of the system depicted in Figure 17.1, there are multiple sensors used to collect details and monitor the baby continuously. Microphone sound sensor is to detect the noise made by the baby. This sensor is active when sound intensity reaches a certain threshold. The sensitivity limit of sound sensor can be adjusted with a potentiometer in

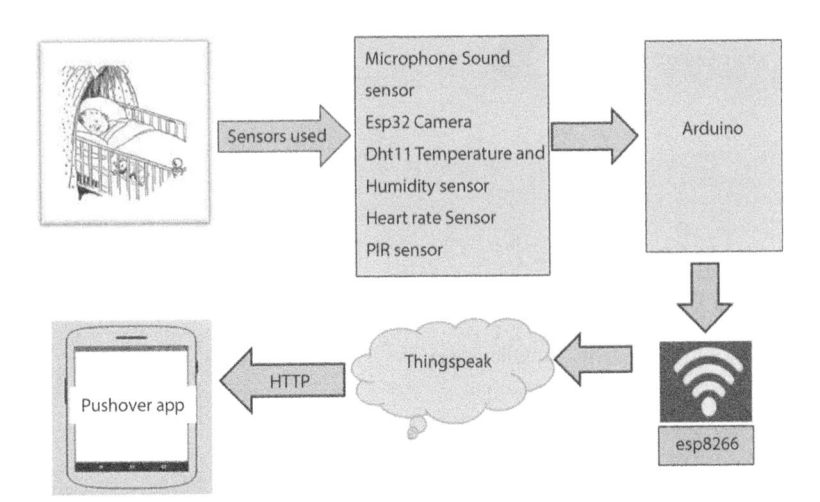

Figure 17.1 Block diagram 1 – Infant stage.

action. DHT11 Temperature and Humidity sensor is used to measure the allowed temperature and humidity around the baby. Heart rate sensor is employed to measure pulse waves by emitting infrared light from the body surface and detecting the variation in blood flow during heart beats as a change in the amount of light transmitted through the body and motion sensor is used to monitor drastic movements of the baby. Threshold values for each sensor are set which, when exceeded, alerts the parents. ESP32 camera is present to live monitor the condition of the baby. All the data collected by the sensors are then processed by Arduino UNO and sent to esp8266 Wi-Fi module. Then the data gets stored in ThingSpeak and using HTTP protocol the alert notifications are sent to parents using Pushover app.

A toddler is supposed to wear this wrist band all the time, which will help parents to ensure the kid's safety. From Figure 17.2, it can be observed that GSM GPS module is used in the band to help parents to track the toddler. GSM allows microcontrollers to have wireless communication to other devices. It can be used to Send, receive or delete SMS messages and also add, read and search the contacts in the SIM Card. Ultrasonic sensor is also used, which helps the toddler maintain social distance during this pandemic. The data is then processed by Arduino UNO and the live location of the toddler is sent to the parent via SMS once every 20 seconds.

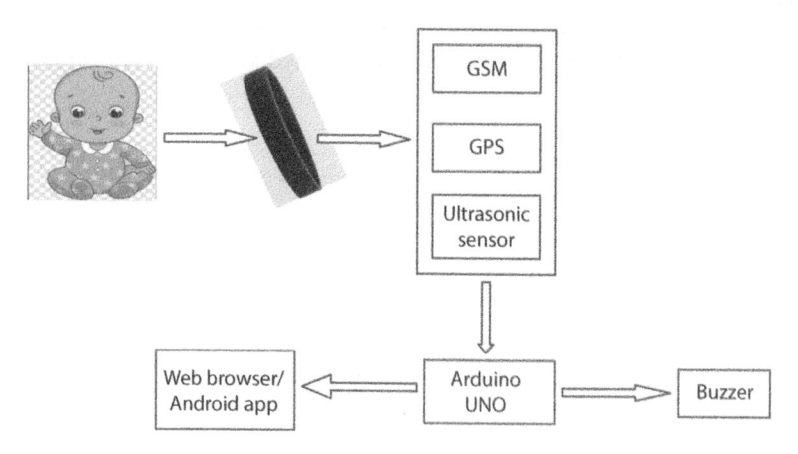

Figure 17.2 Block diagram 2: Toddler stage.

Parents can click on the link received on SMS and view the current location of the baby on Google maps. Also the proximity of the child with another person is kept in track. If that goes below 30 centimeters, a buzzer sound is made to alert people around.

17.3.2 Design Approach

The design of the proposed system uses sensors to capture the ambience of thebaby and record the readings continuously and store it to the cloud. In case of any anomaly detected, the system notifies the parent or guardian immediately. The data collection, storage and process are done in real time. The assemblage of sensors includes temperature and humidity sensor (DHT 11), heart rate sensor, ESP 32 Camera module, PIR sensor, ultrasonic sensor, GSM GPS modem. For monitoring of the infant, threshold values for normal temperature, humidity, heart rate are fixed and if the detected values exceed the threshold values, a message is sent to the parent through Pushover app, informing them about the abnormality. HTTP protocol is used for cloud integration using ThingSpeak. Sound detection sensor is used to detect the cries of the infant. As soon as the sound sensor detects a cry, a DC motor is powered to make the cradle swing, thus soothing the baby. For monitoring of the toddler, once every 20 seconds, the live location of the toddler is updated to the parent in the form of SMS to their mobile phone. Also if anybody comes in close proximity to the toddler (less than 30 centimeters), an alert notification is sent to the parent; along with it buzzer sound is made to alert the people around.

17.3.3 Software Analysis

Proteus software is used for the software simulation and analysis. Two different circuit simulations are made for the infant and toddler stage and the results are analyzed.

Figure 17.3 is the first circuit at the infant stage and is interfaced with all the sensors for ambience monitoring. Figure 17.4 is the second stage of the circuit with the ultrasonic sensor for distance measurement and Figure 17.5 is the wrist band simulation with GPS and GSM modules.

Sensors interfacing:

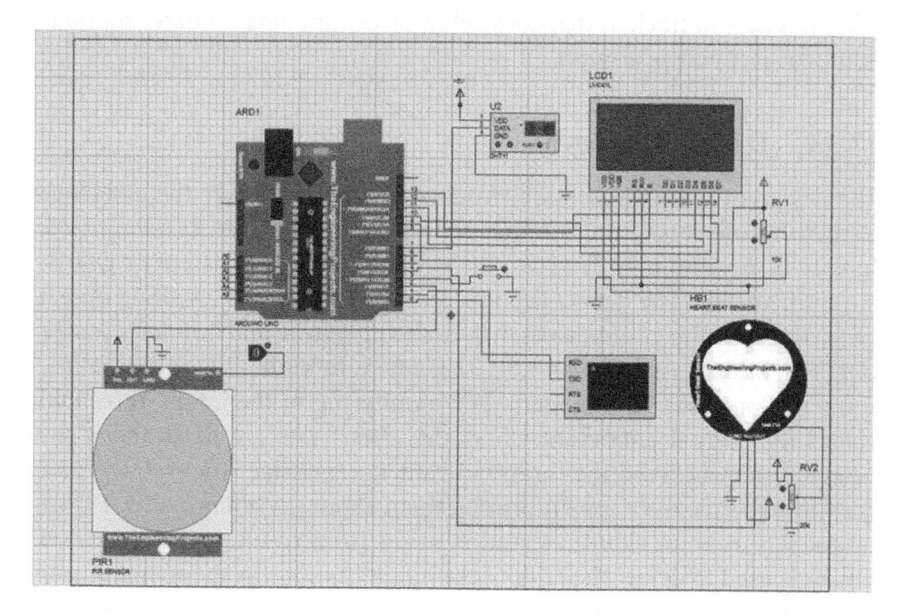

Figure 17.3 Proteus simulation circuit 1.

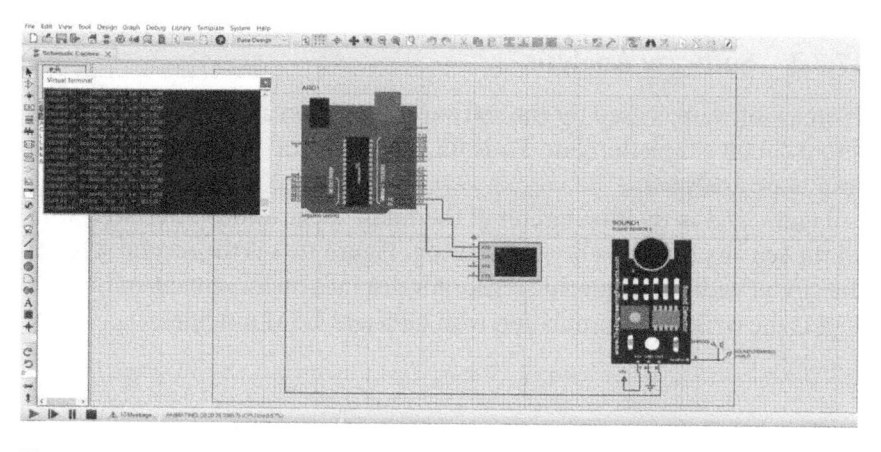

Figure 17.4 Proteus simulation circuit 2.

For Toddler band:

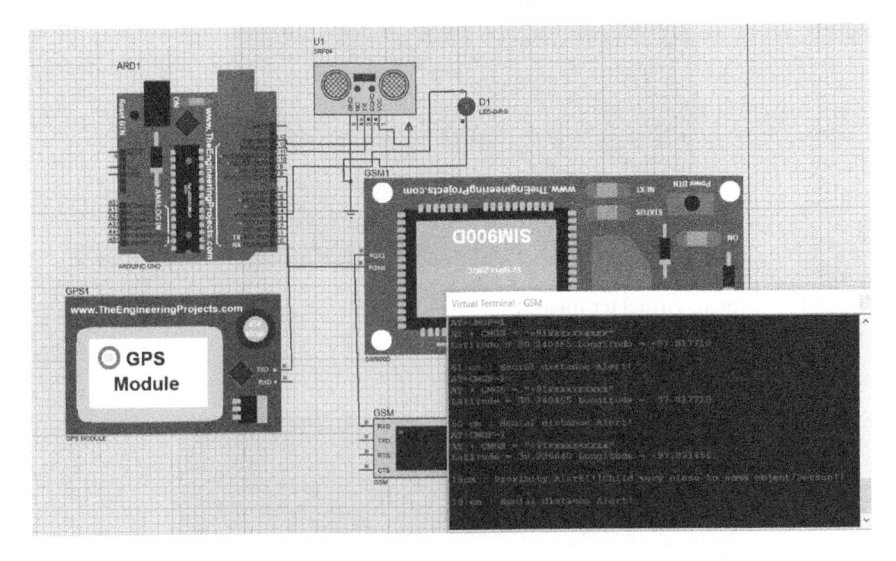

Figure 17.5 Proteus simulation circuit 3.

Proteus Simulation outputs:

Figure 17.6 is the Proteus simulation output that senses the temperature, humidity and if motion is detected or not. Figure 17.7 is the GPS, GSM output display along with proximity alert. Figure 17.8 is the output display of the sound sensor.

Figure 17.6 Proteus virtual terminal 1.

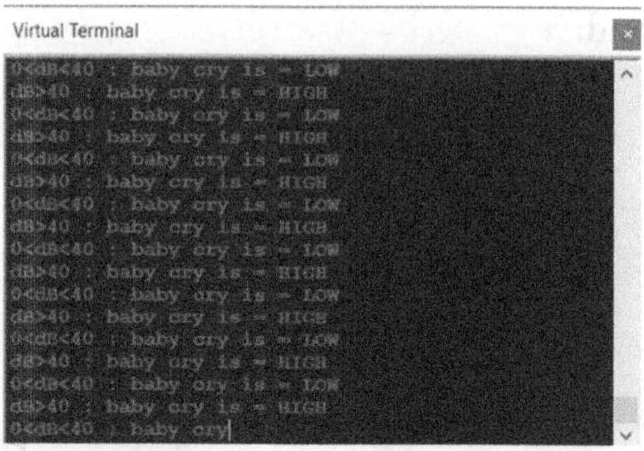

Figure 17.7 Proteus virtual terminal 2.

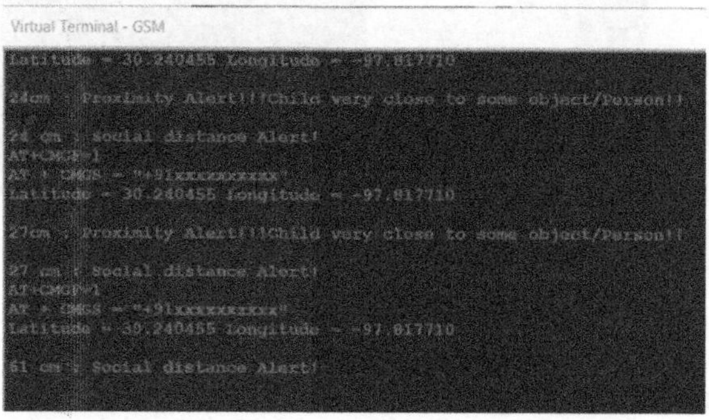

Figure 17.8 Proteus virtual terminal 3.

17.3.4 Hardware Analysis

For monitoring the infant, temperature sensor DHT11, heart rate sensor, sound sensor collects the continuous real-time respective values of the baby which is stored into ThingSpeak cloud using ESP8266 Wi-Fi. The PIR sensor is fixed at the corners of the infant's cradle to detect if the baby tends to fall out of the cradle or attempts to get down from the cradle. From ThingSpeak using ThingHTTP we transfer the values passing the threshold condition to 'React App' in ThingSpeak to send the alert messages to Pushover application to notify the user. This monitoring system also automates the movement of the cradle. When the baby's cry is detected from the sound sensor passing above the threshold decibels, it turns on the motor of the cradle and turns it off when the baby stops bawling, which is determined through the sound sensor. ESP32 camera is used for 24x7 surveillance. For monitoring of the toddler, the wrist band is developed. This encompasses a GPS and GSM module to track the live location of toddler continuously and send it once in every 20 seconds to the parent's mobile. Parents will receive a link in SMS, by clicking the link, the live location of their child will be shown using Google maps. The band also has an ultra-sonic sensor to help the toddler maintain social distancing in the time of Covid-19 and also to prevent the baby from any abuse by creating a buzzer sound to alert people and also alerting the parent by sending a message and location. The added advantage of the system is the passive data collection with which it performs all these tasks without any intervention.

17.3.4.1 Experimental Setup

I. Sensors Setup
Arduino is chosen as the controller of the system. The temperature sensor, soundsensor and heart rate are connected to the analog pins and PIR sensor, motor driver module are connected to the digital pins.

II. ESP826 Wi-Fi Interface
ESP8266 Wi-Fi functionality check is performed and the system is connected to a Wi-Fi using ESP8266. Figure 17.9 is the hardware setup with sensors and camera module for ambience monitoring, Arduino for control and processing and Wi-Fi module to enable cloud interfacing.

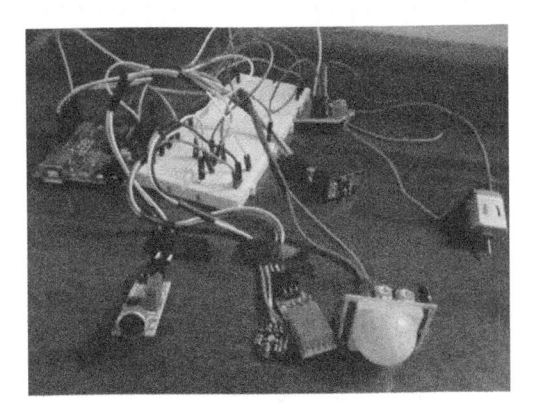

Figure 17.9 Hardware setup.

III. Cloud interface

Through esp8266 Wi-Fi the sensor values are passed into a ThingSpeak channel through API keys and the sensor values are stored in it.

IV. Pushover Application

The first step is creating an account in Pushover. Advantage of the application is that anynumber of devices can be added and the notification can be sent and also this app is very well encrypted as this is signed-in through email address and API token keys.

V. ThingHTTP

Creating an HTTP for every sensor values is the second step. With the Pushover app credentials and API keys we create the HTTP to print the required alert text for the respective sensors. We create the HTTP using an authorized username and password by using the api.pushover.net as host.

VI. React App Interface

Similar to ThingHTTP, we create React for every sensor. This interface helps in fixing the threshold value for respective sensors and when to deliver the notification to the user.

VII. Camera Surveillance

The 24x7 camera surveillance system captures the baby and video streams it to a particular IP address in which the parent can view the baby's actions anytime andanywhere they want. This esp32 camera stream comes with inbuilt facial recognition. The esp32 camera is bootloaded using FTDI programmer. Figure 17.10 displays the camera interface with the FTDI.

Figure 17.10 ESP32 camera flashing memory with FTDI.

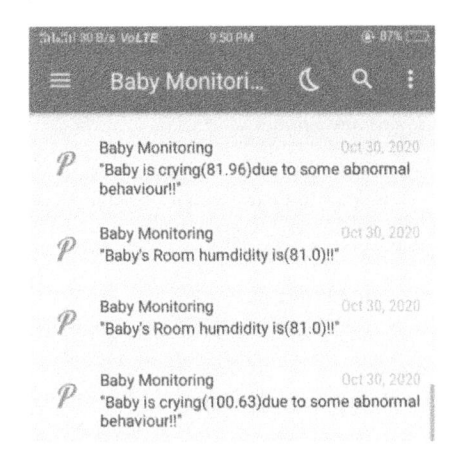

IoT-Based Child Monitoring and Security System 255

Figure 17.11 Pushover APP alert notifications sent to user.

VIII. User Notification

The Pushover is programmed to send notifications in regular intervals to the user's registered mobile phone. Figure 17.11 is the sample image of the notifications being sent.

IX. GSM GPS Configuration

Figure 17.12 is the GPS and GSM module interface with the Arduino Uno. The serial monitor output of the GPS location is displayed in Figure 17.13.

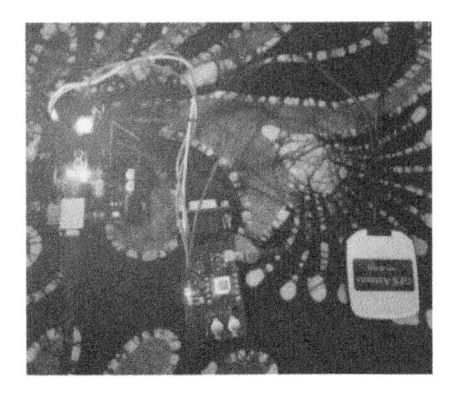

Figure 17.12 GSM GPS module.

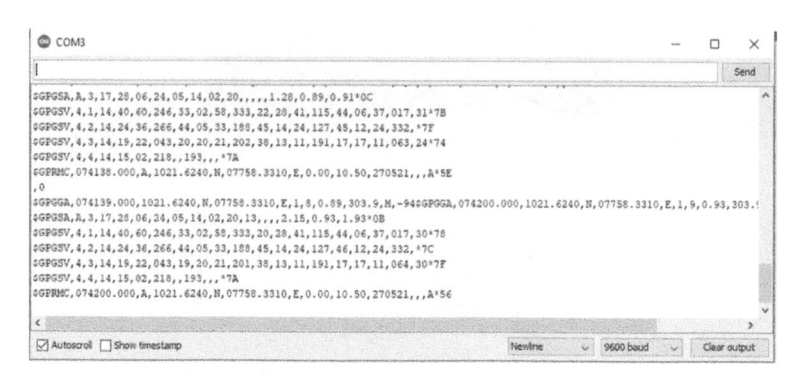

Figure 17.13 GPS output in Arduino serial monitor.

17.4 Result and Analysis

ThingSpeak Cloud Dashboard:

Figures 17.14 and 17.15 display the ThingSpeak dashboard. The data from the sensors are sent via Wi-Fi module in real time and the data is displayed graphically in the dashboard where it can be accessed using authorized user name and password.

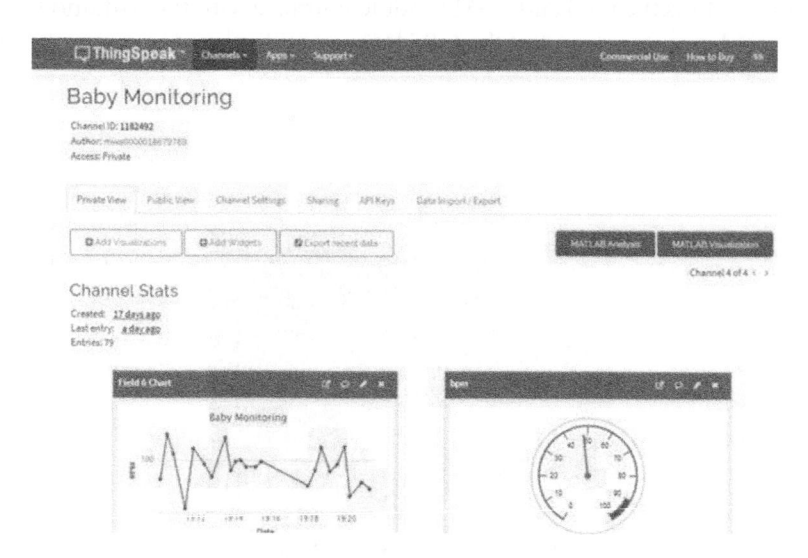

Figure 17.14 ThingSpeak dashboard.

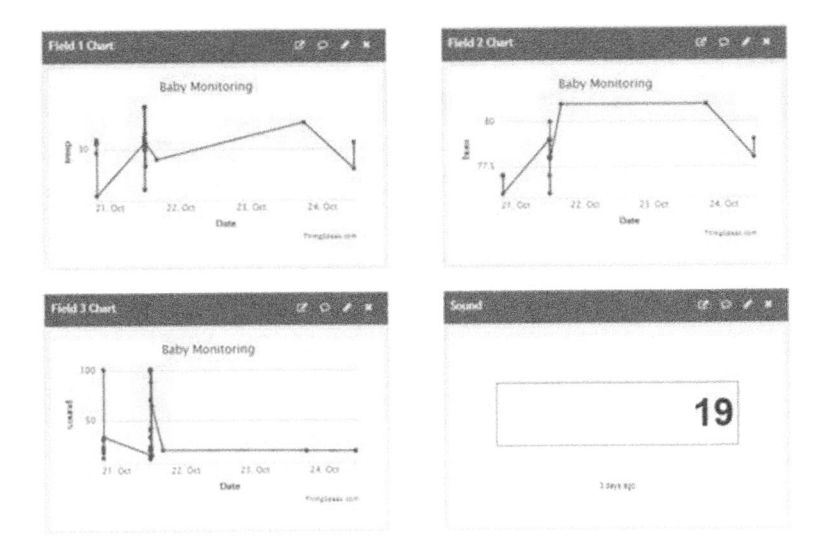

Figure 17.15 ThingSpeak dashboard.

Output Video Stream:

Figure 17.16 is the sample image of the live 24x7 video surveillance. The video is streamed directly to the parent's mobile phone. Figure 17.17 is the sample image where the system uses face recognition algorithm to detect the presence of strangers and alert the parents about the same.

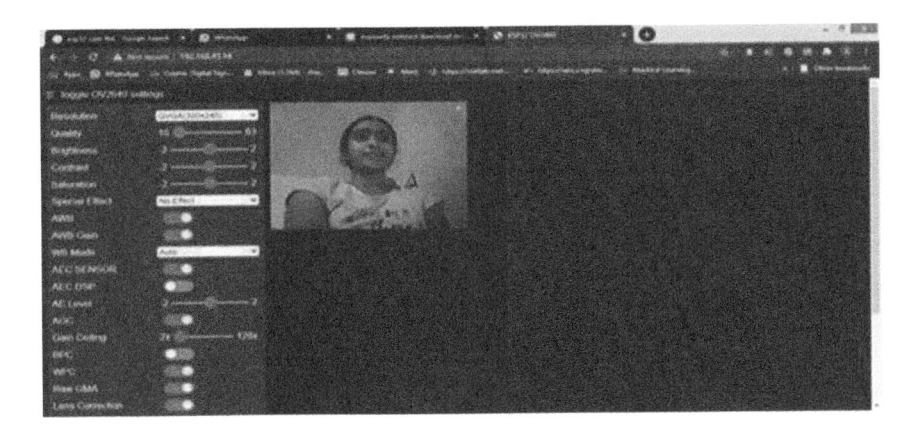

Figure 17.16 Camera surveillance.

Figure 17.17 OVA260.inbuilt facial detection.

Locations received by parents as SMS:

Figure 17.18 is the sample image of the SMS message received via GSM every 20 seconds and Figure 17.19 is the Google map location available after clicking the link.

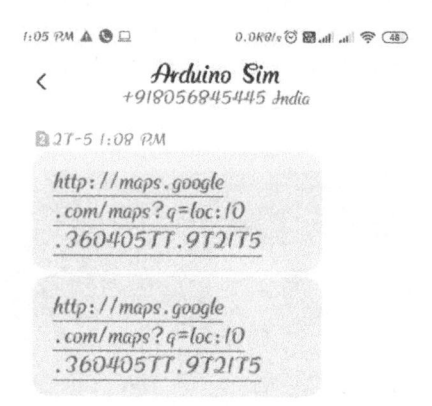

Figure 17.18 GSM message received.

Figure 17.19 Google map location display when the link is clicked.

17.5 Conclusion and Future Enhancement

17.5.1 Conclusion and Inference

Recent advancements in sensor technologies, IoT and wireless communications enabled the innovation of a new-generation baby healthcare monitoring device with wearable electronics. The constant recording of the ambience of their baby and analysis of the same facilitates the parents to understand the overall health condition of the baby and take prior actions for improvement, if necessary. The Sensor variables send alarms in real time to an Android mobile terminal via the application. Also, it is possible to observe the history of sensor data, allowing an analysis at a later stage also. With the help of the band worn by the toddler we can locate the baby and alert parents in case of any threats; it also helps kids maintain social distancing. Furthermore, the total expenditure spent in this baby monitoring is Rs.4000, making this system affordable and easy to use. This system is also energy efficient and can improve the quality of parenting.

17.5.2 Future Enhancement

The future scope of the project includes providing vaccine reminders for the baby. The reminders for common vaccines like polio, hepatitis, etc., can be set in the smartphone application in accordance with the date of birth of the baby.

When the baby cries, he/she can be temporarily distracted by the automatic music or toys placed in the cradle.

Weight monitoring can be done by the sensors placed on the crib. The periodical measurements can be recorded, which may be useful for medical analysis.

Access to a doctor (family doctor information acquired from the parents already) can also be provided remotely in case of queries and emergencies for the baby.

References

1. Doss, Pradeep, S. N. Siva, and Madhu Sudhan. "Non invasive health monitoring system for infant using IOT." In *2019 1st International Conference on Innovations in Information and Communication Technology (ICIICT)*, pp. 1-3. IEEE, 2019.
2. Badgujar, Dipali, Neha Sawant, and Dnyaneshwar Kundande. "Smart and Secure IoT based Child Monitoring System." (2019).
3. Shasna, M., Kerala Mathilakam, Mohsina Mohamed Kabeer, Nedheela N. Nazar, and Nighila Ashok. "Infant cradle monitoring system." (2019).
4. Mandke, Shilpa, Komal Kudave, Rakshanda Labde, and J. W. Bakal. "IOT based Infant Health Monitoring System." *International Journal of Engineering and Technology (IRJET)* (2018).
5. P. Nandhini and K. Moorthi. "A Study on Wearable Devices for the Safety and Security of a Girl Child and Women". (2018)
6. Lai, Chinlun, and Lunjyh Jiang. "An Intelligent Baby Care System Based on IoT and Deep Learning Techniques." *International Journal of Electronics and Communication Engineering* 12, no. 1: 81-85, 2018.
7. Jabbar, Waheb A., Hiew Kuet Shang, Saidatul NIS Hamid, Akram A. Almohammedi, Roshahliza M. Ramli, and Mohammed AH Ali. "IoT-BBMS: Internet of Things-based baby monitoring system for smart cradle." *IEEE Access* 7 : 93791-93805, 2019.
8. Amol Srivastava, BE Yashaswini, Akshit Jagnani, Sindhu K, "Smart Cradle System for Child Monitoring using IoT", *International Journal of Innovative Technology and Exploring Engineering (IJITEE)* ISSN: 2278-3075, Volume 8, Issue 9, July 2019.

9. Guo, Fang, and Jia Yong Duan. "Design and implementation of intelligent baby room monitoring system based on the internet of things." In *Applied Mechanics and Materials*, vol. 513, pp. 2484-2486. Trans Tech Publications Ltd, 2014.
10. Levy, M., Deepali Bhiwapurkar, Gokul Viswanathan, S. Kavyashree, and Pawan Kumar Yadav. "Smart cradle for baby using FN-M16P Module." *Perspectives in Communication, Embedded-systems and Signal-processing-PiCES* 2, no. 10 : 252-254 (2019) .
11. Kumar, Shubham, and Er Chhavi Gupta. "Child Safety Wearable Device." (2020).
12. Saadatian, Elham, Shruti Priya Iyer, Chen Lihui, Owen Noel Newton Fernando, Nii Hideaki, Adrian David Cheok, Ajith Perakum Madurapperuma, Gopalakrishnakone Ponnampalam, and Zubair Amin. "Low cost infant monitoring and communication system." In *2011 IEEE Colloquium on Humanities, Science and Engineering*, pp. 503-508. IEEE, 2011.
13. Rajesh, G., R. Arun Lakshman, L. Hari Prasad, and R. Chandira Mouli. "Baby monitoring system using wireless sensor networks." *ICTACT Journal on Communication Technology* 5, no. 3 (2014).
14. Morales, Laura Valentina Martinez, Karen Ximena Currea Piratova, and Monica Andrea Rico Martinez. "Real-time temperature and audio baby monitoring using IoT technologies." In *2019 Congreso Internacional de Innovación y Tendencias en Ingenieria (CONIITI)*, pp. 1-6. IEEE, 2019.
15. Kaur, T., M. Mittal, and H. Singh. "The baby monitoring room prototype model using IoT." *Int J Adv Res Sci Eng* 7, no. 6 (2018).
16. Joshi, Madhuri P., and Deepak C. Mehetre. "IoT based smart cradle system with an Android app for baby monitoring." In *2017 International Conference on Computing, Communication, Control and Automation (ICCUBEA)*, pp. 1-4. IEEE, 2017.
17. Symon, Aslam Forhad, Nazia Hassan, Humayun Rashid, Iftekhar Uddin Ahmed, and SM Taslim Reza. "Design and development of a smart baby monitoring system based on Raspberry Pi and Pi camera." In *2017 4th International Conference on Advances in Electrical Engineering (ICAEE)*, pp. 117-122. IEEE, 2017.
18. Hussain, Tanveer, Khan Muhammad, Salman Khan, Amin Ullah, Mi Young Lee, and Sung Wook Baik. "Intelligent baby behavior monitoring using embedded vision in IoT for smart healthcare centers." *Journal of Artificial Intelligence and Systems* 1, no. 1: 110-124, 2019.
19. Gare, Harshad Suresh, Bhushan Kiran Shahne, Kavita Suresh Jori, and Sweety G. Jachak. "IOT Based Smart Cradle System for Baby Monitoring." (2019).
20. Priyanka, M. Nandini, S. Murugan, K. N. H. Srinivas, T. D. S. Sarveswararao, and E. Kusuma Kumari. "Smart IOT Device for Child Safety and Tracking." *International Journal of Innovative Technology and Exploring Engineering (IJITEE)* 8, no. 8 (2019).

18

IoT-Based Plant Health Monitoring System Using CNN and Image Processing

Anindita Banerjee[1], Ekta Lal[2] and Berlin Hency V.[2*]

[1]School of Electronics Engineering (SENSE), VIT University, Chennai, India
[2]SENSE, Vellore Institute of Technology, Chennai, India

Abstract

India is a global agricultural powerhouse and farmers are the backbone of it. Here, most of the economic income comes from agriculture. However, an ever-increasing incidence of suicides among farmers is registered. These are attributable to the losses incurred due to low price for crops, loss of crops due to diseases or natural calamities. Additionally, the labour charges are also huge. Production damages due to bugs and viruses are approximately 15% and 35% of the entire harvest. Improper maintenance and protection of crops leads to more infections and affects overall productivity.

This paper presents a simple plant monitoring system made for disease detection and analysis. This system will help the farmer identify what type of diseases their plants are being affected with. We have created a web application where, with the help of neural network classification, the image is processed, and the status of the leaf is analyzed. The system also prescribes medicines for the diseases of the plants. The farmer uploads a picture of his crop and instantly gets the results. The web application is available in three different languages, Hindi, English and Tamil.

Keywords: IoT, plant disease detection, Arduino, CNN, deep learning, leaf disease, neural network training, image processing

---

*Corresponding author: berlinhency.victor@vit.ac.in

18.1 Introduction

Agricultural plants are the main factors on which the Indian economy heavily depends [13]. A large amount of energy is wasted in agriculture. This energy can be utilized in basic survival in cooking, lighting and space heating by the poor people in rural areas. Therefore, we need proper management of this energy in the agriculture sector. Precise use of this energy can help with better profits and vital supplies for the farmers. Healthy plants contribute to profitable crop production that satisfies the national demands of food, fuel, and fibre provision. Therefore, detecting plant diseases at primary stages is necessary. Monitoring the condition of the plant is crucial for successfully detecting plant diseases and hence, the successful cultivation of crops.

Plant health is strongly associated with human health at the most basic level. Agriculture produces medicines, food and fibre for human needs. It is also responsible for the livelihood of millions of farmers. Better crop production leads to better pay for the product as well as better health of consumers. India losses approximately Rs 55,600 crore worth of crops every year due to pollution, pests, weather and the diseases spread in the crops. A large number of farmers go bankrupt every year due to the destroyed crops. For a nation like India, where food is scarce, this is a big issue. Our system aims to help the user check at regular intervals easily if the crop is healthy or not. If the crop is suffering from any disease, the system will notify the user of the same. It will tell the user about the disease name, cause and cure of the same. Thus, it will help the farmer cure the disease before it affects the crop, helping with larger produce and incomes. The system works in three languages, Hindi, English and Tamil, helping it reach a large population. Also, there is no installation cost and it does not need much setup, making it accessible for poor farmers.

Using the convolution neural network and IoT together for the agriculture sector has given rise to new technologies and opportunities. This monitoring system is an appreciable system that instantly notifies concerning the crop's health and indirectly helps in increasing the national economy.

The main idea of this automated plant health monitoring system is to monitor the health of each plant in the field in real time [15]. It detects if the plant is healthy or not by checking the condition of its leaves. This system uses an enormous database of various leaves with commonly seen diseases. The images of the plant to be analyzed are captured and compared with the database of leaves. Then the system identifies the ailment that the plant is suffering from. It provides the user with information about the disease, cause and cure. It is capable to diagnose almost six different variants of major diseases in plants. If the discolouration is due to natural reasons, it tells the user that the crop is healthy and also the cause of discolouration. This automated system aims to reduce the number of crops destroyed due to infections.

18.2 Literature Survey

S. Dey, E. M. Amin and N. C. Karmakar [1] introduce and analyse a paper-based chip-less RFID device for condition measures in plants. A Resonator in this system is made with an oblong loop influencing in addition to AN inter-digital condenser. To prove its effectuality, an exploratory analysis of the same is done. An abundant amount of backscattered signal is laid by the resonator for the wireless measure inside the microwave waveband. Once it is encountered with some of the UWB antennas and a VNA, it gives the correct leaf condition response of the plant.

S. D. Calisgan, V. Rajaram, S. Kang, A. Risso, Z. Qian and M. Rinaldi [2] have exhibited a VOC detector of power zero for the initial process. This VOC is supported by micro-mechanical switches. Otherwise, from progressive active chemical sensors, the system given here uses a very passive, chemically sensitive switch supported cantilever and switch-based readout mechanism to sight VOCs prodigious a decided concentration free by not healthy crops. When focused on VOCs, the chemical compound bi-material beam diverts down and stimulates the mechanical switch because of strain elicited because of chemicals immersion within the layers of the polymer.

Y. Liu, K. Akram Hassan, M. Karlsson, O. Weister and S. Gong [3] state health, well-being, and luxury is closely associated with an indoor climate. Observance and administration are current for several regions, from government offices to peopled homes. Their preceding analysis revealed a vigorous leaf wall mechanism that successfully scales back the concentrations of stuff and VOC and balances greenhouse emission accumulation for an inside atmosphere. However, crop care was impossible due to geographic conditions and high cost.

Yafei Wang, Frank Bakker, Rudolf de Groot, Heinrich Wörtche [4], in their paper, have provided a scientific summary concerning questionable ecosystem services. This paper contributed significantly to the study of factors of home atmosphere like air quality, etc.

Rowlandson T, Gleason M, Sentelhas P, Gillespie T, Thomas C, Hornbuckle B. [5] have examined the connection between leaves conditions and plant diseases. They claim that due to stagnant water on leaves and fruits, the diseases in plants have increased.

Dadshani, S., Kurakin, A., Amanov, S. et al. [6] proposed that the stagnant water on leaves of plants display pressure. This study gives insights into pressure tolerance in plants at the time of water scarcity. They have successfully calculated moisture percentage and thus worked on better crop management.

M. I. Pavel, S. M. Kamruzzaman, S. S. Hasan and S. R. Sabuj [7] have designed an IoT and Raspberry Pi-based device to collect plant pictures and identify the diseases. This information is collected over a time frame in MySQL.

Kacira M and Ling PP. [8] have introduced a system managed by a computer that contains a motor-controlled turntable. This system comprises a motionless sensing device that calculates the microclimate condition difference between numerous crops. It considers the conditions required for plants for their growth and stores the values in the cloud.

Zeidler Conrad, Zabel Paul, Vrakking Vincent, Dorn Markus, Bamsey Matthew, Schubert Daniel, Ceriello Antonio, Fortezza Raimondo, De Simone Domenico, Stanghellini Cecilia, Kempkes Frank, Meinen Esther, Mencarelli Angelo, Swinkels Gert-Jan, Paul Anna-Lisa, Ferl Robert J. [9] give an outline in their paper of many years of observing plant health.

Harshitha, Jainita R, Apoorva M Gowda and Rekha [10] in their paper introduce image processing techniques for malady and detection; show the share of diseases; style and build-out of an affordable automatic system to produce ideal setting conditions for higher production of the crop.

Dhandapani, Vignesh, S. Remya, T. Shanthi, and R. Vidhy [11], in their project, deal with image processing techniques for locating diseases of the plants. Their project entirely revolves around disease detection through leaf images. However, they have not mentioned any medications for the same.

Tejas G. Mahale, Sanath S. Shetty, Salman B. Bandri, Gyanchandra Gupta and Prof. Gitanjali Korgaonkar [12] have proposed a custom handmade device to control crop health using a pan and tilt method. They have used a tiny colour zoom camera and RGB colour sensors. Leaf colour sensors examine the sensing element outputs to provide data on the status of plant tissue health in a non-destructive manner.

Dash, Jyotirmayee, Shubhangi Verma, Sanchayita Dasmunshi and Shivani Nigam [13] focus on plant health watching supported by NDVI (Normalized distinction Vegetation Index) calculation that helps in noticing the distinction between the healthy and unhealthy plants by scheming their NDVI values. The pictures of the crop area unit taken from a NIR camera, linked with a Raspberry Pi. Python is used for coding on Raspberry Pi. So, through the VNC viewer package, the outcome area unit shared with the user helps them distinguish between healthy and unhealthy plants.

A. M. Ezhilazhahi and P. T. V. Bhuvaneswari [14] state that, currently in India, about 70% of the people are in the agriculture system. Agricultural including harvesting of plants, fruits and crops on a large scale. This harvested product should be disease-free. Thereby, diseases in plants might cause a lot of economic losses. The most standout cause of harm is plant diseases. Farmers experience a lot of difficulties in the detection and control of these diseases with the naked eye.

P. Subashini, Abhishek, Sourav and Anurag [15] have studied agriculture powers the economy of the Republic of India. They have examined the needs of the plant for healthy growth. They have proposed software that gathers data about the surroundings. Thus, using different sensors, treating the soil and atmospheric conditions according to the needs of the plants.

Siddagangaiah and Srinidhi [16] state that monitoring plant health is crucial for a nation like ours. They have researched plant fitness measures for the betterment of biodiversity.

Pravin A, Prem Jacob T, Asha P. [17] have addressed the water scarcity problem in Indian agriculture. They have proposed a system using the internet that limits water wastage and proper utilisation of water. This system calculates the humidity requirements of the plants with the help of some sensors. Agriculture is the spine of our country; most humans rely on agriculture for their sole income.

M. S. Farooq, S. Riaz, A. Abid, K. Abid and M. A. Naeem [18] studied that due to the larger population, demand for meals is increasing. Drastic weather changes and water shortage is also on the rise. Industries are replacing agricultural lands, particularly in big cities. Because of the extent of vehicles, the degree of air pollution is growing with each day. Most of these problems should not wreck the spine of our country. Apart from these above problems, the actual venture exists in the cultivation of plants that offers extra yield; with the on-hand space, new methods, modern strategies have to be followed. The perfect solution is to change our domestic terrace or outside into a greenhouse. Substances like glass or plastic keep heat inside. It may additionally be bloodless weather; however, the inexperienced residence will provide first-rate heat surroundings through which vegetation can grow and flourish. It is seen that the yield of a greenhouse is substantially excessive due to the managed environment.

This does now not happen in the ordinary field. An introduced benefit of a greenhouse is the safety of plant life from environmental pollution due to its closed shape.

Kaur, Gaganpreet, Sarvjeet Kaur, and Amandeep Kaur [19] state awareness of plant leaf illnesses is a very salient factor in the agricultural environment. The prior identification of plant leaf ailments can assist in stopping the losses that farmers face due to more than a few plant leaf diseases. The plant leaf illnesses in photo processing are detected via observing the patterns of leaf snapshots at a positive duration of time. Observation and identification of leaf ailments is no longer an effortless assignment to do manually due to the fact it requires a lot of time, money, effort, etc. So, it is better to pick out the ailments through an automatic device in picture processing. There are various photo processing methods reachable for the detection of plant leaf ailments which consist of some simple steps such as photo acquisition, photograph pre-processing, picture segmentation, function extraction and photo classification.

Nurhasan, Usman, Arief Prasetyo, Gilang Lazuardi, Erfan Rohadi, and Hendra Pradibta [20] propose a Deep Flow Technical (DFT) hydroponics system using IoT. Hydroponics is the cultivation of flowers via water usage except for the use of soil by focusing on the success of dietary desires for crops. This system counts the crop booming factors. They have used various built-in sensors with Raspberry Pi Hydroponics. For the healthy growth of the fauna, all the essential factors are taken care of.

18.3 Data Analysis

Compared to the different image recognition techniques, deep learning–based image recognition technology does not require extracting distinct features.

In this paper, we have built an image classification model [21] using a combination of IoT and deep CNN. For the same, we have used deep learning techniques such as Keras and TensorFlow.

18.3.1 Convolutional Neural Network

Convolutional Neural Network or CNN holds a multiplex network and can execute convolution operations [22]. As depicted in Figure 18.1 below,

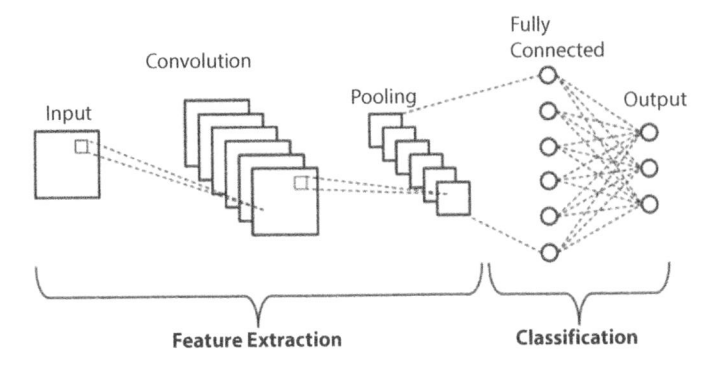

Figure 18.1 The basic structure of CNN.

the model contains an input layer, convolution layer, pooling layer, full connection layer and output layer. CNN is the most used model in the area of deep learning. The cause is the enormous model size and aggregate knowledge brought by the structural characteristics of CNN, which facilitates CNN to play an edge in image recognition. Similarly, the benefits of CNN in computer vision tasks have heightened the expanding admiration of deep learning.

18.3.2 Phases of the Model

The model will operate in two phases:

1. **Training detector:** Here, we have focused on loading our dataset from our disk, trained a model (using Keras and TensorFlow) in our dataset, and then serialized our detector to the disk.
2. **Deployment:** Once we trained our detector, we then moved to start the detector, performing dry/green leaf detection, and then assorting each leaf as green or dry.

Figure 18.2 below explains the phases operating in the model. The phase 1 and phase 2 include the train detector and the apply detector stages respectively.

18.3.3 Proposed Architecture

The proposed architecture is shown in Figure 18.3 below. This architecture shows the entire process proposed in this work in brief.

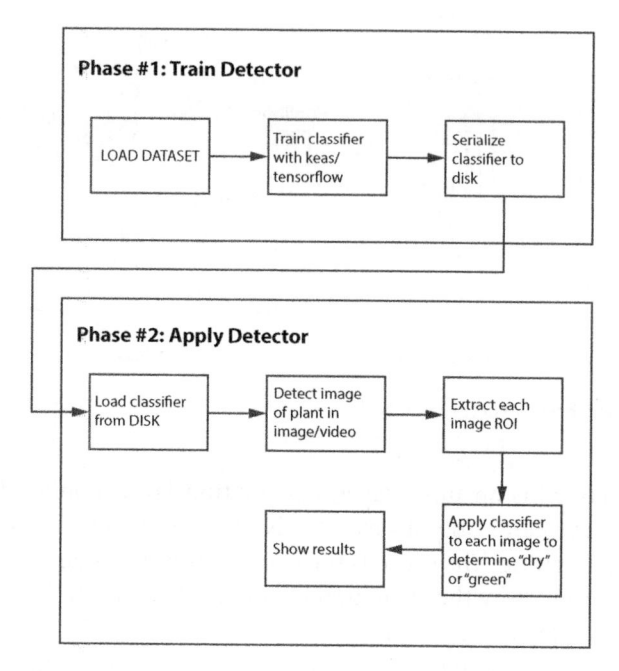

Figure 18.2 Phases of model training.

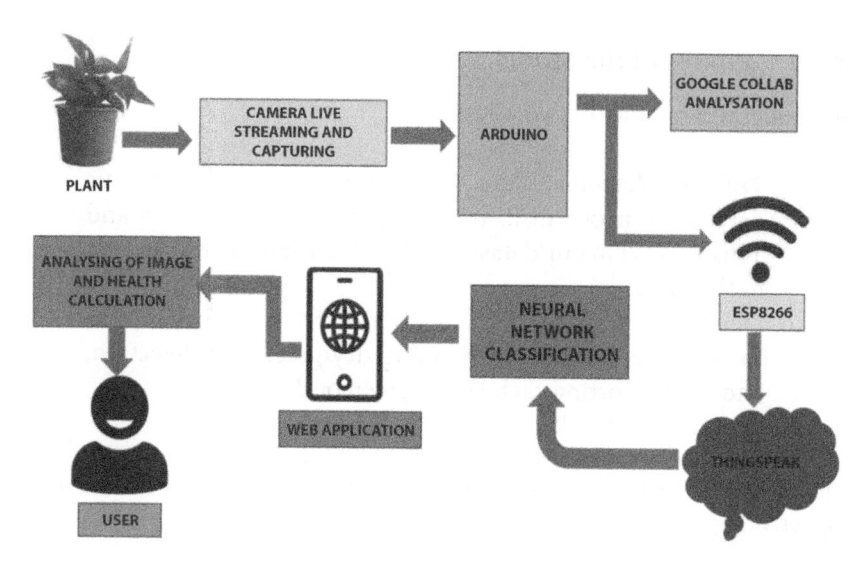

Figure 18.3 Flowchart of proposed model.

18.4 Proposed Methodology

Internet of Things (IoT) advances along with convolutional neural network and can be used in agricultural cultivation to enhance the quality of agriculture. This paper includes various features like detection of plant diseases at an early stage and recommending medications accordingly. In this system, to decide if the plant is ordinary or infected, a computerized framework is used.

18.4.1 System Module and Structure

In this system, a Python-based web application is built using the convolutional neural network (CNN). This web application takes the input image from the user and then analyses it. The training of our model used an enormous database of photographs of plants with various diseases. Plant recognition and disease detection is executed by the MTCNN package, which is a Convolutional Neural Network. This app also prescribes required medication if the plant has ailments. The application is available in three languages. The design of the proposed system uses an ESP32 OV2640 Camera to capture the plant that is to be monitored. It records the readings continuously and stores them in the cloud. We have also used Google Colab and Tinkercad to analyze whether the plant is green or dry to predict if it requires any additional water supply. For the web API, we have used ThingSpeak, which allows our Google Colab code and Tinkercad code to communicate with each other. Each API key has a Read code and Write code. Here, we put our Write code on Google Colab and read API key on Tinkercad. Google Colab will write values to the web API whether or not a plant is detected. The Tinkercad code reads the value from the API and converts it to physical means of detecting whether the leaf is dry or green. We achieve this by transmitting to ThingSpeak not only the default host address, but along with it, we send additional data (character "x", followed by "d/g" depending on the result). Our Tinkercad circuit can read the value of this transmitted data and can extract the last two digits. The results are finally stored in the ThingSpeak cloud using the ESP8266 Wi-Fi module. We have made the structure very affordable so that every person can make use of it. Moreover, the device's basic setup is quite effortless to establish, plus it can be introduced quickly by everyone.

18.4.2 System Design and Methods

We have used Arduino UNO and an ESP8266 Wi-Fi module to create our system. An ESP8266 Wi-Fi module is pretty inexpensive and hence very economical. The framework is associated with Google Colab and serial communication with Arduino alongside the Wi-Fi module. Arduino captures all the outputs produced by the Google Colab and stores them all in string format.

Side by side, a Python-based CNN web application is created. We have used Keras to train our model. It is an open-source library, and anyone can make use of this powerful library. After shaping the CNN model, automatically, an URL or Uniform Resource Locator gets generated using the IP address of the data server. Here, the URLLIB of Python is used to transmit an HTTP request using the previously created URL. Flask, an API of Python, is used to implement the web application. A camera captured multiple-coloured images of the leaves. These images were then sent to the database. Consequently, the database stored all the required values and images and analysed them to give results.

18.4.3 Plant Disease Detection and Classification

We have implemented CNN, a class of deep neural networks, to identify plant diseases. We were able to diagnose almost six different variants of major diseases in plants. In the upcoming sections, the entire procedure of training the model to detect plant diseases is detailed. We have divided the complete mechanism into multiple vital subsections.

18.4.3.1 Dataset Used

An appropriate and huge dataset is vital at each stage of the computer vision technique. We took all the images necessary for the model training purpose from multiple sources in numerous languages from the web. Further, we classified these images into thirty distinct groups. Twenty-eight classes represent the visually diagnosed plant conditions by observing the leaves.

We added a group of healthy leaves to this dataset to clearly distinguish healthy leaves from infected ones. For accurate classification, we supplemented a group of images with the added background to the dataset. Ultimately, the system prepared itself to distinguish the leaves from the surrounding. At this stage, the rejection of all replicated images by an advanced Python script implementing the comparing mode took place.

The resulting action implied enriching the dataset with extended images. This paper aims to train the system to acquire the traits that distinguish one group from the others. Consequently, when we used a higher number of augmented images, the probability to learn relevant features increased. The ultimate dataset contained 51,680 training images and 4,250 validation images. Section 18.4.3.3 contains the augmentation process in depth. The table given below presents the dataset used for training the plant disease detector.

18.4.3.2 Preprocessing and Labelling Methods

The web images used for the dataset were all of the different sizes, resolutions and properties. Hence, all the pictures were made of equal dimensions and preprocessed to extract the accurate features. To highlight the region of interest, we cropped all the leaf images manually into a square shape. During this procedure, we rejected the leaf images with dimensions less than 600 pixels. The only accepted photographs for feature extraction were the ones that had all the essential information required. To ensure a reduced training period, we resized all the images to 256 by 256 format.

It is essential to classify all the images thoroughly to build a legitimate detecting system. After collecting the pictures, we segregated them all and then labelled them with the appropriate disease names. At this stage, the elimination of all the replica and non-important images took place.

18.4.3.3 Procedure of Augmentation

The sole purpose of introducing augmentation is to include an insignificant amount of deformity in the images. This step helps not only to improve the dataset but also minimizes overfitting. Overfitting occurs when the model learns the details and noises in the data so much that it worsens its performance. For augmenting, we applied affine transformation, perspective transformation, and image rotations on the images. Figures 18.4, 18.5 and 18.6 below demonstrate the transformations implicated in the augmentation process. Figure 18.4 depicts the results of the Affine transformation. Figure 18.5 shows images obtained from perspective transformation, and Figure 18.6 reflects the simple rotation of the input image.

18.4.3.4 Training Using CNN

In this paper, we have built an image classification model from the dataset described in section 18.4.3.1. We have trained the model using deep

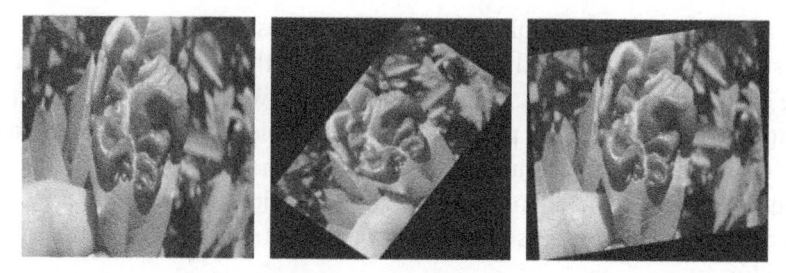

Figure 18.4 Affine transformation of images.

Figure 18.5 Perspective transformation of images.

Figure 18.6 Rotation transformation of images.

CNN algorithm. There are numerous popular deep learning frameworks available. For this research, we have used the Keras API of the TensorFlow Python library, which is the fastest and the best framework.

The Keras library in Python makes it moderately manageable to build a CNN. CNN is a unique kind of deep neural network technique used to analyze visual data. On the other hand, Keras is an easily accessible, powerful open-source Python library tool. Keras is the interface for the TensorFlow library. With the help of this library, we have productized our deep model on the web using very few lines of code.

18.4.3.5 Analysis

The standard procedure used to build this model included dividing our data into training and test dataset. We trained the model using the training dataset. For prognostication, we used the test set. After getting the accurate results for the trial set and our model's prognosticated outcomes, we calculated a rough estimation of the system's efficiency. We analyzed by training 4,250 original images with 51,680 database images.

For testing the precision of our predictive system, we followed a ten-fold cross-validation method. We repeated this procedure after each thousandth iteration during the training period. In Table 18.1, we have mentioned all the details of the dataset images used from each specified group. Section 18.5 details the analysis conclusions.

18.4.3.6 Final Polishing of Results

Fine-tuning is used to enhance the results by creating minute alterations. It is beneficial to intensify the efficiency of the system [23]. Fine-tuned training operations are usually simple, quick and effortless. This dataset has thirty diverse categories. Using experimental modifications, we were able to accomplish a plant disease detection system, which is further exhibited and illustrated in section 18.5.

18.4.4 Hardware and Software Instruments

For the training and testing purpose of our IoT-based model, we have used a single PC. To train the CNN, we used GPU mode instead of the normal mode. This machine worked roughly around six to ten hours for every iteration. We have detailed the basic properties of this machine in Table 18.2 below.

18.5 Results and Discussion

We have obtained the conclusions presented in this section by training all the images, including the original and augmented ones. To improve the accuracy of the results, we have tried to extend our dataset as much as possible.

After fine-tuning, the overall accuracy achieved was about 96.42% and roughly around 95.67% without fine-tuning. Figures 18.7 and 18.8 display

Table 18.1 Database used for training.

Group	Quantity of primary images	The total quantity of images: primary plus enlarged	Quantity of validation images
1. Undiseased leaves	426	2,952	322
2. Pear decline phytoplasma	265	2,124	152
3. Pear Leaf spot	211	1,057	110
4. Pear Viral diseases	225	1,190	182
5. Pear Phytoplasmal disease	190	2,110	59
6. Peach Leaf Curl	101	1,273	79
7. Peach Bacterial Spot	122	1,235	175
8. Apple Erwinia amylovora	245	2,258	211
9. Apple Venturia	189	2,150	153
10. Apple Marssonina leaf blotch	135	1,340	116
11. Apple scab	163	1,960	163
12. Grape wilt	264	2,109	114
13. Grape mites	223	1,094	230

(Continued)

Table 18.1 Database used for training. (*Continued*)

Group	Quantity of primary images	The total quantity of images: primary plus enlarged	Quantity of validation images
14. Grape Phylloxera	201	1,680	148
15. Grape Grey Mold	239	2,696	198
16. Orange sooty mold	211	1,678	108
17. Orange tip or Marginal Leaf Burn	209	2,211	106
18. Orange citrus Greening	298	1,890	93
19. Orange cigar Leaf Curling	178	1,560	112
20. Raspberry leaf spot	256	2,134	122
21. Raspberry spur blight	287	2,341	199
22. Raspberry yellow rust	198	1,066	85
23. Soyabean brown spot	180	1,500	72
24. Soyabean frogeye leaf spot	216	1,345	180
25. Soyabean bean pod mottle virus	135	1,171	76
26. Cucumber mosaic virus	181	1,099	152

(*Continued*)

Table 18.1 Database used for training. (Continued)

Group	Quantity of primary images	The total quantity of images: primary plus enlarged	Quantity of validation images
27. Squash bacterial wilt	201	2,058	211
28. Squash downy mildew	106	1,722	114
29. Tomato leaf mold	209	1,442	96
30. Background images	1,235	1,235	112
TOTAL	7,299	51,680	4,250

Table 18.2 Fundamental features of the machine.

Machinery components	Features
1. System in-built memory	4 Gb
2. Processor of the system	Intel(R) Core (TM) i3-7020U CPU @ 2.30GHz 2.30 GHz
3. Graphics	Intel(R) HD Graphics 620
4. Operating system	Windows 20H2 64 bits

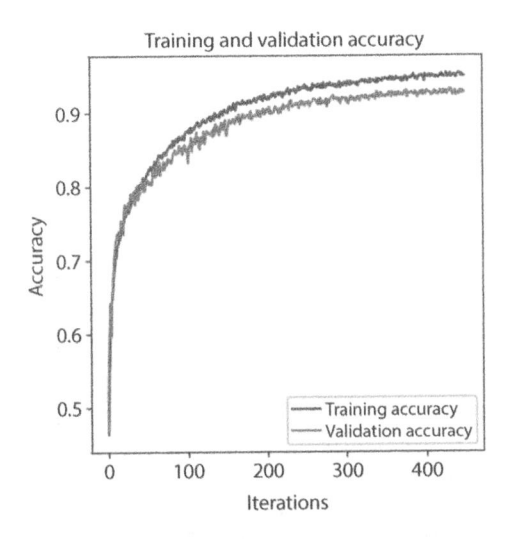

Figure 18.7 Training and validation accuracy.

the graphs of accuracy and loss, respectively. We clicked a snapshot of the training model after every twenty-thousandth repetition. The loss curves rapidly diminished through training iterations.

Moreover, we tested each group of the trained model separately. Each image was inspected individually by us. Figures 18.9 and 18.10 represent the graph of testing accuracy for all thirty groups.

Figures 18.9 and 18.10 clearly show that the model's accuracy is directly proportional to the number of images used. The pictures with background proved to segregate the leaf images better from their surroundings. Eventually, our system proved to be more efficient than other disease-detecting techniques.

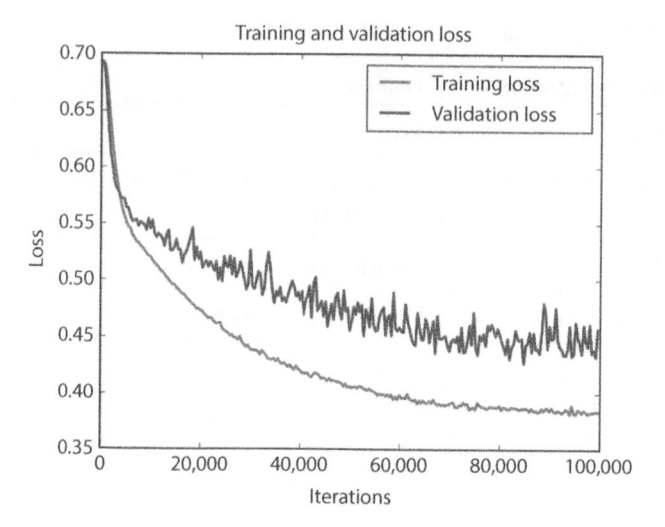

Figure 18.8 Training and validation loss.

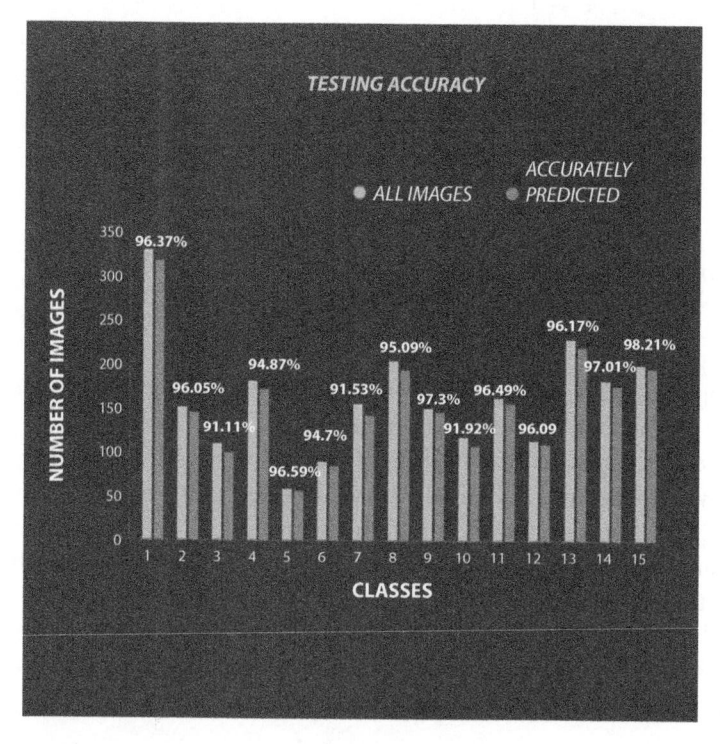

Figure 18.9 Graph of testing accuracy (first 15 classes).

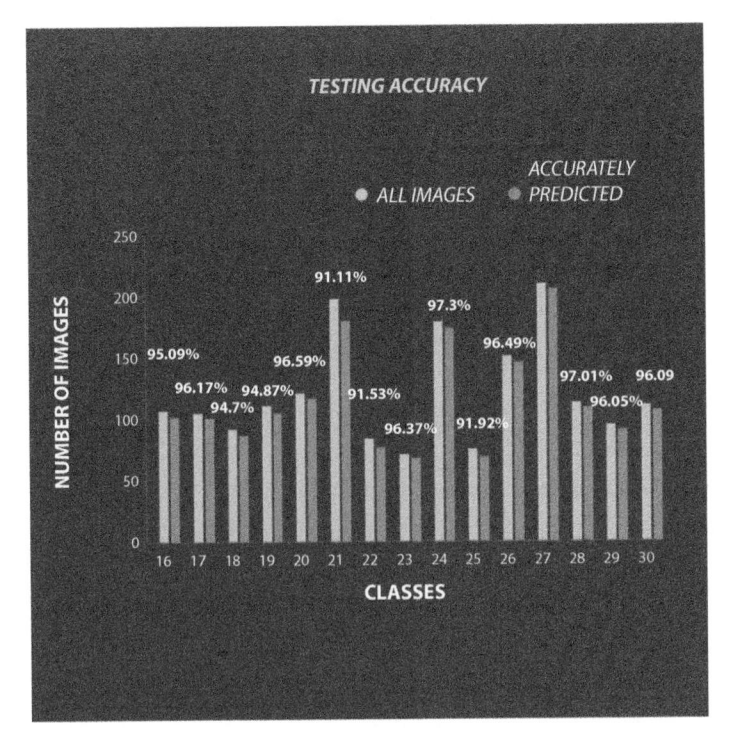

Figure 18.10 Graph of testing accuracy (last 15 classes).

We have used ThingSpeak to generate API keys. Every API key has a Read code and Write code. Here, we embed our Write code on Google Colab and read API key on Tinkercad. Figures 18.11 and 18.12 show our ThingSpeak write key and read key, respectively.

The Tinkercad code reads the values from the API and converts them to physical means of identifying whether the leaf is dry or green. Figures 18.13 and 18.14 given below show the serial monitor outputs.

Google Colab writes values to the web API whether or not a plant is detected. Figures 18.15 and 18.16 given below show the Google Colab outputs.

Figures 18.17, 18.18, 18.19, 18.20 and 18.21 show the web page created for convenience of users to detect the diseases of their crops.

Figure 18.11 ThingSpeak write API key.

Figure 18.12 ThingSpeak read API key.

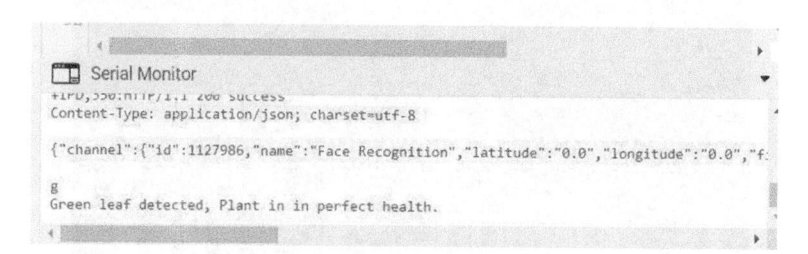

Figure 18.13 Tinkercad serial monitor output for a green leaf.

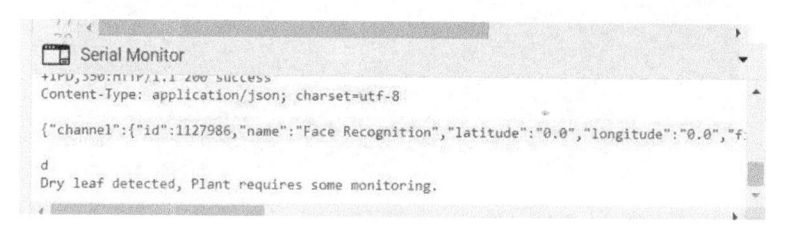

Figure 18.14 Tinkercad serial monitor output for a dry leaf.

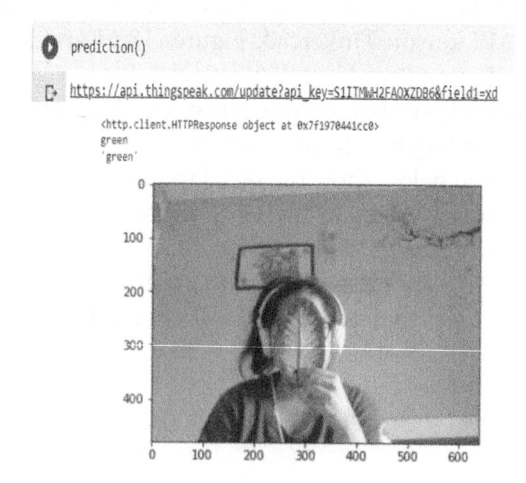

Figure 18.15 Google colab output for a green leaf.

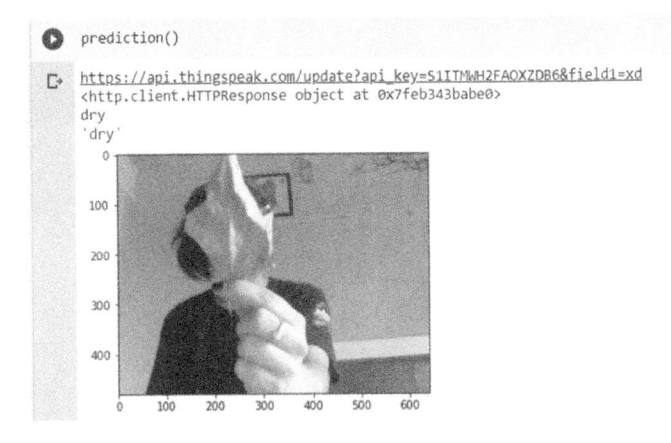

Figure 18.16 Google colab output for a dry leaf.

Figure 18.17 Index page of the web app.

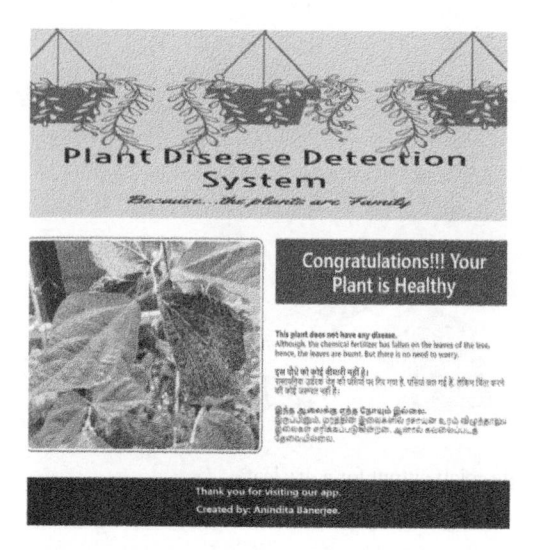

Figure 18.18 Healthy leaf page of the web application.

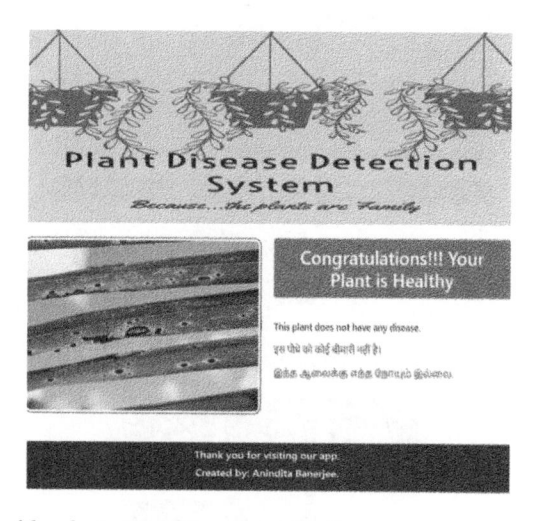

Figure 18.19 Healthy plant page of the web application.

IoT-Based Plant Health Monitoring System   285

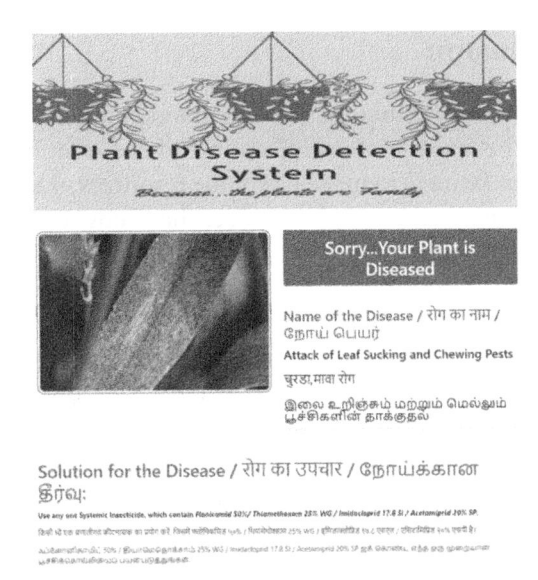

Figure 18.20 Diseased plant page of the web application.

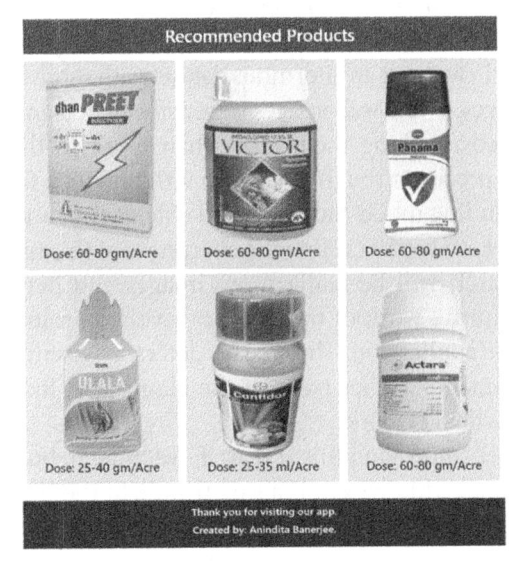

Figure 18.21 Medications prescribed for the plant diseases.

18.6 Conclusion

In this paper, we have ensured to use only the required amount of energy and resources. The appropriate medication and the amount necessary for each disease helps farmers to make use of the resources in the best possible way to fight the disease and not let it affect the produce. This helps them with greater profits and more resources for their daily needs.

The research in the domain of plant disease detection is scanty despite having so many computer vision processes. In this paper, we have attempted successfully to detect as many as six major plant diseases from their leaves using deep learning. Our model is competent in segregating the leaf images, which is the region of interest from the background. It also distinguishes visually diagnosable healthy leaves from infected ones. We have described every step of this research in detail, starting from gathering data to training and preprocessing and eventually polishing the results. Lastly, we tested our model multiple times before being published to ensure an accuracy level of 100%. We have implemented this model into a web application so that needy farmers can make use of this system. This application does not require any expertise and is usable by anyone. It is a convenient, fast and inexpensive solution to all the needs of a farmer.

For this research, we built an enormous dataset with more than seven thousand original images. These web images were reformatted and retransformed to match our needs, making them more than fifty-two thousand images. Our system has a precision roughly between 90% and 98% for various tested groups. The concluding model has an overall efficiency of around 96.25%.

For future work, to further refine the sharpness of this model, an expansion of this research will be using more images and better augmenting techniques. The future work of this project revolves around converting it into a smartphone application. The disease detector system will not be limited to only leaf images. Other parts of the plants also, like fruits, vegetables, stems, etc., will also work.

Thus, by further improving this research, we hope to build a sound system for the peasants. This system is a tiny tribute to the Indian farmers, to prevent farmer suicides. Using this system, we also aim to motivate more of the younger generation to pursue farming as a career.

References

1. Dey, Shuvashis, Emran Md Amin, and Nemai Chandra Karmakar. "Paper based Chipless RFID Leaf wetness detector for plant health monitoring." *IEEE Access* 8 (2020): 191986-191996.

2. Calisgan, Sila Deniz, Vageeswar Rajaram, Sungho Kang, Antea Risso, Zhenyun Qian, and Matteo Rinaldi. "Micromechanical Switch-Based Zero-Power Chemical Detectors for Plant Health Monitoring." *Journal of Microelectromechanical Systems* 29, no. 5 (2020): 755-761.
3. Liu, Yu, Kahin Akram Hassan, Magnus Karlsson, Ola Weister, and Shaofang Gong. "Active plant wall for green indoor climate based on cloud and Internet of Things." *IEEE Access* 6 (2018): 33631-33644.
4. Wang, Yafei, Frank Bakker, Rudolf De Groot, and Heinrich Wörtche. "Effect of ecosystem services provided by urban green infrastructure on indoor environment: A literature review." *Building and Environment* 77 (2014): 88-100.
5. Rowlandson, Tracy, Mark Gleason, Paulo Sentelhas, Terry Gillespie, Carla Thomas, and Brian Hornbuckle. "Reconsidering leaf wetness duration determination for plant disease management." *Plant Disease* 99, no. 3 (2015): 310-319.
6. Dadshani, Said, Andriy Kurakin, Shukhrat Amanov, Benedikt Hein, Heinz Rongen, Steve Cranstone, Ulrich Blievernicht *et al.* "Non-invasive assessment of leaf water status using a dual-mode microwave resonator." *Plant Methods* 11, no. 1 (2015): 1-10.
7. Pavel, Monirul Islam, Syed Mohammad Kamruzzaman, Sadman Sakib Hasan, and Saifur Rahman Sabuj. "An IoT based plant health monitoring system implementing image processing." In *2019 IEEE 4th International Conference on Computer and Communication Systems (ICCCS)*, pp. 299-303. IEEE, 2019.
8. Kacira, Murat, and P. P. Ling. "Design and development of an automated and Non–contact sensing system for continuous monitoring of plant health and growth." *Transactions of the ASAE* 44, no. 4 (2001): 989.
9. Zeidler, Conrad, Paul Zabel, Vincent Vrakking, Markus Dorn, Matthew Bamsey, Daniel Schubert, Antonio Ceriello *et al.* "The plant health monitoring system of the EDEN ISS space greenhouse in Antarctica during the 2018 experiment phase." *Frontiers in Plant Science* 10 (2019): 1457.
10. Harshitha, Jainita R, Apoorva M Gowda, Rekha, "Plant health monitoring and controlling of environment in a Conservatorium," 2018, International Research Journal of Engineering and Technology (IRJET), Volume 05, Issue 07, July 2018.
11. Deepika, P., and S. Kaliraj. "A Survey on Pest and Disease Monitoring of Crops." In *2021 3rd International Conference on Signal Processing and Communication (ICPSC)*, pp. 156-160. IEEE, 2021.
12. Tejas G. Mahale, Sanath S. Shetty, Salman B. Bandri, Gyanchandra Gupta, Prof. Gitanjali Korgaonkar, "Plant Health Monitoring Using Wireless Technology," *International Journal of Scientific and Research Publications*, Volume 6, Issue 9, September 2016, 720 ISSN 2250-3153 www.ijsrp.org
13. Jayaraj, Devu, R. P. Aneesh, P. S. Sooraj, S. Arya Lekshmi, and P. Deenath. "Smart Agro: IOT Based Rice Plant Health Monitoring System."

14. Khan, Jihas. "Vehicle network security testing." In *2017 Third International Conference on Sensing, Signal Processing and Security (ICSSS)*, pp. 119-123. IEEE, 2017.
15. Subashini, P., and Sourav Abhishek. "Anurag (2019)." *Real Time Plant Health Monitoring System Using Sensors and Clouds. IRJCS: International Research Journal of Computer Science* 6: 189-192.
16. Siddagangaiah, Srinidhi. "A novel approach to IoT based plant health monitoring system." *Int. Res. J. Eng. Technol* 3, no. 11 (2016): 880-886.
17. Pravin, A., T. Prem Jacob, and P. Asha. "Enhancement of plant monitoring using IoT." *International Journal of Engineering and Technology (UAE)* 7, no. 3 (2018): 53-55.
18. Farooq, Muhammad Shoaib, Shamyla Riaz, Adnan Abid, Kamran Abid, and Muhammad Azhar Naeem. "A Survey on the Role of IoT in Agriculture for the Implementation of Smart Farming." *IEEE Access* 7 (2019): 156237-156271.
19. Mangla, Monika, Deepika Punj, and Shilpa Sethi. "A review paper on techniques to identify plant diseases." *Annals of the Faculty of Engineering Hunedoara* 18, no. 2 (2020): 177-182.
20. Nurhasan, Usman, Arief Prasetyo, Gilang Lazuardi, Erfan Rohadi, and Hendra Pradibta. "Implementation IoT in System Monitoring Hydroponic Plant Water Circulation and Control." *International Journal of Engineering & Technology* 7, no. 4.44 (2018): 122-126.
21. Sladojevic, Srdjan, Marko Arsenovic, Andras Anderla, Dubravko Culibrk, and Darko Stefanovic. "Deep neural networks based recognition of plant diseases by leaf image classification." *Computational intelligence and neuroscience* 2016 (2016).
22. Chowdhury, Muhammad EH, Tawsifur Rahman, Amith Khandakar, Mohamed Arselene Ayari, Aftab Ullah Khan, Muhammad Salman Khan, Nasser Al-Emadi, Mamun Bin Ibne Reaz, Mohammad Tariqul Islam, and Sawal Hamid Md Ali. "Automatic and Reliable Leaf Disease Detection Using Deep Learning Techniques." *AgriEngineering* 3, no. 2 (2021): 294-312.
23. Behmann, Jan, Anne-Katrin Mahlein, Stefan Paulus, Heiner Kuhlmann, Erich-Christian Oerke, and Lutz Plümer. "Calibration of hyperspectral close-range pushbroom cameras for plant phenotyping." *ISPRS Journal of Photogrammetry and Remote Sensing* 106 (2015): 172-182.

Bibliography

Al-Hiary, Heba, Sulieman Bani-Ahmad, M. Reyalat, Malik Braik, and Zainab Alrahamneh. "Fast and Accurate Detection and Classification of Plant Diseases." International Journal of Computer Applications - IJCA, www.ijca-online.org/archives/volume17/number1/2183-2754. Accessed 10 Nov. 2022.

Aoki, Takayuki & Tanaka, Fumio & Suga, Haruhisa & Hyakumachi, M. & Scandiani, Mercedes & O'Donnell, K.. (2012). Fusarium azukicola sp. Nov., an exotic azuki bean root-rot pathogen in Hokkaido, Japan. Mycologia. 104. 1068-84. 10.3852/11-303.

Bock, C. H., P. E. Parker, A. Z. Cook, and T. R. Gottwald "Visual Rating and the Use of Image Analysis for Assessing Different Symptoms of Citrus Canker on Grapefruit Leaves - PubMed." PubMed, 1 Apr. 2008, pubmed.ncbi.nlm.nih.gov/30769647.

Costa, Corrado & Schurr, Uli & Loreto, Francesco & Menesatti, Paolo & Carpentier, Sebastien. (2019). Plant Phenotyping Research Trends, a Science Mapping Approach. Frontiers in Plant Science. 9. 1933. 10.3389/fpls.2018.01933.

Elena, Santiago & Fraile, Aurora & García-Arenal, Fernando. (2014). Evolution and Emergence of Plant Viruses. Advances in virus research. 88. 161-91. 10.1016/B978-0-12-800098-4.00003-9.

Fankar Armash Aslam, Hawa Nabeel Mohammed, Jummal Musab Mohd, Munir, Murade Aaraf Gulamgaus, Prof. P. S. Lokhande, Case study and report in "Efficient Way of Web Development Using Python And Flask", International Journal of Advanced Research in Computer Science, ISSN No.0976-5697, Volume 6, No. 2, March-April 2015.

Jian, Lin, and Zhu Bangzhu. "Neural network ensemble based on feature selection." In 2007 IEEE International Conference on Control and Automation, pp. 1844-1847. IEEE, 2007.

Karmokar, Bikash Chandra, Mohammad Samawat Ullah, Md Kibria Siddiquee, and Kazi Md Rokibul Alam. "Tea Leaf Diseases Recognition Using Neural Network Ensemble." International Journal of Computer Applications - IJCA, www.ijcaonline.org/archives/volume114/number17/20071-1993. Accessed 10 Nov. 2022.

Korotaeva, Daria, and Maksim Khlopotov. "CEUR-WS.org/Vol-1975 - Supplementary Proceedings of AIST 2017." CEUR-WS.org/Vol-1975 - Supplementary Proceedings of AIST 2017, 7 Nov. 2017, ceur-ws.org/Vol-1975.

Liu, Jun, and Xuewei Wang. "Plant Diseases and Pests Detection Based on Deep Learning: A Review - Plant Methods." BioMed Central, 24 Feb. 2021, plantmethods.biomedcentral.com/articles/10.1186/s13007-021-00722-9.

Paulus, Stefan & Dupuis, Jan & Mahlein, Anne-Katrin & Kuhlmann, Heiner. (2013). Surface feature based classification of plant organs from 3D laser-scanned point clouds for plant phenotyping. BMC bioinformatics. 14. 238. 10.1186/1471-2105-14-238.

Riker, Albert Joyce, and Regina Stockhausen Riker. "Introduction to research on plant diseases. A guide to the principles and practice for studying various plant-disease problems." Introduction to research on plant diseases. A guide to the principles and practice for studying various plant-disease problems. (1936).

Rumpf, T., A-K. Mahlein, U. Steiner, E-C. Oerke, H-W. Dehne, and L. Plümer. "How to Access Research Remotely." How to Access Research Remotely, www.cabdirect.org/cabdirect/abstract/20103316634. Accessed 10 Nov. 2022.

Sharma, Parul, Yash Paul Singh Berwal, and Wiqas Ghai "Performance Analysis of Deep Learning CNN Models for Disease Detection in Plants Using Image Segmentation." Performance Analysis of Deep Learning CNN Models for Disease Detection in Plants Using Image Segmentation - ScienceDirect, 18 Nov. 2019, www.sciencedirect.com/science/article/pii/S2214317319301957.

Wang, X. & Zhang, M. & Zhu, J. & Geng, Shengling. (2008). Spectral prediction of Phytophthora infestans infection on tomatoes using artificial neural network (ANN). International Journal of Remote Sensing - INT J REMOTE SENS. 29. 1693-1706. 10.1080/01431160701281007.

West, Jonathan S., Cedric Bravo, Roberto Oberti, Dimitrios Moshou, Herman Ramon, and H. Alastair McCartney. "Precision Crop Protection- the Challenge and Use of Heterogeneity | Ebook | Ellibs Ebookstore." Precision Crop Protection - the Challenge and Use of Heterogeneity|Ebook| Ellibs Ebookstore, www.ellibs.com/book/9789048192779/precision-crop-protection-the-challenge-and-use-of-heterogeneity.

Zhou, Chi, H. B. Gao, Liang Gao, and W. G. Zhang "International Journal of Computer Networks and Applications (IJCNA)." International Journal of Computer Networks and Applications (IJCNA), www.ijcna.org/abstract.php?id=871.

19

IoT-Based Self-Checkout Stores Using Face Mask Detection

Shreya M.[1], R. Nandita[1], Seshan Rajaraman[1] and Berlin Hency V.[2*]

[1]B.Tech Electronics and Communications, Department of SENSE, VIT University, Chennai, India
[2]Department of SENSE, VIT University, Chennai, India

Abstract

Currently, the world is witnessing a second wave of the Covid-19 pandemic, and the situation is getting worse day by day. Simple protocols like minimising human contact and wearing a mask outdoors are proving to be good measures to control the spread of the virus. We saw a huge rise in the demand for daily items and due to a lack of availability, large numbers of people gather without taking any precautions to stock essentials. This has led to the spread of the virus to a great extent. In self-checkout stores, the shopping experience is completely automated and there is no physical presence of the shop owner. The automation enables the customers to pick their goods, scan and make payments by themselves without the intervention of the owner or a cashier. In such stores there is a high chance of people not following Covid protocols. So, there is a need for a system that maintains an allowed threshold of people inside the store at any one time, thus minimizing the potential dangerous human contact at all possible cases. We propose an IoT-Based Self-Checkout Store Using Mask Detection. The primary goal of this project is to create a safe environment for the consumers who visit the shop, by keeping a check on the number of customers present at the store and ensuring that each and every customer is following the protocol of wearing a mask. The system consists of two parts, the face mask detection and the customer count. For the mask detection part, deep learning algorithms like CNN are used to generate a model that helps detect a mask, and for the customer count part, a threshold value is set, which gives us the maximum number of people allowed inside the store at a time. The PIR sensors detect the entry and exit of customers and help regulate the count below the threshold. So once the face mask detection of the customer is complete

---

*Corresponding author: berlinhency.victor@vit.ac.in

and the number of people present inside the store is checked, the system takes the decision of either allowing the customer inside or asking him or her to wait. This project is designed to provide a solution to the current real-world problem using minimally efficient technology with high accuracy.

Keywords: Internet of Things (IoT), mask detection, self-checkout stores, CNN, RaspberryPi, sensors

19.1 Introduction

This project aims to provide complete automation of a day-to-day aspect of ordinary people, i.e., shopping in stores, which is also seen as a threat to the safety of consumers during the pandemic. The proposed system is easy to implement and is designed with efficient technology so as to not compromise the safety of vulnerable age groups of customers. The system is proposed by use of Raspberry Pi SBC and Pi camera to input the real-time video/image capture of the customers to enter the self-checkout store in order to pass the condition of mandatory presence of properly worn mask, which is implemented using deep learning models. Sensors are used to continuously detect the movement of customers in and out and transmit data to the single board computer in order to carry out the process designed.

19.2 Literature Review

19.2.1 Self-Checkout Stores

Firstly, we need to understand what a customer looks for when he or she shops at a self-checkout store, and this was clearly elaborated in [1]. There is a need of a faster and more reliable service and an "easier-to-use" system that can be understood by anyone across different age groups. It was noted that a customer who has a satisfactory overall experience attributes it to the retailer responsible for the superior service and hence this improves retail patronage intentions.

The Study of Success Factors in Supermarket Self-checkouts [2] dives deep into some factors that may affect the preference for self-checkout stores over traditional checkout. Although the fact remains that, the former is less time consuming and more efficient, the author points out that if the customer does not appreciate the self-checkout technology and still uses it, it may create a bad experience in some way or the other and hence may affect the store's rating.

The research [3] highlights the issues faced by the existing physical cashier system and how self-checkout stores can help eliminate these effects. Firstly, it mentions that the speed of the cashier's workflow depends on the length of the line at the station and also the physical difficulties faced, like posture problems, by the cashier throughout the day. It also mentions the long waiting time that can be eliminated by using self-service technology. It also mentioned that the older generation may not jump at this idea as they would perceive it as extra work for the customer. But this is outweighed by all the other positives that the self-checkout stores offer.

Empirical evidences from a study [4] concentrates on the technology Readiness and likelihood to use Self-Checkout Services (SCS) using smartphone in Retail Grocery Stores and how different factors are correlated to one another. The process of a SCS is as follows: the customer scans the barcode of the product using a self-checkout app. If the scan fails, they can manually enter the barcode. Once the scan is complete, they can use inbuilt banking interfaces in the app. After a successful transaction the receipt is sent as an email. They found that the Indian market was moderately ready to adopt the new technology and also concluded that if the users were comfortable using their technology, then there is a high possibility of them to consider it useful and start using it.

The security concerns in such stores call for a fraud prevention system which was elaborated in [5] that focuses on the retailer's perspective of running a self-checkout store without encountering any kind of loss. The system developed earlier would take in the barcode (either by scan or manually) and identify the item. Once the item is placed on the belt for further processing, the item's weight will be cross checked with the one that the system identified through the barcode to ensure that no extra item is taken by the customer without it being accounted for. But a common loophole in such systems is that a customer can scan a lower priced item of the same weight and clearly get away with an expensive item. So, to smartly overcome this loophole, an additional visual screening is developed to measure the height, colour, shape, etc., of the item and hence improve security.

19.2.2 Face Mask Detection

A main aspect of this project is detecting whether the customer is wearing a mask or not and for this [6] proposed a faced mask recognition system based on discarding masked regions and deep learning-based features in order to address the problem of the masked face recognition process.

Face mask detection and masked face detection are two different tasks; in this paper they have carried out the second, where a pre-trained deep Convolutional neural network (CNN) is applied to extract the best features, i.e., to recognize a face based on the eyes and the forehead regions. This method involves heavy sampling and training of data.

A new approach in [7] combined the algorithms of image super-resolution and classification networks (SRCNet) and developed a new facemask-wearing identification method. This method is mainly for mask detection in low-quality images. The idea of image super resolution is for reconstructing high-quality images from low-resolution images. Therefore, high-performance algorithms like SRCNet have been introduced to improve efficiency.

A system that uses deep learning techniques in addition to an alarm system in distinguishing facial recognition and recognizing if the person is wearing a mask was explained in [8]. The system mentioned is a RPi-based real-time face mask recognition that alarms and captures the facial image. In this paper, they are acquiring images and annotating the images which is to classify the images using text to help the DL model detect. This is adding metadata to the dataset, further performing the image processing

In [9] the researchers applied methodology of identification of the data sets and then the model, to investigate pictures as masked and unmasked using the Inception v3 model, which is an image detection algorithm used for machine learning. This paper improves the conventional systems of face recognition. To resolve this, the identification is carried out on faces that are not noticeable. Here the data sets have been analyzed to make specific algorithms feasible. They have also used the mathematical formula for different layers.

A deep learning model called facemasknet which can work with still images and a live video stream was discussed in [10]. Previously, binary classification has been performed to detect face masks. This uses three-class classification, the region of interest (ROI) is extracted and then the mask detector is applied. This classification is uncommon, which increases the data to be trained significantly but also provides an efficient application.

A design of a real-time DNN-based face mask detection system which used a single-shot multibox detector and MobileNetV2 was proposed in [11]. The accuracy seemed like a good measure, as the classes were balanced. Precision gave the measure of positive predicted values. Recall helped the classifier to find all positive samples and f1 score indicated the measure of test accuracy. These evaluation metrics were chosen because of their ability to give best results in a balanced dataset.

A masked faces detection model [12] in the Wild with LLE-CNNs Robust features play an important role in detecting masked faces, especially when most facial regions are occluded by masks. These results may point out a future direction of developing robust and effective face descriptors with the assistance of the MAFA dataset.

A face mask detection with a warning system for preventing respiratory infection using the Internet of Things was developed and discussed in [13]. The result showed that the accuracy of the developed system is 91.60% for mask detection. When we talk about the system evaluation, the mean value around 4.51 has a standard deviation 0.51 and a mean value 4.62 has a standard deviation 0.49 by experts and users, respectively.

A masked face detection using the Viola Jones Algorithm [14] is a progressive approach for less time-consuming masked face detection that consisted of different stages and were analysed. As compared to others, calculating the distance of a person from the camera is more robust and correct. Eye line detectors now are easier for us to implement; however, it contributes to detections that are fake within inadequate quality images. Eye feature detection is reliable for identifying eyes on the face. Facial component discovery can be sturdy along with time-consuming steps due to many regions. The face-eye-based multi-granularity model achieves 95% recognition accuracy.

19.3 Convolution Neural Network

Generalizing across typical images, a person's face may be at any distance from the camera, there could be multiple irrelevant objects, the person's face may be a bit tilted, people may have various different features on their faces, the masks themselves may be different in each case, etc. Thus, we can come to a definite conclusion that a handcrafted, simple rules-based model would not be effective at all in this case. There are simply too many probable variations to hard code. This is where Deep Learning and specifically Convolutional Neural Networks came into the picture. Neural Nets are complex learning algorithms which try to learn a simpler, purpose-oriented representation of the data given to them. Convolutional neural networks take it one step further and employ convolution to focus on certain aspects of the image, thus making it viable to use for image classification and object detection tasks.

In essence, a neural network is basically a model full of tuneable parameters which can learn complex representations. Any neural network has two stages, primarily: Training, and testing. In the training stage, training data consisting

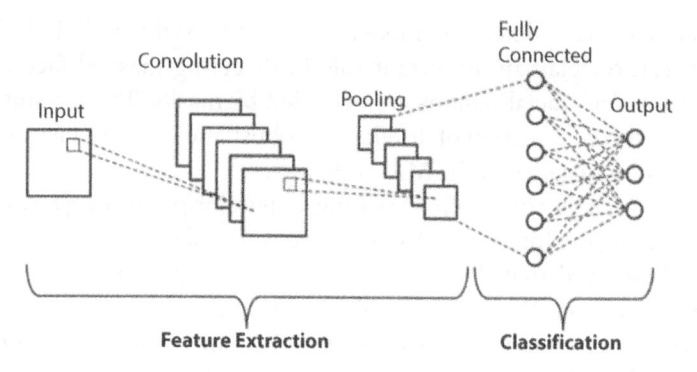

Figure 19.1 A simple ConvNet.

of the inputs (images of people wearing masks) and expected outputs (yes – 1 or no – 0) are fed into the network, and the network learns the patterns in the data which can help it to achieve the expected output. A convolutional neural network contains two parts (Figure 19.1): the convolutional layers in the front and the fully connected (a.k.a. Dense) layers in the back.

The input layer (consisting of the image matrix captured through the Rpi camera) is fed into a convolutional layer first, which is basically a matrix of weights that is convolved with the input. This matrix is referred to as a "kernel" or a "filter". After this, an integer, called the "bias" is added to the whole convolved output matrix.

The Figure 19.2 depicts the process of convolution in images. You can see a 3x3 kernel (filter) being convolved over a 6x6 input. Here, the kernel basically slides over the entire image from top left to bottom right, and the

Figure 19.2 Convolution operation.

corresponding numbers are multiplied, added, and the result is entered into the corresponding position of the output matrix. This output matrix is referred to as the feature map, because it represents all the features that were focused on and gathered by the convolution process. But in a CNN, there would be multiple kernels, even as many as 64 kernels convolving an image.

One important thing to notice here is that the pixels in the middle of the picture are focused on more than the pixels in the corner. This leads to the model focusing more on the objects in the center and less on the objects near the corners. In order to avoid this, we add more pixels around the boundary of the image, with a value of 0. Thus, the actual corner pixels will get more focus during the convolution process. This process of adding pixels to the boundary is termed as padding.

The filter matrix and the bias are tunable parameters, and they are adjusted so that the network's output is as close as possible to the expected output during training. Thus, over time the network learns to discern the important features in the input image. After convolution, a non-linear function called the **activation function** is applied over the output.

In Figure 19.3, an entire CNN is shown. There are multiple batches of convolution and pooling layers, followed by the Dense layers. One observation to be made while looking at each layer's outputs is that, as you go deeper into the network, the layers focus on more sophisticated parts of the image. The first layers start with enhancing edge features, and as they move in, the layers start to focus and enhance groups of pixels to entire objects, provided the training is done properly. The output that we will get through

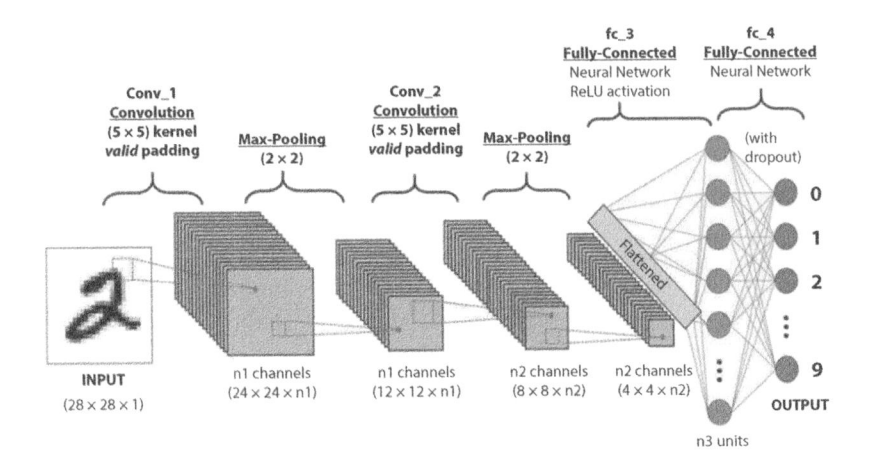

Figure 19.3 An entire CNN.

the CNN model for this system is either 1 (indicates presence of mask) or 0 (indicates absence of mask). Based on this output we further have certain conditions like the capacity of the room (explained in later sections) that need to be fulfilled to let the customer enter the store.

19.4 Architecture

In the architecture diagram of the proposed system (Figure 19.4), we aim to achieve a closed loop operation of face detection and people counter. We start off by initialising a threshold value 'n' which indicates the maximum number of people allowed inside the store at any point of time. This helps us reduce the spread of the virus by avoiding overcrowding inside the store.

The PIR sensors present at the entry and exit doors, will give us an indication of how many people enter and exit the store, respectively, and this will be updated in the variable P. so the variable P always indicates the number of people present inside the store. Now we check if the number of people 'P' present inside the store is less than the threshold value 'n'. If yes, then the face mask detection will take place and if the mask is detected then, we open the doors and the PIR sensor detects a person entering, so the value of P goes up by unity. If the number of customers is equal to the threshold 'n', the customer will be asked to wait for his or her turn. If the face mask is not detected, the customer will be asked to wear a mask and try again.

The flow in Figure 19.5 depicts the processes included in developing the detector model and testing it using a RPi camera module. Once we train

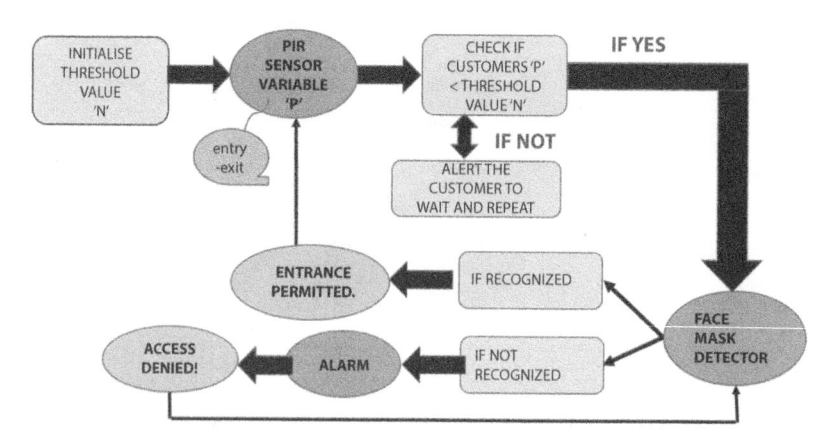

Figure 19.4 Architecture of proposed system.

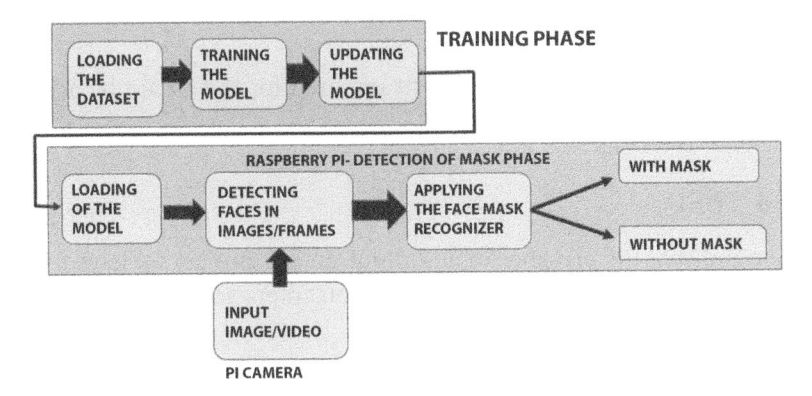

Self-Checkout Stores with Face Mask Detection 299

```
TRAINING PHASE
LOADING THE DATASET → TRAINING THE MODEL → UPDATING THE MODEL

RASPBERRY PI- DETECTION OF MASK PHASE
LOADING OF THE MODEL → DETECTING FACES IN IMAGES/FRAMES → APPLYING THE FACE MASK RECOGNIZER → WITH MASK / WITHOUT MASK
                            ↑
                    INPUT IMAGE/VIDEO
                    PI CAMERA
```

Figure 19.5 Face mask detection process.

and develop the model using CNN, we get the image/video feed using the Pi camera attached to the RPi and pass it on to the face mask detector.

19.5 Hardware Requirements

19.5.1 PIR Sensor

It is a sensor which detects the human motion at a range of 5m to 12m from the sensor. In our project we have used this sensor to measure the number of people entering and exiting to properly monitor the capacity.

19.5.2 LCD

It is used to display a particular output with the help of light modulated properties of liquid crystals.

19.5.3 Arduino UNO

It is an open-source microcontroller board based on the ATmega328P. With the help of the digital and analog input/output pins this board can be used to interface various components and circuits.

19.5.4 Piezo Sensor

Used to measure the physical quantities from the surrounding.

19.5.5 Potentiometer

Used to provide an amount of current required for the circuit by adding/reducing the resistance in the circuit.

19.5.6 LED

Used to indicate the presence of an output. In our project we have used 2 LED's, each one of them glows when a person enters and exits respectively.

19.5.7 Raspberry Pi

A Raspberry Pi is a mini version of a computer at the size of a credit card which performs all the basic functions of a computer.

19.6 Software

19.6.1 Jupyter Notebook

It is one of the programming software that supports Python, R.

19.6.2 TinkerCAD

It is a 3D modelling program that supports visual code blocks and electronic circuits. In simple terms our hardware circuit can be reconstructed virtually with the help of TinkerCAD.

19.7 Implementation

The project consists of two parts: Face Mask detection and People counter. To test the working of the code and concept and due to lack of hardware, the face mask detection was done based on the CNN concept using Python programming on the Jupyter Platform. To train the model we have two folders (with mask and without mask). These two folders consist of around a thousand pictures of people wearing a mask and people with bare face, respectively. The algorithm learns the characteristics using these pictures as dataset and generates a model that can accurately detect the presence of a mask. This model is then imported in the Test.py file and is applied on the images it captures from the webcam of the laptop to detect the presence of a face mask. This code can be easily used on the Raspbian interface and the images can be captured using the Pi camera connected to the Raspberry Pi.

19.7.1 Building and Training the Model

We first import all the libraries that are required like Keras, SkLearn, etc. Keras is a numerical computing library used mainly for deep learning. SkLearn is a machine learning library used with Python programming. Building the Neural Network: we first add 2 pairs of Conv2D and MaxPool layers to extract the features from the image dataset. Conv2D is used to introduce 100 filters with size 3x3 to learn the traits of the image of height and width 150 and having 3 channels (RGB). We use the 'Relu' activation function, which is to be applied after convolution and it is a mathematical function defined as

$$Y = x; \text{ if } x \geq 0$$

$$Y = 0; \text{ if } x < 0$$

The MaxPool layer helps apply a 2x2 filter over the output of the previous convolution layer to get the maximum of the elements in that pool window. Flatten() is used to flatten the output vector of the above layers. For example, an output shape of 1x1x1x2048 from the previous layers will be flattened into a 2048 length vector. The Dropout() function is used to randomly switch off certain neurons at each pass. Say the probability is 0.5, only 50% of the neurons will accept the input and give an output. This helps make sure that no one neuron gains priority over the others and the weights are distributed across each neuron. Since the purpose of the neuron is to identify a different feature, this function serves as a regularization layer which helps reduce overfitting. The next two layers are dense layers of 50 - neuron output with activation function 'Relu' and a 2-neuron output Dense layer, which is activated by the softmax activation function. Softmax is essentially the activation function used for multi-class classification. We then use model.compile() with various parameters to configure the model for training. Now that we have the model ready, we need a dataset to **train and test the model** on.

We first store the path of the training and testing datasets in TRAINING_DIR and VALIDATION_DIR. Since we have a limited dataset to train, we use the ImageDataGenerator() function to augment the data in different ways like changing the rotation angle, height and width of the image, zooming in, flipping the image, etc. This helps us get a wide range of images to train the model on which helps increase the accuracy of the model. We apply this ImageDataGenerator() function on the images present in the training dataset and create batches of size 10 with each image resized to dimensions 150 x 150 as this is the required input size of our CNN model that we developed earlier.

ModelCheckpoint() is used to save the model at a particular location. Since we set monitor to 'val_loss' and save_best_only as 'True', every time we get a model with loss lower than the previous model, the current model gets saved at the given path. Using model.fit_generator() we pass in the training data set (training_dir) to the model to learn the features and test it on the validating dataset (validation_dir). The epoch value is nothing but one iteration of training process. We use callback to store the model generated at the path specified in ModelCheckpoint() and give details about the accuracy, loss etc.

19.7.2 Testing The Model

We first load the model with lowest loss that was developed during our training session. We define the labels i.e., 0 as without mask and 1 as with mask. This will be used to name the box frame drawn around the face in the image. We also mention the color of the box frame as red for without mask and green for with mask in the 'color_dict' dictionary.

Now to test the model, we assigned the number of people (n) present in the store as 12. But in the real implementation this value will be obtained from the PIR sensors present at the entry and exit doors.

We run a while loop which begins by accessing the webcam of the laptop and creates an object called webcam. OpenCV already contains many pre-trained classifiers for face, eyes, smile, etc. So we import one classifier "haarcascade_frontalface_default.xml" and store it in variable classifier. Webcam. read() returns 2 values, one is a Boolean value (rval) which tells us if the image was captured successfully, and if it is "true", the captured image gets stored in variable 'im'. We resize 'im' and apply the classifier defined before and the output is stored in variable 'faces'. Now for every face 'f ' in faces we resize it to make it suitable to input to our model. The results from the model are stored in variable 'result'. The label 0/1 (without/with mask) is extracted using the np.argmax() function. This label is used to determine the color of the box frame and label the frame in the live window.

The image is displayed in the live window and the 'n' value is checked using the 'if loop' against a threshold value that indicates the maximum number of people allowed inside the store. The If loops works as follows:

1. The number of people inside the store is less than the threshold and the label is 1 (i.e., mask is detected) the doors will open and the n value is incremented.
2. The number of people inside the store is equal to the threshold and the label is 1, the customer will be asked to wait and the entry doors will remain closed until one person exits from the exit door.
3. Label is 0 (i.e., mask is not detected) then the doors will remain closed and the customer will be asked to wear a mask and try again.

This runs in a while loop till the 'ESC' button is pressed, upon which the loop is stopped and the webcam is turned off and image display windows are closed.

19.8 Results and Discussions

Let's assume the number of people already present in the store is 14. So, N=14. Practically, the values of the number of people entering and exiting are obtained from the PIR sensors placed at the entry and exit doors. But since we are simulating the scenario, we enter those values ourselves using an input function (Figure 19.6).

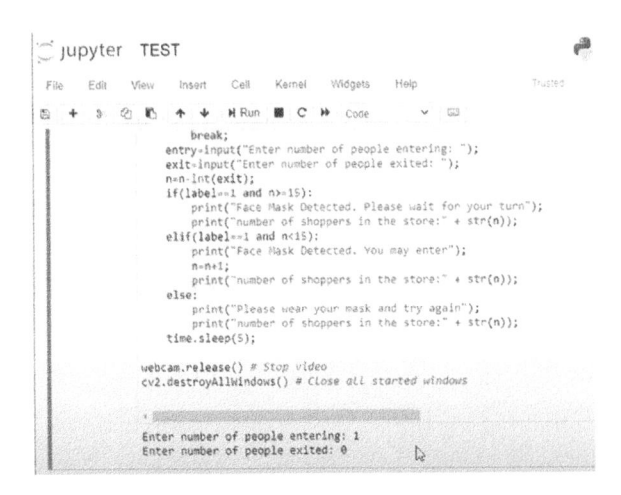

Figure 19.6 Entering the PIR values.

The following cases were assumed:

CASE 1: NO FACE MASK DETECTED
As in Figure 19.7, the number of people inside the store (N) is 14, but since a face mask wasn't detected, the person will not be allowed to enter the store.

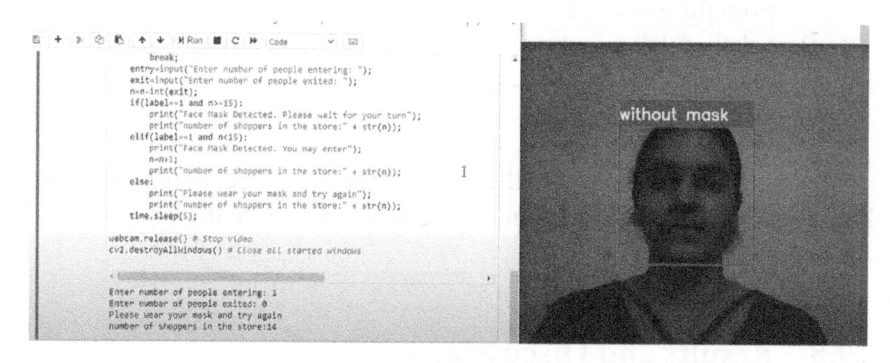

Figure 19.7 No mask detected.

CASE 2: FACE MASK DETECTED
Face mask is detected and since N is less than 15, the person is allowed to enter and N is incremented by 1, hence, N=15 (Figure 19.8).

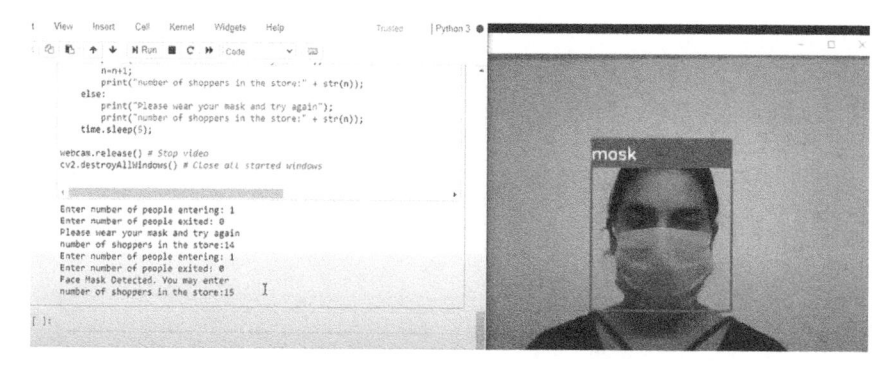

Figure 19.8 Face mask detected and N<15.

CASE 3: FACE MASK DETECTED BUT N=15

Although a face mask is detected, the number of people present inside the store at the moment N=15, thus we ask the customer to wait till someone exits the store (Figure 19.9).

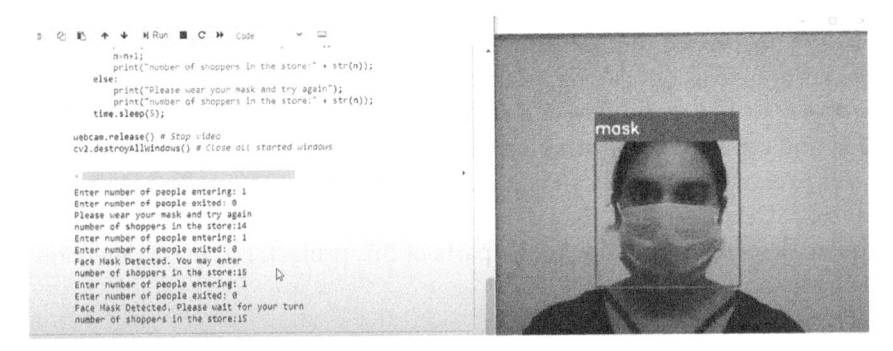

Figure 19.9 Face mask detected but N=15 (threshold).

CASE 4: FACE MASK DETECTED AND 1 PERSON EXITS

In Figure 19.10, the face mask is detected and one person exits, hence making N=14, which is less than the threshold, so the person is allowed to enter. Now N=15 again.

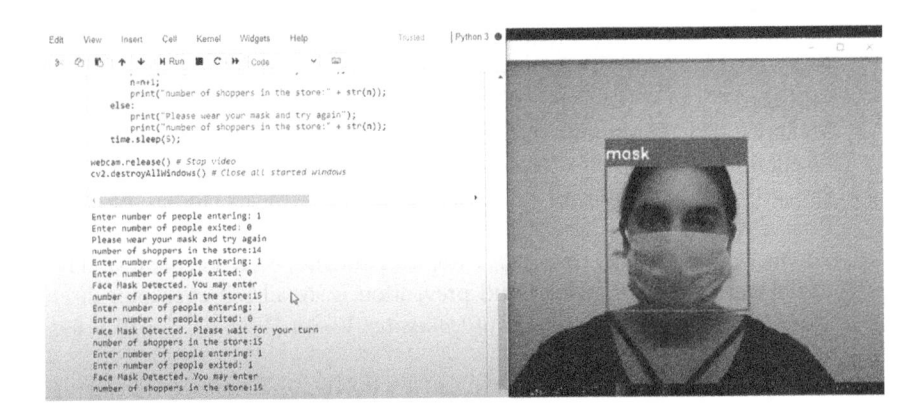

Figure 19.10 Face mask detected and one person exits the store.

19.9 Conclusion

Technology has had an immense effect on our surroundings, and its development in human safety is especially noticeable during the pandemic period. The project is completely eco-friendly and economical as it does not cause any harm to the environment. With the help of mask detection along with capacity monitoring the shop owner can assure the safety of the customers as only a certain number of people who are wearing masks properly will be allowed in at a time. This idea of combining mask detection and capacity monitoring must be implemented in all possible public areas to avoid unwanted spreading of the Covid virus. With the help of the efficient Deep learning algorithm CNN (Convolutional Neural network) the implementation of one of the parts of this project, i.e., mask detection is made simpler.

References

1. Fernandes, T., & Pedroso, R. (2016). The effect of self-checkout quality on customer satisfaction and repatronage in a retail context. *Service Business*, 11(1), 69–92. doi:10.1007/s11628-016-0302-9
2. Russell J. Zwanka "A Study Of Success Factors in Supermarket Self-checkouts". *International Journal of Management Information Technology and Engineering (IJMITE)* ISSN (P): 2348-0513, ISSN €: 2454-471X, Vol.7, Issue 4, Apr. 2019, 7-18.
3. Hassan, H., Sade, A. B., & Rahman, M. S. (2013). Self-service Technology for Hypermarket Checkout Stations. *Asian Social Science*, 10(1). doi:10.5539/ass.v10n1p61
4. Mukerjee HS, Deshmukh GK, Prasad UD. Technology Readiness and Likelihood to Use Self-Checkout Services Using Smartphone in Retail Grocery Stores: Empirical Evidences from Hyderabad, India. *Business Perspectives and Research*. 2019;7(1):1-15. doi:10.1177/2278533718800118
5. Bobbit, R., Connell, J., Haas, N., Otto, C., Pankanti, S., & Payne, J. (2011). Visual item verification for fraud prevention in retail self-checkout. *2011 IEEE Workshop on Applications of Computer Vision (WACV)*. doi:10.1109/wacv.2011.5711557
6. Hariri Walid (2020). "Efficient Masked Face Recognition Method during the COVID-19 Pandemic." DOI: 10.21203/rs.3.rs-39289/v1
7. Qin, Bosheng; Li, Dongxiao (2020). "Identifying Facemask-Wearing Condition Using Image Super-Resolution with Classification Network to Prevent COVID-19". *Sensors* 2020, no. 18: 5236. DOI: 10.3390/s20185236

8. S. V. Militante and N. V. Dionisio (2020) "Real-Time Facemask Recognition with Alarm System using Deep Learning" *2020 11th IEEE Control and System Graduate Research Colloquium (ICSGRC), Shah Alam, Malaysia,* 2020, pp. 106-110, DOI: 10.1109/ICSGRC49013.2020.9232610.
9. Priyanka S, Manikantha K, Kavita V. Horadi (2020) "Deep Learning Model for Identification of People with or without Masks During Pandemic" *European Journal of Molecular & Clinical Medicine.* ISSN 2515-8260, Volume 07, Issue 08, 2020.
10. Madhura Inamdar and Ninad Mehendale, "Real-Time Face Mask Identification Using Face masknet Deep Learning Network" (2020). Social Science Research Network. DOI:10.2139/ssrn.3663305
11. "A real time DNN-based face mask detection system using single shot multibox detector and MobileNetV2". *Sustainable Cities and Society*, Volume 66, March 2021, 102692.
12. "Detecting Masked Faces in the Wild with LLE-CNNs" Conference: *2017 IEEE Conference on Computer Vision and Pattern Recognition (CVPR)* DOI: 10.1109/CVPR.2017.53
13. "Face Mask Detection and Warning System for Preventing Respiratory Infection Using the Internet of Things". Vol. 9 No. 9 (2020): Volume 9, Issue 9 ISSN :2320-0790.
14. "Masked Face Detection using the Viola Jones Algorithm: A Progressive Approach for less Time Consumption". *December 2018 International Journal of Recent Contributions from Engineering Science & IT (iJES)* 6(4):4 DOI: 10.3991/ijes.v6i4.9317

20
IoT-Based Color Fault Detection Using TCS3200 in Textile Industry

T. Kalavathidevi[1], S. Umadevi[2*], S. Ramesh[1], D. Renukadevi[1] and S. Revathi[1]

[1]Department of Electronics and Instrumentation Engineering, Kongu Engineering College, Perundurai, Erode, Tamil Nadu, India
[2]Centre for Nanoelectronics and VLSI Design, School of Electronics, VIT University, Chennai, Tamil Nadu, India

Abstract

In the past, measurements for detection of color have been rigid to perform and expensive to handle. As well as in color display monitors, computer printers, plotters, paint, textiles, cosmetics, medical analysis, dental matching, etc., color sensors are often used in semiconductor (LED) sorting and testing, industrial sorting, and identification, and in label management as well. In other words, it is important to understand about the color sensing element, as it works for enormous applications in modern technology and industries. An in-depth analysis of color areas, transformations, and color management systems is usually required. This paper discusses the use of the TCS3200CS color sensing element (I2C sensing element Color Grove), which is an inexpensive hardware based on Arduino solution that can capture, process, and manage the color of any non-self-luminous object. In this way, the sample color can be sensed and the proportion of each color element color in it can be estimated. With the use of a proximity sensor, the length of the fabric that has faulty color is also detected. The formulated system is based on combining RGB color sensors and photo detectors to perceive color on colored garments.

Keywords: Internet of Things, sensors, color capturing, preprocessing

---

Corresponding author: umadevi.s@vit.ac.in

20.1 Introduction

Modern measurement and detection techniques are influenced by technology developments in the electronic sector. The color may be determined by the exploitation of the tools. Colors are classified as primary and secondary in physics. The three primary colors are RGB (Red, Green, and Blue).

Figure 20.1 represents a part of the textile industry. The textile business is primarily involved with the design, production, and distribution of yarn, fabric, and apparel. Materials could also be either natural or artificially manufactured products of the industry. These industries attribute the most important to color when choosing clothing colors. Detection of color in fabric plays a critical role in both the paper and textile industries. It is considered to be a main parameter in determining the sale volume of a product, and it is a primary consideration for product acquisitions. It is a standard objective of color technology to manage and reproduce a color within a collection of similar conditions. However, several of the factors that enable color control in textile supply chains are not optimal, which could result in a longer path from product conception to a client, with related price efficiency. Sensors for Color detection offer solutions for color intolerance, and dimension across a variety of applications like color feedback management in solid-state illumination and RGB backlight systems.

This industry requires a real-time response and improved method management to minimize waste. Miserably, modern color segmentation strategies rely on gradual, complicated, and massive-sign processing

Figure 20.1 Part of the textile industry.

techniques that cannot be implemented in compact architectures. As an end result, low-fee, compact, and brief coloration sensors should be developed for several industrial, agricultural, and strength-saving applications. Conventionally, automatic machines are used for specifying the share of shade elements within the material coloration. Manufacturers determine those combinations relying on the variety of sun shades hooked up in the product. This approach is restricted to the advent of a very hooked up quantity of sun shades of happiness. The work is focused on three primary colors, Red, Green, and Blue, with a color sensor element TCS3200 used to determine the given shade. Arduino Uno R3 is used as the controller chip in this technique.

The sensing element detects the color and transmits that information to the chip. As per the programming code, the input from the sensing building block provides the output. Sensor output is represented by dissimilar intensities of key colors, which, when diverse, give the color detected. The problem of the textile unit in color fault detection is addressed with the proposed technique of detection of color fault through color sensor and its length by IR sensor.

20.2 Literature Survey

Errors caused due to human negligence are avoided [13] by the use of an automated system by color-based sorting using a color sensor. Experimental result shows that the gadget is automated and reduces the labor paintings and it is far utilized in airport baggage separation device.

Basil plants were fertilized [7] with 0, 2.5 mM and 10 mM nitrogen (with different NO_3^-/NH_4^+ ratios), and then monitored using a low-power technique based on an optical leaf meter and a low-cost RGB sensor interfaced with an Arduino UNO board. Fluorescence spectral technique and UV methods [5] were used to study the sensing behavior of the probe.

The color sensor can notice the intensity [9] of the mild reflected from the product which then integrate it with the frequency of the wavelength. This is used to provide the essential value in the RGB space. The use of RGB records and changing equations, the crimson colors are given in tiers so that it's far simpler to determine precisely in which it belongs inside every color area. By the given results of the deviation and the ranges of the deviation, sorting can be accomplished.

Measuring the intensity of light [2] transmitted by the solution is the another research study where authors determined the concentration of the solution using a calorimeter. This study used the samples of dyes of both damaging and risk-free. Photodiode sensors use the light source to detect light that passes through the sample. The photodiode sensor try to change the light into an output voltage. The output voltage is amplified by using a voltage amplifier circuit. The Arduino confirms the output voltage, which is then displayed on the LCD.

The voice recognition [3] is used to perform color detection recognition with the prototype device. According to 80% of blind people, the prototype of color detection can perform well, while 70% of the juvenile population reports that the prototype can perform color detection.

The possibility of estimating the pH through the analysis of the color of the pH paper is suggested [6]. It isn't a subjective approach of judging the degree of pH visually however it needs some quantitative facts. By means of the optical sensor involved in this cram, it can be able to gain the value of the RGB colors with a small measuring device.

Software library called Open CV (Open Source Computer Vision Library) and Microsoft visual studio 2010 [14] is used as a development podium with a influential function. It makes the programmer's work more efficient and flexible and supports multiple development applications.

TCS3200CS [4,1] is a color sensor which captures, processes, and manages the colors in the fabric with the controller namely Arduino.

A cost-effective and simple resolution [9] for providing a sorter automaton using sensors. By measuring the reflected wavelength from the artifact, the color sensor detects its frequency. An example of colored candy sorting was used in the investigation of the machine.

RGB sensor values were acquired from the field surface using the RGB sensor. The densities of the weeds are estimated with the neural network model [8]. This method proved to be accurate for assessing the weed density in the field at 83.75 percent.

The research work [15] planned an innovative reflective color sensing system for controlling and monitoring the color of the in paper and textile processes. The system is created by means of a solid-state RGB sensor and an elegant signal procedure put into practice on microcontroller architecture. Several demos and results of the simulation are presented to demonstrate the significance of sensors and correctness in measurement.

Computer vision-based [10] approaches with application in textile industry to detect fabric defects was discussed. Various methods proposed are histogram-based approaches, color-based approaches, image segmentation-based approaches, frequency domain operations, texture-based defect detection, sparse feature-based operation, image morphology operations, and recent trends of deep learning.

Biologically stirred processing architecture [12] to distinguish and categorize fabrics with reverence to the weave pattern yarn color (fabric color) and (fabric texture) was proposed. Red Green Blue (RGB) color descriptor based on rival color channels simulating the single opponent and double opponent neuronal purpose of the brain is incorporated in to the texture programmer to extract the yarn color feature values. Support vector machine classifier is involved to train and test the proposed bioinspired algorithm.

Fabric fault [11] in the jacquard fabric is detected based on the number of color yarns employed in the color channels can be extracted from an arbitrary jacquard fabric. Images of each color channel are patterned are generated with a test input fabric. Preprocessing the fabric images is done to eliminate the noise. The characterization of the pattern is done with a comparison method involving

20.3 Methodology

Figure 20.2 shows the block diagram of the IoT-based color sensing system. In this proposed system, associate information is provided about the color of textiles within the thread color section. In any scenario, the user will receive a message with the corrective measures that need to be taken. It also detects the length of the color that is woven into the fabric. Upon successfully connecting to the web, the method looks for instructions on an online server and uploads the information through Wi-Fi. During this process, we tend to perform encoding into the microcontroller unit following the initial estimated values of multiple sensors intended for normal conditions.

Three sensors used in the process are the color sensor, IR sensor, and Proximity sensor. They are used in the detection of the color defect area and the length of the fabric. Based on the detection the data is continuously sent to the Blynk app all the way through the Wi-Fi module. As per the detection, the servo motor is turned to stop the running

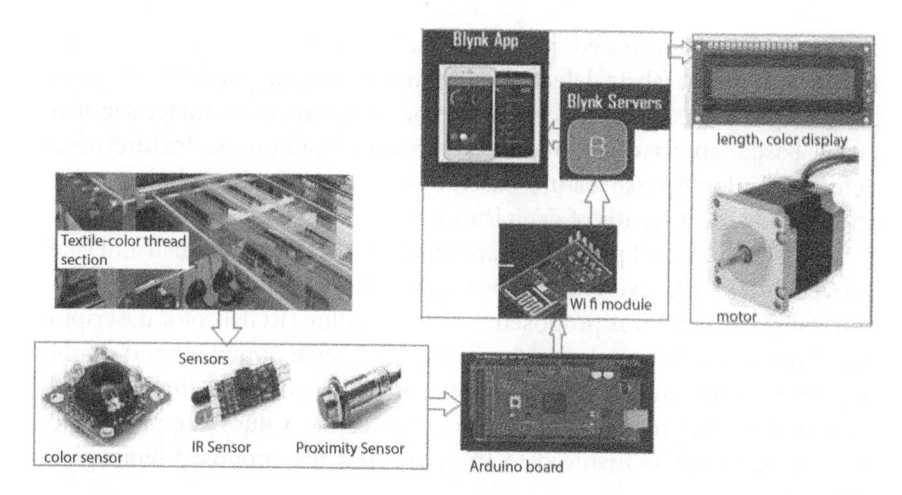

Figure 20.2 Block diagram of the proposed model.

process. After the correction, the servo motor is mechanically OFF. The accuracy of those sensors was instituted to be high and also the final response was acceptable. Controlling the entire system is handled by Arduino boards.

The following is a short description of the system's main components.

20.3.1 Sensor

A programmable color light-to-frequency converter is the TCS3200 sensor. The microcontroller enables this module to detect colors. Based on the wavelength, it can detect a wide range of colors such as Red, Blue, and Green. TCS3200 provides frequency scaling so that the readings of the sensor can be optimized. White LEDs illuminate the surface of objects whose color needs to be detected. By calculating the light intensity reflected by the object, reflection intensity can be determined. The converter produces a frequency proportional to the intensity that can be used by the microcontroller to predict the color of the object. Block diagram of TCS3200 is given in Figure 20.3. It uses configurable silicon photodiodes and an A/F converter. The array of 8x8 photodiodes is composed of 8x8 photodiodes each. Through its digital input and digital output pins, the TCS3200 can be directly interfaced with any microcontroller. A photodiode can be

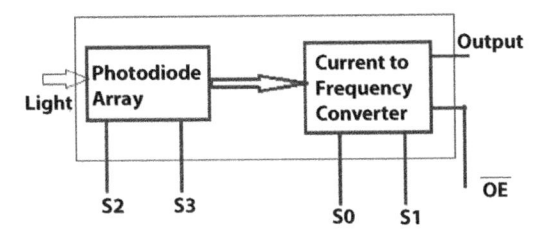

Figure 20.3 Block diagram of TCS3200.

activated by the pins S2 and S3. The frequency scaling factor can be determined by S0 and S1.

A basic TCS3200 sensor interface circuit is used for this project. It robotically detects the red colors with the help of photodiode arrangements. It displays RGB color intensity values on the serial monitor window along with names of the colors. Meanwhile, in the Arduino serial monitor window, a red LED blazes in the RGB LED. Likewise, the left over two colors are perceived and found to (green and blue) glow in the RGB LED.

20.3.2 Microcontroller

A microcontroller board originally designed to support ATmega328 (datasheet) could be called the Arduino UNO. This board features 14 digital input and output pins. By reading the sensor information, the microcontroller translates it into serial values which are then sent to the ESP8266

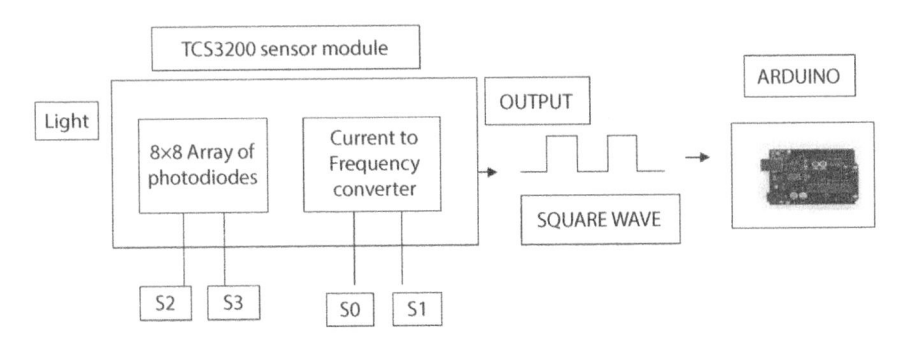

Figure 20.4 Microcontroller working with TCS3200 color sensor.

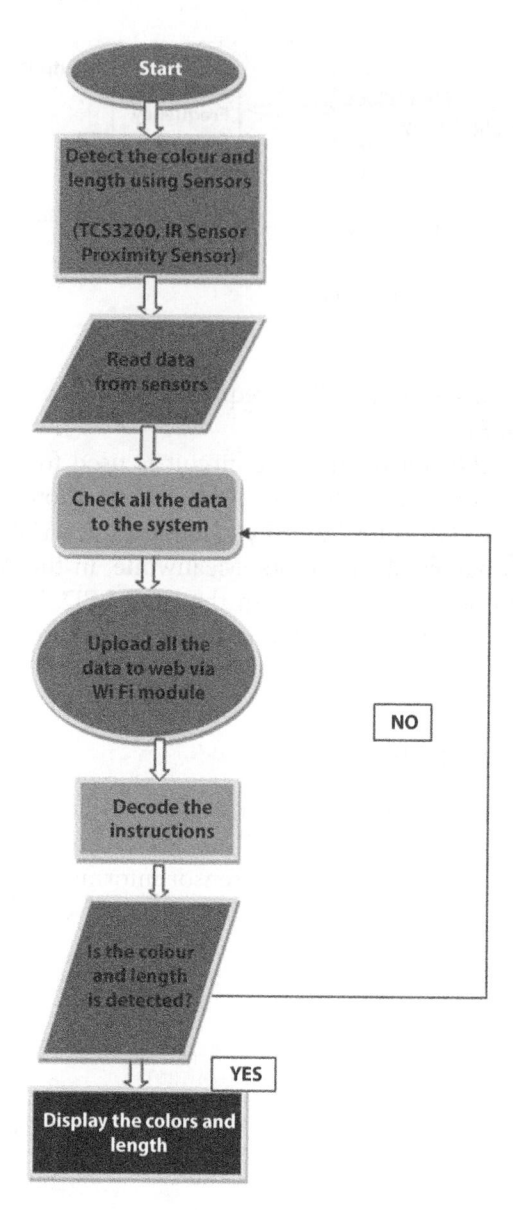

Figure 20.5 Flowchart of our proposed model.

which is a Wi-Fi module via the TXD pin. Furthermore, it obtains the serial data sent by the ESP8266 Wi-Fi component through the RXD pin. Using the ESP8266 Wi-Fi element, the microcontroller transmits information from the net server to the relevant I/O device and executes the desired task on the device which is shown in Figure 20.4. The Software for programming Arduino Uno Board 1 is available in Arduino IDE. Arduino UNO boards come with a preprogrammed boot loader that enables to upload of code in them. The Arduino IDE is set initially to use the correct COM port and connect the Arduino board to the PC. Compile the program/sketch (TCS3200.ino). This project is compiled with Arduino IDE 1.6.4 and uploaded to the internet. The workflow is discussed in the flowchart shown in Figure 20.5.

20.3.3 NodeMCU and Wi-Fi Module

NodeMCU is initially based on a Wi-Fi unit (ESP8266) that converses easily through the microcontroller by establishing serial message. Communication in Serial form is a method of contact between the (ESP8266) Wi-Fi element and the microcontroller. Associating with a web service provider, the web is accessible once connected for data access. Information received from the specified web page is transmitted to the microcontroller via this module. Figure 20.6 shows how the output from the microcontroller is uploaded to a particular web sheet.

20.3.4 Servomotor

To detect and correct color, the motor must automatically activate or deactivate. Servo motors control angular position precisely with DC motors.

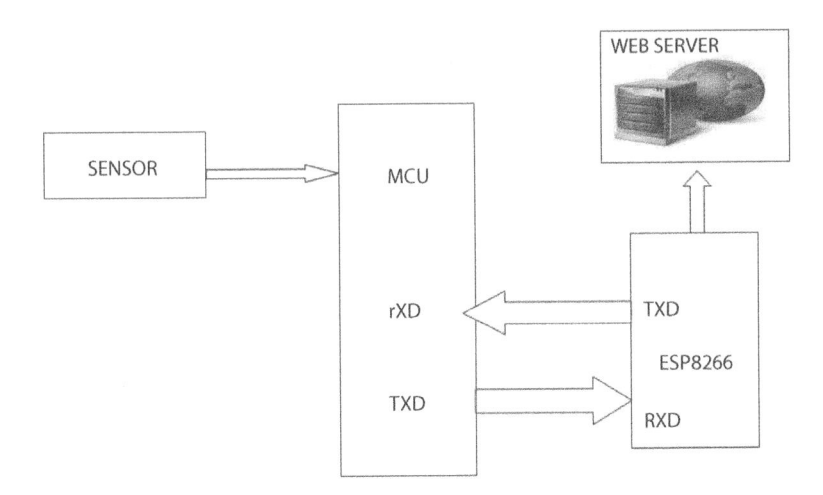

Figure 20.6 Function of the microcontroller and Wi-Fi module.

The gears gradually lower the speed of DC motors. In general, servo motors are designed to operate between 90° and 180°. The system uses an appropriate motor along with a position feedback tool. Servo motors are controlled by pulse width modulation. By applying a pulse width to the motor for a fixed duration of time, the servo motor's angle can be controlled. Once the color fault is detected, the position of the servomotor is used to stop the mechanism.

20.3.5 IoT-Based Data Monitoring

IoT reveals exquisite software in improving the progression efficiency for cloth producers. With employees appearing tasks driven by means of commands acquired through linked devices, there is extra perception into the time it takes to complete the job orders, filter out the old stocks and lock them in that instant. Given that all the facts are recorded in real time, managers can access the information quickly. IoT can assist the producers in production asset tracking and tracking place tracking in addition to monitoring of parameters inclusive of good, performance, bottlenecks, anomalies, and further. The technologies that connect with IoT may be through RFID, Wi-Fi, and Bluetooth. In this work, IoT using the Blynk model is developed for executing the online monitoring of fault and the length.

20.3.6 IR Sensor

The IR sensor emits a light that detects one or more objects in the environment. The IR sensor is capable of sensing the heat created by an article in much the same way it detects motion. Most of the articles emit some type

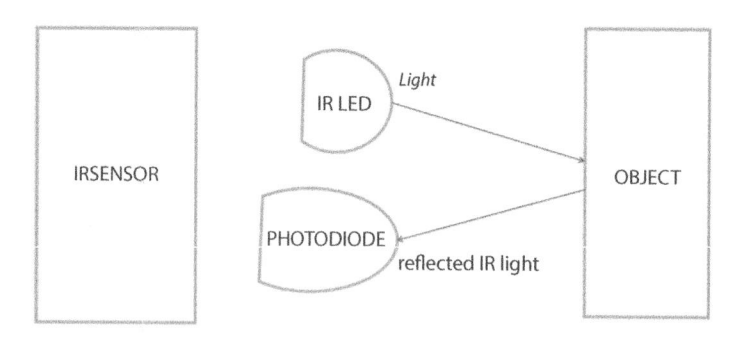

Figure 20.7 Working principle of IR sensor.

of warm radiation in the infrared range. Figure 20.7 shows the maker is IR sensor and the receiver is an IR photodiode. The IR photodiode is dedicated to the IR light transmitted by IR drivers. Figure 20.7 shows that the fundamental working guideline of the IR sensor. At the point when the IR source creates emission, it arrives at the item and a portion of the emission that reflects the IR collector. In light of the force of the gathering by the IR collector, the yield of the sensor characterizes.

20.3.7 Proximity Sensor

A capacitive proximity sensor is a gadget that can identify or detect the presence of objects nearby without requiring actual tactile contact. A proximity sensor can detect a defect in a yarn automatically, or determine where on the cloth a defect occurs.

Figure 20.8 illustrates the working of the Capacitive Proximity Sensor Capacitive proximity sensors, which consist of a high-frequency oscillator and two metal electrodes for sensing surfaces. The electrostatic field of the electrodes changes in response to an object approaching the sensing surface, changing the oscillator's capacitance. This method has been found to be an excellent choice for the industrial environment in the detection of color fault.

20.3.8 Blynk

Blynk is an app that displays output on a smartphone. This Internet of Things solution was designed for Blynk. It allows remote control of hardware, displays sensor data, stores data, visualize it, and can do many other cool things. Figure 20.9 shows how Blynk Software works in this proposed model. The first step is to launch the Blynk app. Using Blynk, create a user account for login. If the password is correct, it goes to the next step. If there

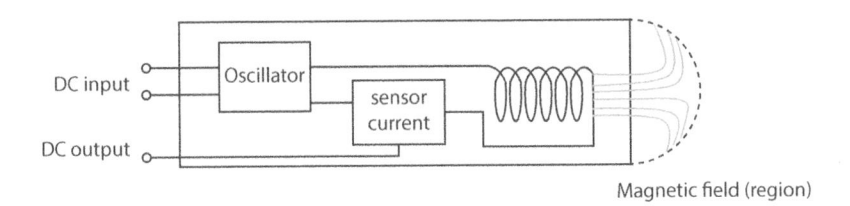

Figure 20.8 Working principle of capacitive proximity sensor.

is a problem with the password, proceed to the first step (Create an account) and then open the new project. Now, Blynk app screen is opened in the background. Smartphone notification is sent when the color is detected in the current executing project.

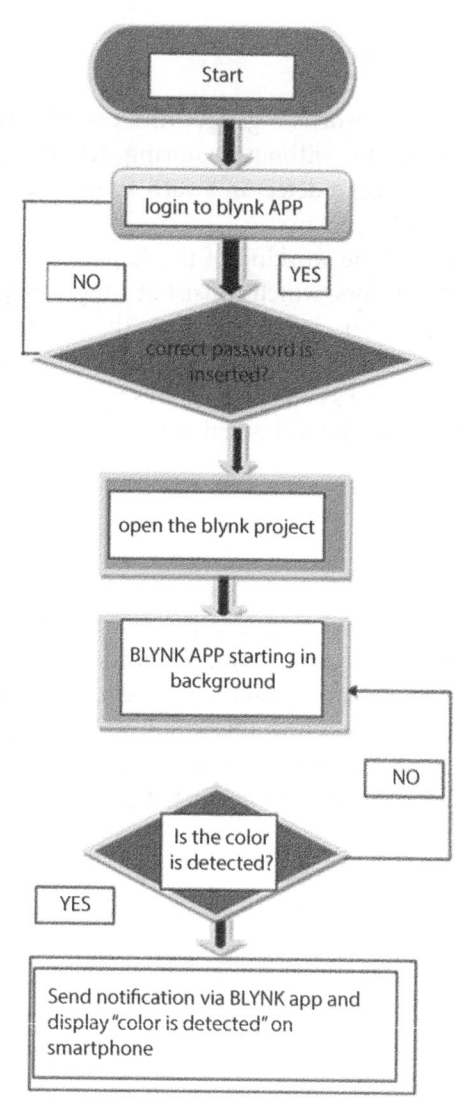

Figure 20.9 Process that describes the working of Blynk.

20.4 Experimental Setup

A color detector can sense or detect colors in the fabric or threads. A color detector utilizes an external means that emits light to detect the color of the mirrored light from an article by using lightweight photodiodes. The material's correct color can be determined from the color sensors. With the sensors discussed above, an Arduino color detector application was developed that has the capability of detecting when a color change occurs in the fabric. The TCS3200 color sensor is the main color fault detector. The capacitive proximity sensor is used to measure the length of time in which a color has changed. The Arduino UNO was used as the controller in the project.

A prototype of the IoT color detection system is shown in Figure 20.10, using a TCS3200 sensor module. IoT is used to display the results via the Blynk application on a smartphone. Color values like red, blue, and green were displayed in Blynk. The Blynk application displays the color values red, blue, and green, depending on the color detected.

The pin configuration for the prototype system is shown in Figure 20.11. The real-time data in this experiment was processed by a central microprocessor. It then checks for any knowledge limit violations after collecting the sensors' knowledge. As a result of a violation, a predefined call is formed, and corrective action is taken. It then sends all the information to the website via a Wi-Fi module. The Wi-Fi module simply passes any information that is not illegal to the knowledge limit violation function. The unit encodes the sensor's knowledge as well for proper detection of information within the Wi-Fi module. Once a message is received from the webpage,

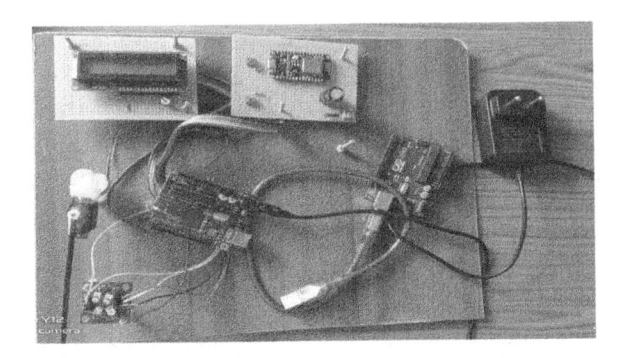

Figure 20.10 Hardware prototype.

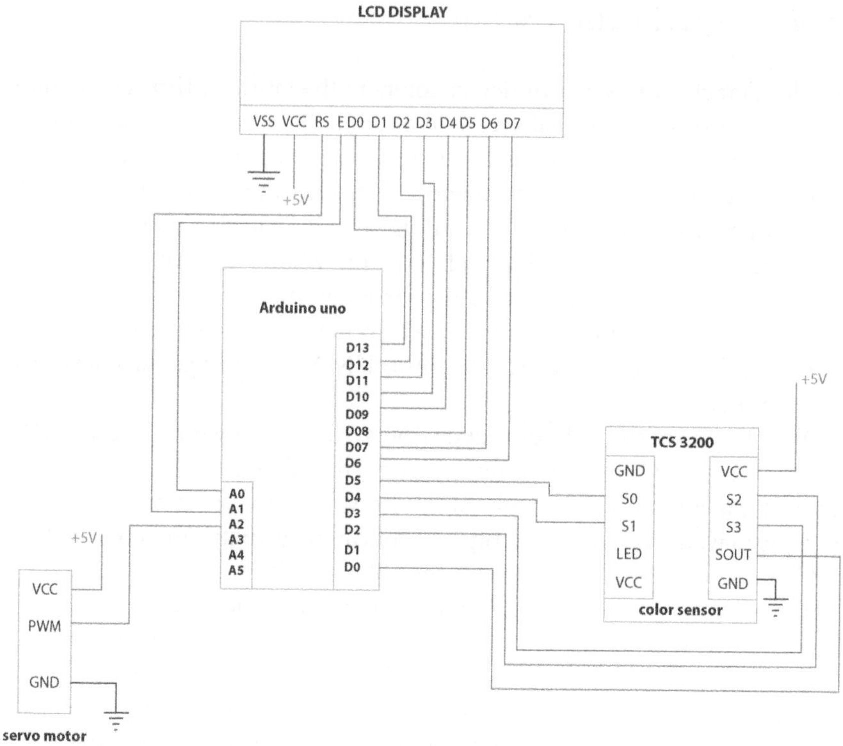

Figure 20.11 Pin configuration of the hardware.

it is transmitted to the central microcontroller via the Wi-Fi module. The message is then decoded and the call is placed.

20.5 Results and Discussion

From the hardware unit (Figure 20.12), the color sensor detects the changes in the color shade from RED, BLUE, or GREEN. The frequency of the defective color and its length are displayed in the LCD and data is continuously sent to the IoT for monitoring. IoT-based data communication is done through the Wi-Fi module using the Blynk app.

In this case, when any red, green, or blue tone is saved for the location before the shading sensor, the ideal shading system turns on and the yield in detecting the color is presented. As an initial step, we place the green shading paper on top of the shading sensor, which detects and turns on

Figure 20.12 Hardware implementation of system.

the drive, and similarly, it is measured for the other two shades, which is shown in Figure 20.13 and Figure 20.14.

Table 20.1 shows the frequency values of various forms of garments. In the Blynk app, frequency values will be displayed when a red object is placed in front of the detector. The same method can be used with blue and inexperienced objects with completely different garments. In case R is the most expensive parameter (in RGB parameters), we will know that it is a red object. At once G is that the most expensive, we all know that we have got an inexperienced object. A blue object appears once B is the most dominant shade.

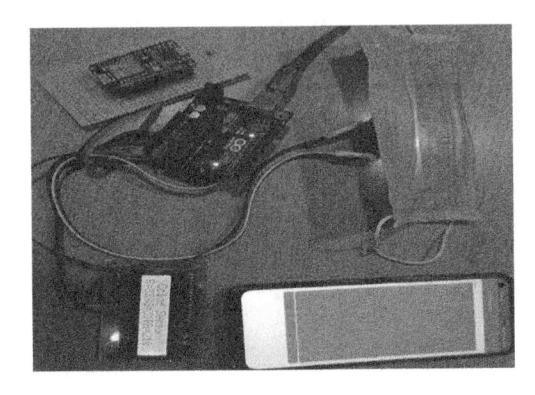

Figure 20.13 Blue color detection.

Figure 20.14 Red color detection.

Table 20.1 Frequency values in Hz of different types of clothes.

S. no.	Type of cloth	Red	Blue	Green
1.	Cotton	275	255	160
2.	Jersey Cotton	249	221	282
3.	Gauda	255	219	175
4.	Nylon	310	195	170

20.6 Conclusion

The purpose is to identify RGB color using TCS3200 sensing element connected to the Internet of Things backbone. In the present scientific scenario, this methodology could become a necessity to make the process easier for the worker in the textile unit. The integrated component, when placed in the fabric area for spinning or weaving or yarning within the textile business, can minimize production loss. The event of IoT-based color detectors employing a TCS3200 sensing element module has reduced these losses. With the proximity sensor, the length is also identified, which allows the workers' role to be much simpler.

References

1. Ch. Shravani, G. Indira, V. Appalaraju, 2019, "Arduino Based Color Sorting Machine using TCS3200 Color Sensor", *International Journal of Innovative Technology and Exploring Engineering*, Volume 8.

2. Yulkifli, P Kahar, R Ramli, 2019, "Development of color detector using colorimetry system with photodiode sensor for food dye determination application", *International Conference on Research and Learning of Physics*, Series 1185.
3. Mohammad Marufar Rahman, Md. Milon Islam, 2020, "Obstacle and Fall Detection to Guide the Visually Impaired People with Real Time Monitoring", *SN Computer Science*, Article 219 Springer.
4. Shubham Kumar, Dasari Karthik, Pradeep Khanna, 2018, Development of a color detection and analyzing system, *International Journal of Research in Engineering and Technology*.
5. Gisu Heo, Ramalingam Manivannan, Hyorim Kim, 2019, "Developing an RGB - Arduino device for the multi-color recognition, detection and determination of Fe(III), Co(II), Hg(II) and Sn(II) in aqueous media by a terpyridine moiety, *Journal of Sensors and Actuators*.
6. Han-Byeol OH, Ji-Sun Kim, and Jae-Hoon, Jun. 2018, "pH Measurement using Red, Green, and Blue Values", *Current Optics and Photonics*, Volume 19, Issue 6, p. 700-704.
7. Massimo Brambilla, Elio Romano, Marina Buccheri, 2020, "Application of a low-cost RGB sensor to detect basil (Ocimum basilicum L.) nutritional status at pilot scale level". *Precision Agriculture*, Springer.
8. M Solahudin, W Slamet and W Wahyu, 2018, "Development of Weeds Density Evaluation System Based on RGB Sensor", *IOP Conference Series Earth and Environmental Science*, Vol. 147, Issue 1.
9. Alaya, M. A., Tóth, Z., Géczy, A, 2019, "Applied Color Sensor Based Solution for Sorting in Food Industry Processing", *Periodica Polytechnica Electrical Engineering and Computer Science*, vol. 63, issue 1, pp.16-22.
10. Aqsa Rasheed, Bushra Zafar, Amina Rasheed, Nouman Ali, Muhammad Sajid, Saadat Hanif Dar, Usman Habib, Tehmina Shehryar, Muhammad Tariq Mahmood, "Fabric Defect Detection Using Computer Vision Techniques: A Comprehensive Review", *Mathematical Problems in Engineering*, vol. 2020, Article ID 8189403, 24 pages, 2020.
11. Yu, X., Hu, J. and Baciu, G. (2005), "Defect Detection of Jacquard Fabrics Using Multiple Color-channel Analysis", *Research Journal of Textile and Apparel*, Vol. 9, No. 1, pp. 21-29.
12. Khan, B., Han, F., Wang, Z. and Masood, R.J. (2016), "Bio-inspired approach to invariant recognition and classification of fabric weave patterns and yarn color", *Assembly Automation*, Vol. 36, No. 2, pp. 152-158.
13. Jakkan D.A., Sudhakar C.B., Vilas J.S., Aslam K.U.. "Color based product sorting machine using IoT".. *Journal of Embedded Systems and Processing*, 4, 3, :25-29, 2019.
14. Huang, Mingfeng, Anfeng Liu, Neal N. Xiong, Tian Wang, and Athanasios V. Vasilakos. "A low-latency communication scheme for mobile wireless sensor control systems." *IEEE Transactions on Systems, Man, and Cybernetics: Systems*, 49, 2, 317-332, 2018.

15. Anupama, P., KV Sathees Kumar, S. Rominus Valsalam, G. Harikrishnan, and V. Muralidharan. "An intelligent reflective colour sensor system for paper and textile industries." *In 2012 Sixth International Conference on Sensing Technology (ICST)*, pp. 481-485, IEEE, 2012.

21

Energy Management System for Smart Buildings

Shivangi Shukla[1], V. Jayashree Nivedhitha[1], Akshitha Shankar[1], P. Tejaswi[1] and O.V. Gnana Swathika[2]*

[1]*School of Electrical Engineering, Vellore Institute of Technology, Chennai, India*
[2]*Centre for Smart Grid Technologies, School of Electrical Engineering, Vellore Institute of Technology, Chennai, India*

Abstract

With increasing availability of wireless networks and home automation systems, the need for energy management in residential and commercial building is also increasing. The primary purpose of EMS is to reduce/manage energy usage optimally, thereby reducing electricity bills while increasing productivity. EMS aims to improve the stewardship of the environment, without compromising on standards of living. The described methodology plans on implementing a real-time monitoring system which analyses the changing graphs of energy usage in the buildings and focuses on minimizing energy wastage in order to implement an energy management system that is efficient in terms of battery energy storage. Considering the well-being of the environment, the system mainly functions on renewable power sources which are cost efficient, reliable in nature and are able to provide an uninterrupted power supply. The proposed plan shall be extended by using the Internet of Things (IoT) to monitor a device's energy consumption as well as its management by frequent interpretation of real-time datasets. An EMS is responsible for optimally scheduling end-user smart appliances, heating systems, ventilation units, and local generation devices. Smart building systems are expected to reach a compound annual growth rate (CAGR) of 30 percent by 2020. The results that we are looking for is to show that the system developed is capable of automatically reducing energy consumption and maintaining steady comfort level required by the occupants in the building.

Keywords: EMS, SEM, SSM, SEMS, PSO, SVR, GHG

Corresponding author: gnanaswathika.ov@vit.ac.in

21.1 Introduction

The ever-increasing power consumption of appliances by consumers is a matter of concern in the energy sector. This increased usage of power brings about instability in the demand and supply ratio of power. In order to cope with this growing imbalance, energy should be managed on the demand side so that the energy is used optimally and the deficiency in the supply side is decreased. This may be done by reducing the power usage during peak hours and shutting off loads which are not having high priority. Thus a smart energy management system should be designed and this system should be portable and efficient enough to take the decision of scheduling the loads efficiently. The primary purpose of EMS is to reduce/manage energy usage optimally, thereby reducing electricity bills while increasing productivity without compromising the user's comfort.

21.2 Literature Survey

Energy management is the need of the hour. With the increase in demand and consumption due to advancements in technology, efficient management of the existing amount of energy has become a necessity in the modern times. Smart buildings require different amounts of energies at different but continuous time frames from varied sources in order to sustain the functioning of their smart devices and to ensure smooth load scheduling operations.

Among all the fields that require energy, buildings and localities require a majority of the generated quantity. According to estimates, buildings are accountable for around 1/3rd of the total energy consumption in the USA. A Micro-Grid Management System is a residential building that has a 2-way communication architecture with different load and supply sources [1].

However, the generating authority must be responsible for many factors such as the price, rate of consumption, limited uptime and downtime, with varying gradient of the energy demand profiles. Additionally, there also a lot of stability and reliability issues with the smart grid. Thus, it becomes very much vital to incorporate optimization techniques, methodologies and smart technologies which are to be answerable to the demand and to sustain the generation without constraints [2].

With the increase in the availability and development of renewable energy sources, energy management systems in smart buildings are being designed and implemented with better and improved results as the days pass by with the advent of smart technologies, algorithms and reliable methodologies. The main aim of all these systems is to reduce the cost of electricity consumption and to improve the load characteristics without compromising the end-user comforts. For example, the main ideology as described by [3] is to design an energy management system that reduces the peak demand as well as the electricity price by following source and load scheduling control algorithm. The source scheduling algorithm works in a way to charge the battery bank (with hybrid car) during low-price electricity hours and acts as a source during high-price electricity hours. The load scheduling algorithm on the other hand shifts the suitable load from peak hours to regular hours thereby reducing demand and price as well as a gap in energy exchange utility band using renewable power supply like PV as primary sources, for DC microgrid loads.

Many techniques and methodologies have been used in recent times to efficiently manage energy in smart buildings. The techniques proposed in [4] provide for utilizing network nodes in a BAS such that low cost is provided. Nodes can be used to determine the location of other nodes within the same block. Information can also be shared between the other nodes during commissioning the process. In addition to this, ranging techniques can be used to track nodes.

Smart buildings not only comprise residential buildings but also other prime buildings such as hospitals, educational institutions, etc. [5] highlights the idea of an Intelligent EMS to harness the benefits of all subsystems in a hospital, and facilitating effective operation, while also ensuring that the operations of the hospital are not affected during extreme conditions such as natural disaster. There is a need to achieve the best healthcare services and at the same time, reduce their costs without affecting the quality. This enhances flexibility and increases environmental sustainability. This is a step towards zero-energy buildings, shifting the focus from diesel-based systems that are still being used in most cases.

HVAC equipment consumes a lot of energy and smart buildings account for a large share in the total energy consumption. With the wide range of energy sources, and advancements in technological and social fronts, it has become a necessity to bring in new energy management systems such as distributed energy sources or hybrid energy sources. Smart Grids are an important advancement in terms of energy management [6].

Smart buildings are a major part of the smart grid architecture. Thus, proper management of these grids are also of utmost importance. They consist of smart buildings, electric vehicles, residential sectors, localities consuming energy along with the incorporation of ICT. Many algorithms and methods have been proposed and implemented to ensure higher grid stability and also to reduce the increased consumption of fuel.

For example, in [7], the authors concentrate on the design of a two-level energy management system for a DC Microgrid that overcomes the difficulties of meeting energy requirements of a multi-source hybrid DC microgrid in the traditional distributed control method.

GHG (greenhouse gas) emissions have led to the increase in use of electric vehicles. But improper scheduling of these vehicles has led to grid failure in many cases. Thus, to effectively reduce the chances of microgrid failures to which these vehicles are connected, many methodologies such as introduction of EVs and renewable power supply sources have been discussed and implemented in [8].

Apart from the usual smart grids, there are islanded grids which are disconnected from the main grid and operate independently with power sources and load. Many EMS also concentrate on increasing its efficiency. [9] proposes a method to satisfy the need for Islanded microgrids, that are used as flexible, adaptive and sustainable smart cells of distribution power systems, operated for technical and economic purposes. Besides, a new economical demand response, dependent on frequency response, is conducted to the EMS to cope with uncertainty. This co-optimizes the microgrid energy resources.

EMS have been in existence for a long time, but there is a need to improve its design and output characteristics to improve stability and reliability.

The primary functions of smart home energy management systems (SHEMS) is to monitor, manage and improve the flow and use of energy. The SHEMS consist of many components such as measuring device, sensing device, ICT, smart devices and EMS [10–13].

Thus, it is of utmost importance to develop advanced EMS using optimization algorithms, smart technologies and other methodologies as discussed in this paper. All of these ideas and implementations aim to decrease cost and increase comfort and also have a constant check on the environment.

21.3 Modules of the Project

The project comprises hardware prototype and the software part (ML model). In order to setup the SEMS there are several stages involved as discussed in Section 21.3.

21.3.1 Data Collection for Accurate Energy Prediction

Data is acquired related to weather in the form of parameters such as temperature, pressure, wind speed, etc., from authorized sites such as Solcast and World Weather Online in order to predict solar power generated based on the collected data. This data is fed to the ML Algorithm for prediction of accurate energy. The data is pre-processed by removing the empty cells and replacing it with the mean of the columns.

Attributes of the dataset:

INPUT:

1. DOY - Day of the Year
2. YEAR
3. MONTH
4. DAY
5. FHOP - First Hour of Period
6. DSN - Distance from Solar Noon
7. TEMP - Temperature
8. SC - Sky Cover
9. VISIBILITY
10. RH - Relative Humidity
11. WS - Wind Speed

OUTPUT:

1. SP - Solar Power

The dataset consisted of 1600 values

ROWS:1600
COLUMNS: 12 (11 inputs + 1 output)

21.3.2 ML Prediction

The best ML prediction algorithm PSO-based SVR method is used to predict the power generated on an hourly/daily basis. Data pre-processing is done initially and then the data is trained and tested in order to predict accurately as explained in Figure 21.1.

The algorithm is chosen based on performance indices such as:

i) MAE - Mean Absolute Error
ii) MAPE - Mean Absolute Percentage Error
iii) RMSE - Root Mean Square Error

21.3.3 Web Server

A web server is designed wherein the user logs into the webpage by entering login credentials and specifies the future date, month and year for which he or she wants to schedule the loads. The user also assigns the priority to the loads on this webpage. Then on clicking 'predict', the predicted power is shown on the screen. The power predicted using ML Algorithm is stored in the web server and this value is accessed by ESP32 main unit, i.e., SEM unit. ESP32 is a microcontroller with integrated W-Fi and Bluetooth. Its CPU is a 32-bit LX6 microprocessor. Further, this value is used by the decisive algorithm embedded in the main unit to schedule the loads.

The below mentioned coding languages are used to design the web server:

DJANGO
NODE JS
HTML
CSS.

21.3.4 Hardware Description and Implementation

The master ESP32, i.e., SEM unit controls the slave SSM, i.e., Load 1 ESP32 and Load 2 ESP32 connected to the two loads by running the decisive load scheduling algorithm in the master ESP32. Initially a Wi-Fi communication is established between the nodes on the SEM unit and on the load side by using wireless mode of communication. **ESP MESH** topology is used to communicate between the master ESP and slave ESPs. Then, SEM Unit

ENERGY MANAGEMENT SYSTEM FOR SMART BUILDINGS 333

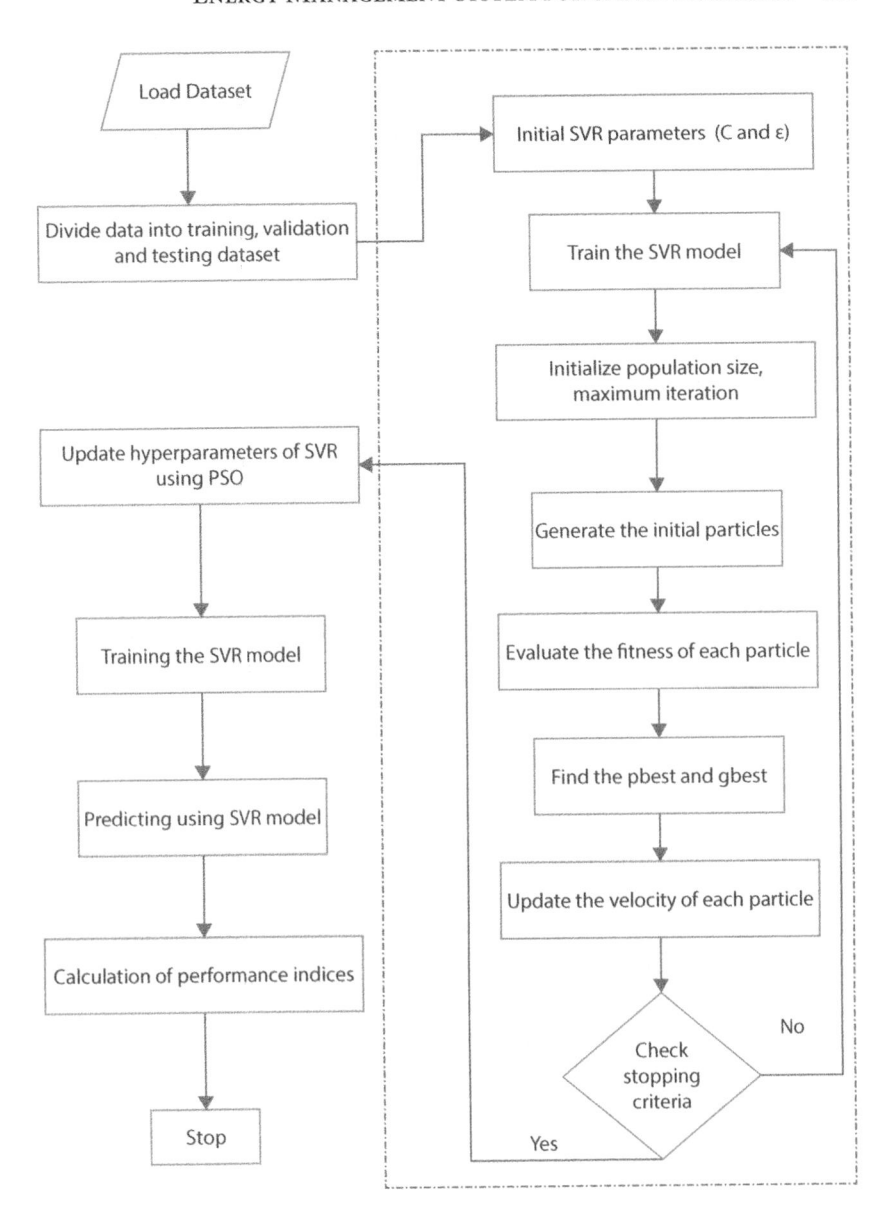

Figure 21.1 PSO-based SVR.

controls the load based on the priority given by the user and with respect to the power available, and the loads are turned ON/OFF using a relay signal received from the SEM ESP32.

1. Data collection from online websites
2. Pre-processing of the data collected
3. Writing and executing proper code for ML Algorithm
4. Creating a server
5. Connecting the server to the webpage
6. Connecting the server to the ML script
7. Storing the results in a file
8. Writing and executing Arduino IDE code
9. Writing and executing SEM (decisive algorithm)
10. Connecting the components forming a proper circuit with a relay and led
11. Uploading the code in the ESP32 modules
12. Ensuring communication between the SEM and SSM ESP32 Modules
13. Analyzing the results.

21.4 Design of Smart Energy Management System

21.4.1 Design Approach

The controlling and monitoring technique of the SEMS mainly comprises the main unit, i.e., SEM unit and the slave units, i.e., smart socket modules which act as load controllers. In order to schedule and control the loads in an optimal manner, the load scheduling algorithm is fed in the main unit. The main unit communicates with the load controllers by using inbuilt Wi-Fi of esp32 module. The main unit thus sends control signals to the load controllers by running load scheduling algorithm based on the user set priority and the power predicted by ML model. At the load side the loads are controlled by the load controller, i.e., SSM units.

21.4.1.1 ML Algorithm

The dataset collected is pre-processed, replacing empty cells with the value of the mean of the columns. The dataset is then divided into training and testing and fitted into the regression model of SVR. The hyper-parameters of the SVR are calculated with the help of the PSO Algorithm.

21.4.1.2 EMS Algorithm

The technique used by this system comprises of a load scheduling algorithm which acts as a central controller. The given system constitutes separate algorithms for the main unit and slave unit in order to use the energy optimally.

The algorithms embedded in EMS system are given as follows:

i) SEM unit(main unit)
- Load scheduling algorithm at the demand side

ii) SSM unit(slave unit)
- Algorithm for receiving control signals from the main unit

<u>i) Load scheduling algorithm at the demand side</u>

The given algorithm is the key algorithm in making decisions and the steps involved in designing the algorithm are explained as follows:

Step 1: Firstly the data of power consumed by the loads is collected in a predefined order. But since we have used DC loads, we have assumed constant power of both the loads.

Step 2: Based on the priority given by the user the load scheduling algorithm functions.

Step 3: Then the SEM unit checks for the given condition:

If Power consumed is greater than MDL, the SEM unit sends a control signal to switch ON the maximum loads with high priorities such that the MDL is not violated and sends a control signal to switch OFF the remaining loads.

The flowchart given in Figure 21.2 describes the load scheduling based on the input priority and the power consumed by the loads.

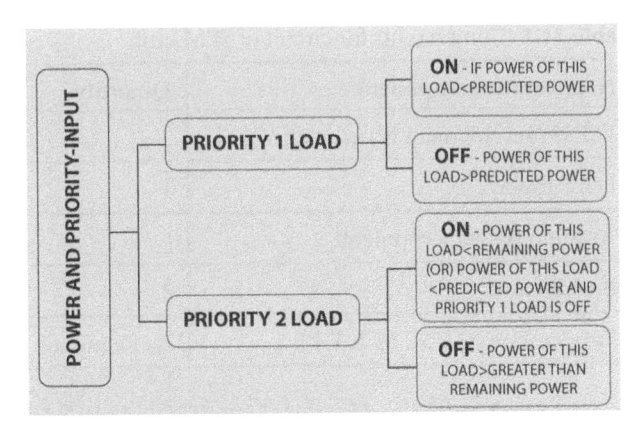

Figure 21.2 Flowchart describing the condition to schedule the loads.

Step 4: On sending the appropriate control signals to the respective loads, the main unit would wait for the next data to come for some seconds. The user may also update the priority of the load during that waiting time. Repeat steps 1-4 for continuing the process.

Before running the decisive algorithm, the user has to set the priorities of loads.

ii) <u>Algorithm for receiving control signals from the main unit</u>
Initially constant power is assumed for both the loads.

Based on the priority set by the user, the load scheduling algorithm functions .A string with 2 characters is communicated between master ESP32 and slave ESP32s.

The first character is for the first load and the second character represents the ON/OFF signal of the second load.
The loads are controlled via a relay
N-ON
F-OFF
String is the combination of N, F such as NN, NF, FN, FF

21.4.2 Design Specifications

The components used for establishing the circuit of SEM unit are shown in Table 21.1.

The components used for establishing the circuit of two SSM units (for two loads) are shown below in Table 21.2:

Table 21.1 Components for circuit of SEM unit.

Name of the component	Quantity
ESP32 Development Board with Wi-Fi	1
LCD display	1
Arduino IDE environment	–
Breadboard	2
Jumper wires (M-M, M-F, F-F)	As Required

Table 21.2 Components for circuit of SSM unit.

Name of the component	Quantity
ESP32 Development Board with Wi-Fi	2
1 Channel 5V Relay Module with Optocoupler	2
LED	2

21.5 Result & Analysis

21.5.1 Introduction

After the communication is established in the experimental setup and all the nodes of the gateway are working fine, the system is tested by first predicting the power, assigning the priority to the loads and then running the decisive algorithm for scheduling the loads. The results achieved are specified in this section.

21.5.2 ML Model Results

In order to predict the power consumed by the load, a power prediction algorithm is designed by using ML technology, namely PSO-based SVR. This process is called power forecasting. For this the data set is collected from an authorized website. Then the solar power generated is calculated using the ML algorithm based on the collected data.

Data acquired is in the form of temperature, pressure, wind speed, etc.

The dataset acquired has to be trained and tested in order to predict the power accurately. The method used in this model is PSO-based SVR. The power consumed is predicted and the value obtained is in the unit of watts.

Figures 21.3 and 21.4 display the output obtained after executing the PSO-based SVR ML algorithm.

21.5.3 Web Page Results

The tests are conducted by using a decisive algorithm which acts as the brain of the hardware prototype and by assigning priority to the loads with different configurations. The priority assigned by the user is used to deal with the customer's comfort while optimizing the energy usage.

Figure 21.3 IDLE showing predicted power value.

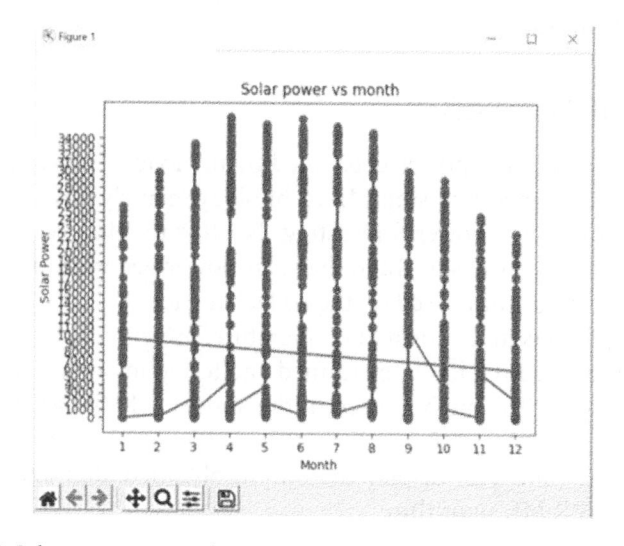

Figure 21.4 Solar power vs. month.

The webpage is developed to display power predicted by machine learning prediction model. The web portal is developed such that only an authorized person can log into the webpage using login credentials as shown in Figure 21.5. On logging in successfully, the user can enter the day, month and year for which he or she wants to predict the data as depicted in Figure 21.6. The user can enter the priority of the loads after which he or she can

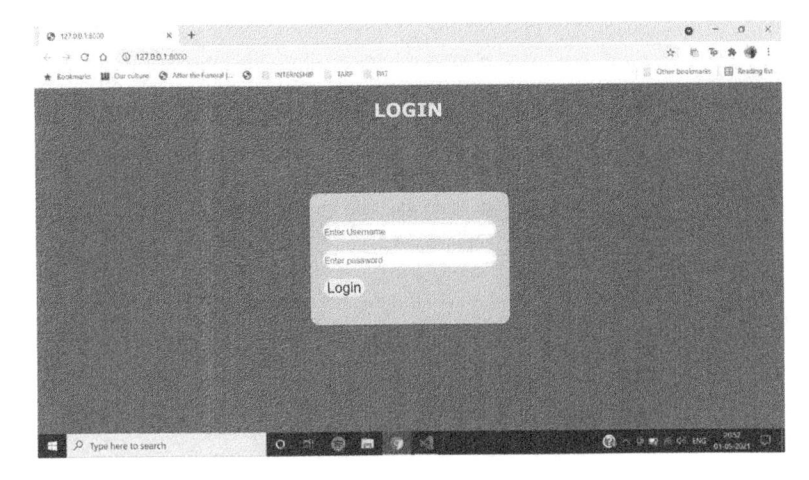

Figure 21.5 Login page of web portal.

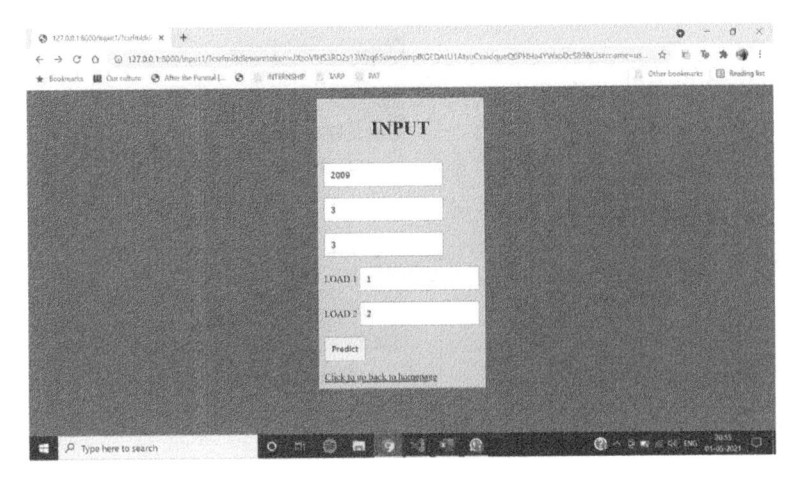

Figure 21.6 Date, month, year and load priority input webpage.

access the predicted power output by clicking 'predict' as shown in Figure 21.7. As shown in Figure 21.4, the graph of solar power vs. month will also be displayed.

21.5.4 Hardware Results

The hardware setup of Smart Energy Management System constitutes a SEM unit on the demand side and SSM units at the load side as shown in Figure 21.8.

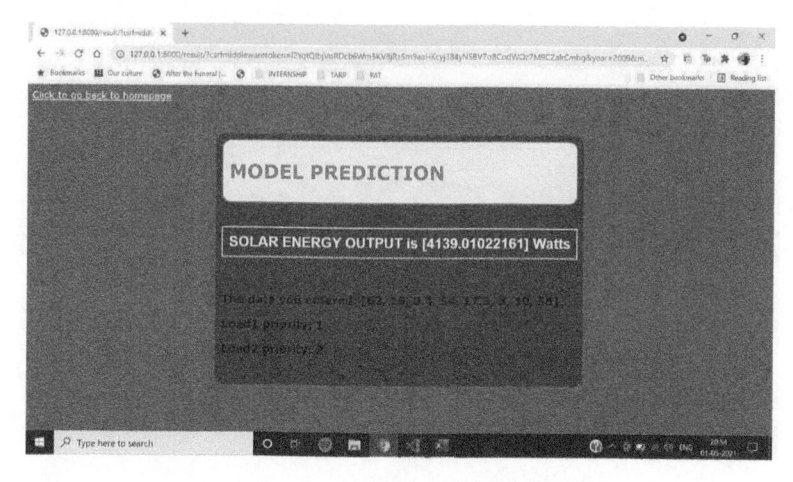

Figure 21.7 Webpage showing the predicted power value.

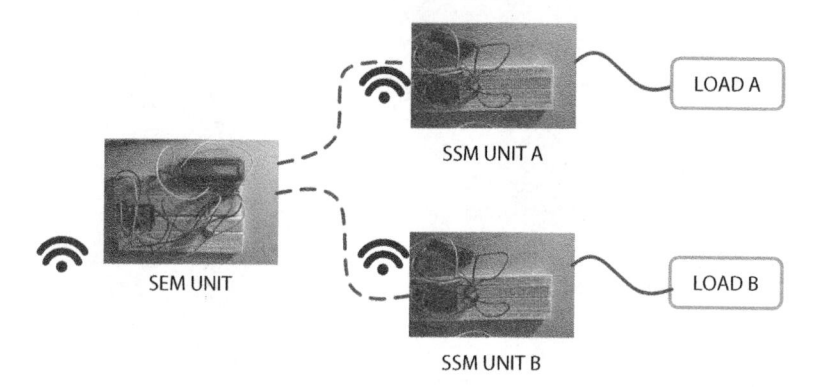

Figure 21.8 Experimental setup of SEMS.

- The topology used in SEMS is ESP Mesh Network topology.
- The loads used in the experimental work are DC Loads (LEDs).

Display Unit interfaced at the user end
Also, an LCD display unit has been used to display the values of electrical parameters like energy predicted by the PSO-based SVR algorithm. It also displays the priority of the loads set by the user. Thus the priority of each load is displayed on an LCD screen on a real-time basis as shown in Figure 21.9.

ENERGY MANAGEMENT SYSTEM FOR SMART BUILDINGS   341

Figure 21.9 Output of LCD screen when load 1 and load 2 are ON.

Communication modules of SEMS

In the given system, three identical esp32 modules with inbuilt Wi-Fi are used for establishing communication:

a) esp32 module in SEM unit (main unit)
b) esp32 module in SSM unit (slave unit)

The SEM unit sends a data request message to the load controller unit associated with the smart socket in the predefined order. This collects the data of power usage by loads connected to the smart sockets. The data received from the load controller is in the string format which is converted to its equivalent decimal form to get the actual value of electrical parameters. On the load side, the load controller unit receives the control signal from the SEM unit and accordingly switches ON/OFF relay.

The load which is turned ON/OFF along with the power is displayed in the Serial Monitor. The command signal transmitted for switching ON the relay is 'n' and for switching OFF the relay is 'f'.

For DC loads:
The different cases for each combination of the communication message shown as follows:

CASE 1: "NF"
☐ The output when load 1 is ON (led glows) and load 2 is OFF (led does not glow) is shown in Figure 21.10(a), the code of SEM unit when load 1 is ON and load 2 are OFF is shown in Figure 21.10(b) and the output when SSM unit receives the signal is shown in Figure 21.10(c).

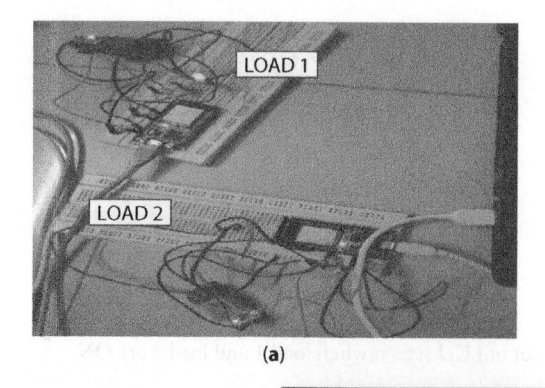

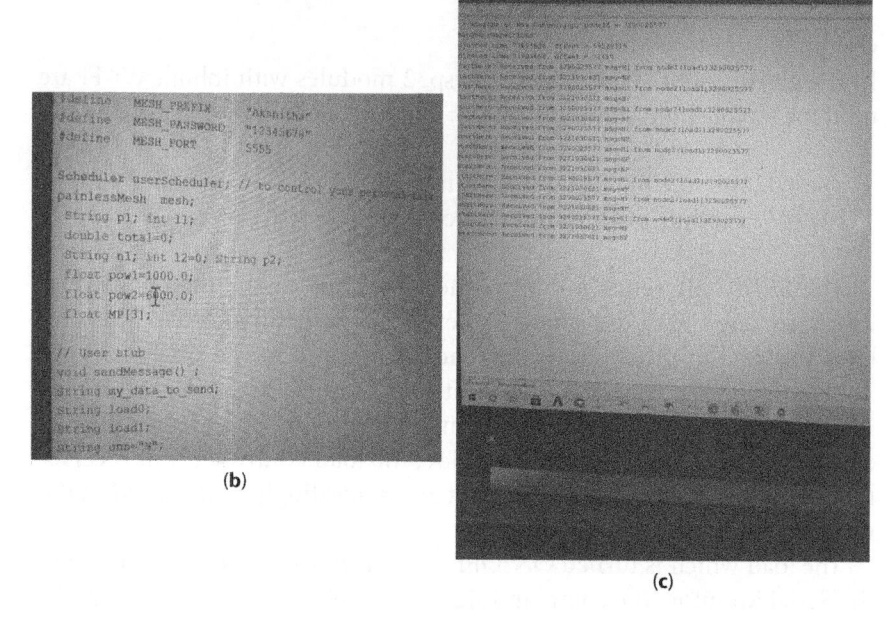

Figure 21.10 Output of (a) Hardware setup, (b) SEM code, (c) Serial monitor output.

CASE 2: "FN"

☐ The output when load 1 is OFF (led does not glow) and load 2 is ON (led glows) is shown in Figure 21.11(a), the code of SEM unit when load 1 is OFF and load 2 is ON is shown in Figure 21.11(b) and the output when SSM unit receives the signal is shown in Figure 21.11(c).

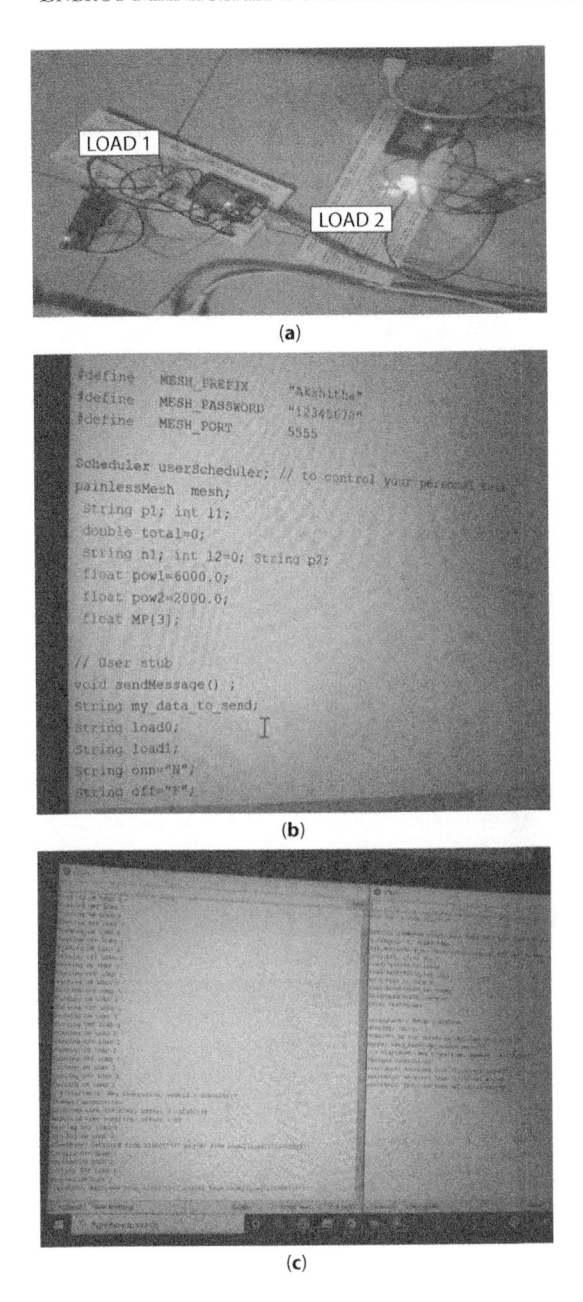

Figure 21.11 Output of (a) Hardware setup, (b) SEM code, (c) Serial monitor output.

344 Integrated Green Energy Solutions Volume 1

CASE 3: "FF"

☐The output when loads 1 and 2 are OFF (both led do not glow) is shown in Figure 21.12(a), the code of SEM unit when load 1 and load 2 is OFF is shown in Figure 21.12(b) and the output when SSM unit receives the signal is shown in Figure 21.12(c).

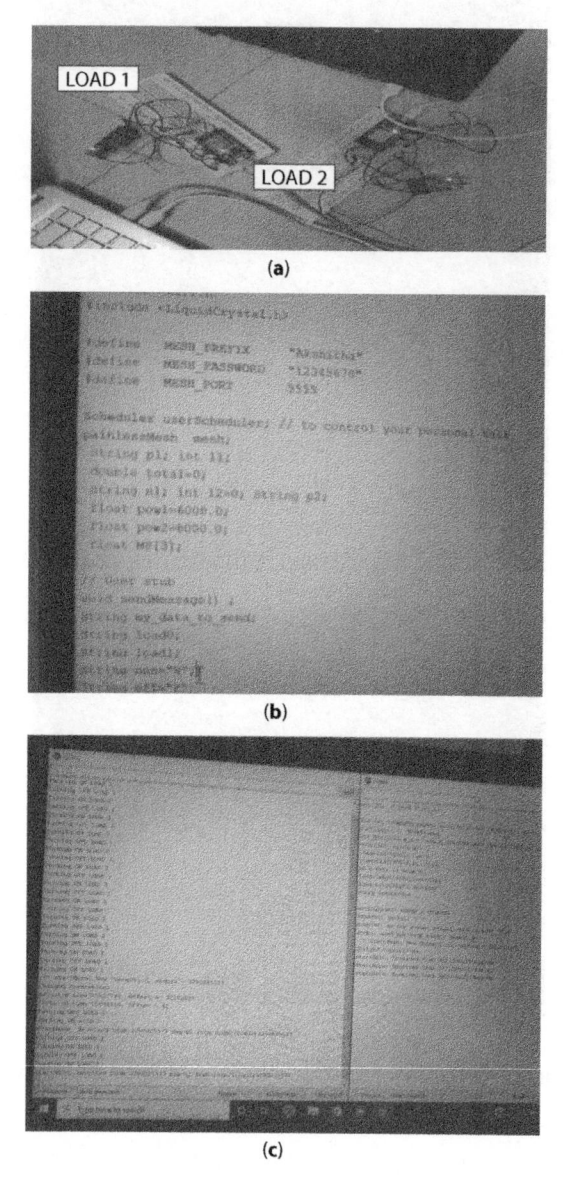

Figure 21.12 Output of (a) Hardware setup, (b) SEM code, (c) Serial monitor output.

CASE 4: "NN"

☐ The output when loads 1 and 2 are ON (both the led glows) is shown in Figure 21.13(a), the code of SEM unit when load 1 and load 2 is ON is shown in Figure 21.13(b) and the output when SSM unit receives the signal is shown in Figure 21.13(c).

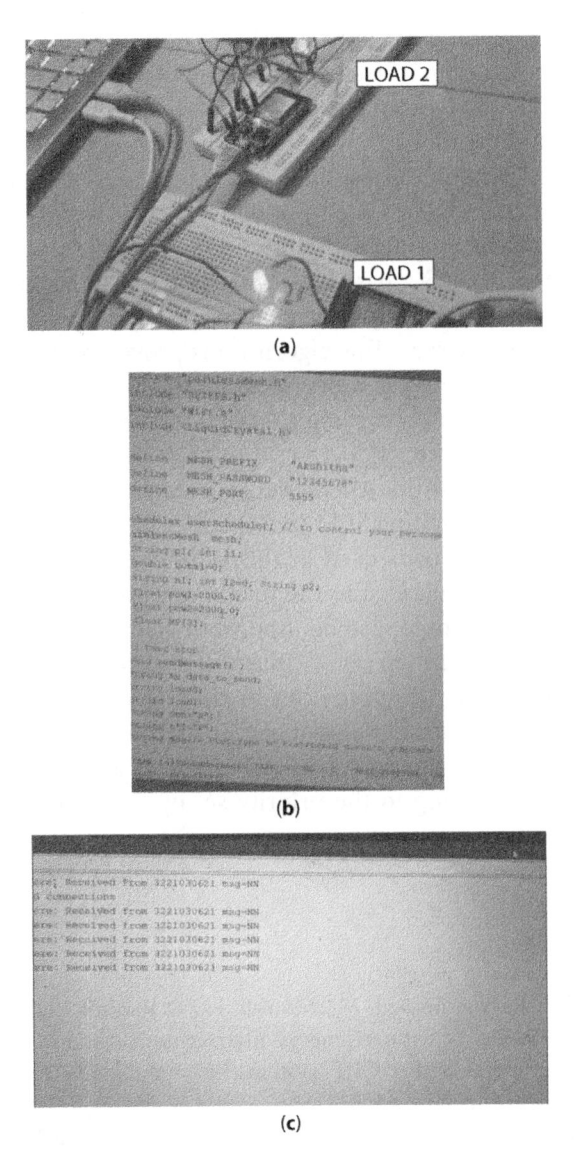

Figure 21.13 Output of (a) Hardware setup, (b) SEM code, (c) Serial monitor output.

21.6 Conclusion

A successful implementation of a SEMS is illustrated in this project. The experimental setup of SEMS is designed and developed which demonstrates the working of decisive load scheduling algorithm embedded in the coordinator unit. The wireless communication is established successfully between SEM unit and load controller (SSM) unit using esp32 inbuilt wifi. The test performed demonstrates scheduling of the two loads based on the priority set by the user and the power predicted by the ML algorithm. During the test, different cases were performed wherein the combinations of loads were switched ON and OFF. These tests were conducted on the hardware prototype which is incorporated with the decisive algorithm in the SEM unit and the outputs with both DC loads and AC loads were successfully attained. The PSO-based SVR machine learning algorithm is able to predict the power consumption value accurately and is able to transfer this value to the webserver. This algorithm is proven to be more efficient than other algorithms and is more accurate for load scheduling. Finally the loads were scheduled according to the priority set by the user and the power consumption value.

This SEM system can be either used as a stand-alone setup or can be integrated with other existing SEM technologies in order to have a greater reach. The number of loads used in this setup has been limited to two, but it can be further increased to the maximum limit set by a single ESP Mesh. An RTC clock could also be used in the circuit to calculate power consumed for a specific interval of time. Other different communication protocols such as TCP, UDP, etc., can be used to communicate between the ESPs. Also, the power consumed can be directly sent from the server to the Arduino IDE also instead of accessing it from a file. Finally the loads were scheduled according to the priority set by the user and the power consumption value.

References

1. Farmani, F., Parvizimosaed, M., Monsef, H., & Rahimi-Kian, A. (2018). A conceptual model of a smart energy management system for a residential building equipped with CCHP system. *International Journal of Electrical Power & Energy Systems*, 95, 523–536.
2. Di Piazza, M. C., La Tona, G., Luna, M., & Di Piazza, A. (2017). A two-stage Energy Management System for smart buildings reducing the impact of demand uncertainty. *Energy and Buildings*, 139, 1–9.

3. Chauhan, R. K., Rajpurohit, B. S., Wang, L., Gonzalez Longatt, F. M., & Singh, S. N. (2017). Real Time Energy Management System for Smart Buildings to Minimize the Electricity Bill. *International Journal of Emerging Electric Power Systems*, 18(3).
4. Meador, J. C., Griffiths, J.C., Sethi, G., Burns, D.W., Floyd, P.D. (2013). Commissioning system for smart buildings.
5. Kyriakarakos, G., & Dounis, A. (2020). Intelligent Management of Distributed Energy Resources for Increased Resilience and Environmental Sustainability of Hospitals. *Sustainability*, 12(18), 7379.
6. Haidar, N., Attia, M., Senouci, S.-M., Aglzim, E.-H., Kribeche, A., & Asus, Z. B. (2018). New consumer-dependent energy management system to reduce cost and carbon impact in smart buildings. *Sustainable Cities and Society*, 39, 740–750.
7. Han, Y., Chen, W., Li, Q., Yang, H., Zare, F., & Zheng, Y. (2018). Two-level energy management strategy for PV-fuel cell-battery-based DC microgrid. *International Journal of Hydrogen Energy*, 44.
8. Ahmad, F., Alam, M. S., & Asaad, M. (2017). Developments in xEVs charging infrastructure and energy management system for smart microgrids including xEVs. *Sustainable Cities and Society*, 35, 552–564.
9. Rezaei, N., Mazidi, M., Gholami, M., & Mohiti, M. (2020). A new stochastic gain adaptive energy management system for smart microgrids considering frequency responsive loads. *Energy Reports*, 6, 914–932.
10. Liu, Y., Qiu, B., Fan, X., Zhu, H., & Han, B. (2016). Review of Smart Home Energy Management Systems. *Energy Procedia*, 104, 504–508.
11. Pawar, P., TarunKumar, M., & Vittal K, P. (2019). An IoT based Intelligent Smart Energy Management System with Accurate Forecasting and Load Strategy for Renewable Generation. *Measurement*, 107187.
12. Pawar, P., & Vittal K. P. (2019). Design and development of advanced smart energy management system integrated with IoT framework in smart grid environment. *Journal of Energy Storage*, 25, 100846.
13. Pawar, P., & Vittal, K. P. (2017). Design of smart socket for power optimization in home energy management system. *2017 2nd IEEE International Conference on Recent Trends in Electronics, Information & Communication Technology (RTEICT)*.

Mobile EV Charging Stations for Scalability of EV in the Indian Automobile Sector

Mohit Sharan, Ameesh K. Singh, Harsh Gupta, Apurv Malhotra, Muskan Karira, O.V. Gnana Swathika* and Anantha Krishnan V.

School of Electrical Engineering, Vellore Institute of Technology, Chennai, India

Abstract

The automobile industry is an ever-evolving industry. For over a century we have seen a paradigm shift in the technology used in building as well as running cars. It is a well-known fact that IC (Internal Combustion) engine cars are highly efficient and have dominated the automobile industry for many decades now. In today's world, every individual either owns an automobile or is dependent on it for their daily commute. The majority of cars being run today are IC engine cars and the main source of energy for them are petrol and diesel. Now, these being fossil fuels are depleting at a much higher rate and are expected to run out soon or even sooner than predicted as the demand or consumption of them is increasing exponentially every single day while petroleum resources are depleting at a much higher rate. Over the years, EVs (Electric Vehicles) have emerged as an alternative to the IC engine vehicles as they curb two major problems—depletion of fossil fuel and heavy contribution to environmental pollution. But the problem with the current EVs is their range. The range of EVs is comparatively much less than that of an IC engine vehicle. Adding to this, there exists no major infrastructure to charge the EVs at any place people want, hence energy management of EVs in the long run becomes a hassle. The current infrastructure of grid-connected charging stations, etc., is not feasible to provide a charging station at every possible location. Hence, to support the growing EV market and scale up the number of EVs on the road a mobile or on-the-go EV charging unit can be handy. This would encourage people to use EVs instead of the traditional IC engine automobile, thereby having a positive effect on the longevity of fossil fuel and on the environment.

Keywords: Electrical vehicle (EV), GPS, OwnTracks, webhook relay, node red, cellular device, IoT

---

*Corresponding author: gnanaswathika.ov@vit.ac.in

Milind Shrinivas Dangate, W.S. Sampath, O.V. Gnana Swathika and P. Sanjeevikumar (eds.) Integrated Green Energy Solutions Volume 1, (349–360) © 2023 Scrivener Publishing LLC

22.1 Introduction

The transport sector is the main cause of air pollution all across the globe as it produces almost a quarter of the total greenhouse gases (GHGs) and especially the road transports which contributes to more than 70 percent of it. With a drastic decline in fossil fuels, the need of the hour is to shift to EVs from traditional IC Engine automobiles. As we are still somewhere in between the shift from traditional IC Engine automobiles to EVs there are no proper facilities for charging the EV everywhere. The sole reason why the electrical vehicle industry is not flourishing is because of consumers' perception of EVs. They do not have the confidence in EVs being a reliable mode of transport, as currently, we do not have omnipresent EV charging stations just like the fuel stations we see on a regular basis.

Electric vehicles play a major role in not only reducing air pollution but also in promoting the use of renewable energy resources. In the coming years, the scalability of the electrically charged vehicle has to be done in order to ensure the availability of resources for the upcoming generation. The scalability of electrical vehicles cannot be done if the infrastructure is not up to the mark. This includes the need for more and more charging stations.

Another major problem in setting up the charging stations which is mostly neglected is the lack of availability of land. Meeting this demand, i.e., building more stations, will lead to more and more deforestation, hence leading to the same problem of climate change.

Hence, one feasible solution to all the problems is the setting up of mobile EV charging stations which will not only solve the problem of air pollution and GHG emission but will also be the most suited solution to the problem discussed above.

22.2 Methodology

Our main goal is to establish a system that could establish contact between the EV owner and the mobile EV charging unit. In order to achieve this the following actions are needed. Mobile chargers will be installed with a cellular tracking system at different locations. It typically uses IoT devices that report the position of charging units based on GPS coordinates and provides either intermittent or real-time location updates. A GPS module incorporated with a cellular device has location-reporting software that queries the GPS devices and shows their location on a map.

Based on the locations as shown in the application the user can call for the nearest charging unit that meets his or her requirements.

For getting this done we will stick to the means of visual programming on node-red.

The following steps are to be performed for the same:

1) Getting the encrypted location data.
2) Pushing the data to the webhook relay.
3) Decrypt Own Tracks location data.
4) Sending location data to the map and visualizing with the help of the world map API.

22.2.1 Design Specifications

1. OwnTracks

OwnTracks is a free and open-source application for iOS and Android (Figure 22.1) that allows you to track your location and send it directly to the required application, which is the webhook here. The mobile screenshot of the OwnTracks mobile application is given below.

The OwnTracks app runs on any smartphone irrespective of the OS, thus improving device interoperability, which is an essential part of any IoT-based solution, and enabling a wide spectrum of customer base. The OwnTracks app runs in the background of an Android or iOS device. It waits for the smartphone to revert and tell it that the device has moved, whereupon OwnTracks sends out a message with its current coordinates.

The HTTP network protocol is preferred over MQTT as we are transmitting the data without using a public IP, henceforth setting up a secure communication line.

The data is backed up on the own tracks logs and can be used for clearing out discrepancies if found. The snapshot of the data is also shared below (Figure 22.2).

The following snapshot is of the data being transported from the OwnTracks mobile application to the bucket end of the webhook.

2. Webhook Relay

Webhooks are used to connect two different applications. When an event happens on the trigger application, it serializes data about that event and sends it to a webhook URL from the active application, which is the Node-Red application here.

Webhook Tokens issued are unique authentication keys. These keys are unique to a user. Hence, unique identification keys are generated for the

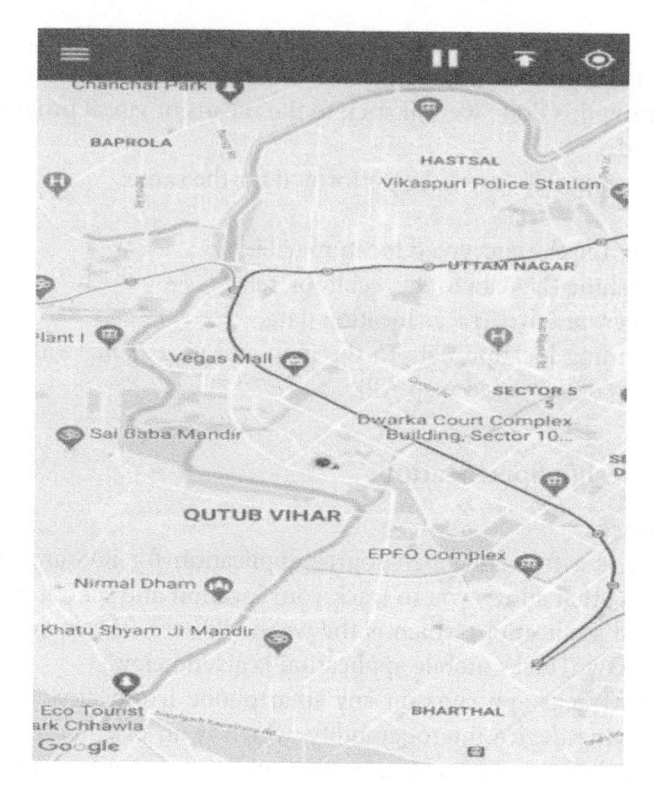

Figure 22.1 Screenshot of the OwnTracks mobile application.

```
04-16 00:28:15.895  8282  8282 I chatty  : uid=10230(org.owntracks.android) identical 49 lines
04-16 00:28:15.978  8282  8282 I chatty  : uid=10230(org.owntracks.android) identical 12 lines
04-16 00:28:15.984  8282  8282 I chatty  : uid=10230(org.owntracks.android) identical 6 lines
04-16 00:28:15.995  8282  8282 I chatty  : uid=10230(org.owntracks.android) identical 8 lines
04-16 00:28:16.044  8282  8282 I chatty  : uid=10230(org.owntracks.android) identical 37 lines
04-16 00:28:16.110  8282  8282 I chatty  : uid=10230(org.owntracks.android) identical 37 lines
04-16 00:28:16.150  8282  8282 I chatty  : uid=10230(org.owntracks.android) identical 22 lines
04-16 00:28:16.160  8282  8282 I chatty  : uid=10230(org.owntracks.android) identical 13 lines
04-16 00:28:16.188  8282  8282 I chatty  : uid=10230(org.owntracks.android) identical 4 lines
04-16 00:28:16.210  8282  8282 I chatty  : uid=10230(org.owntracks.android) identical 31 lines
04-16 00:28:16.241  8282  8282 I chatty  : uid=10230(org.owntracks.android) identical 8 lines
04-16 00:28:16.286  8282  8282 I chatty  : uid=10230(org.owntracks.android) identical 29 lines
04-16 00:28:16.306  8282  8282 I chatty  : uid=10230(org.owntracks.android) identical 24 lines
04-16 00:28:16.362  8282  8282 I chatty  : uid=10230(org.owntracks.android) identical 47 lines
04-16 00:28:16.428  8282  8282 I chatty  : uid=10230(org.owntracks.android) identical 35 lines
04-16 00:28:16.495  8282  8282 I chatty  : uid=10230(org.owntracks.android) identical 37 lines
04-16 00:28:16.561  8282  8282 I chatty  : uid=10230(org.owntracks.android) identical 31 lines
04-16 00:28:16.677  8282  8282 I chatty  : uid=10230(org.owntracks.android) identical 37 lines
04-16 00:28:16.744  8282  8282 I chatty  : uid=10230(org.owntracks.android) identical 37 lines
04-16 00:28:16.795  8282  8282 I chatty  : uid=10230(org.owntracks.android) identical 37 lines
04-16 00:28:16.904  8282  8282 I chatty  : uid=10230(org.owntracks.android) identical 28 lines
04-16 00:28:16.961  8282  8282 I chatty  : uid=10230(org.owntracks.android) identical 41 lines
```

Figure 22.2 Snapshot of the transported data.

different transportation vehicles. So this way each vehicle containing the vaccines has got a different identification key.

These webhook logs can even be checked by the person managing the database, who can simply search for the identification key of the vehicle and thus extract the logs of the vehicle with that unique identification key. This way, coordinates can be known of the entire journey.

Now, these details are relayed (Figure 22.3) at the client end and thus to the Node-Red where they can be viewed in a UI.

In webhook relays, operations are divided into three parts:

1) Token end
2) Bucket end
3) Log end

All three of them function independently to relay the information over the web under HTTP protocol as aligned by the OwnTracks application. The token end generates the access key (Figure 22.4) for the client to a particular logbook.

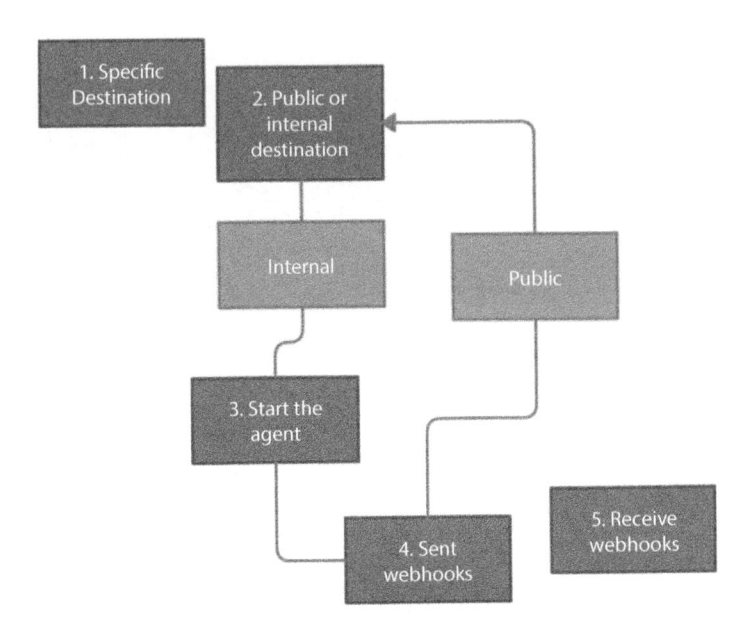

Figure 22.3 Webhook relay functionality.

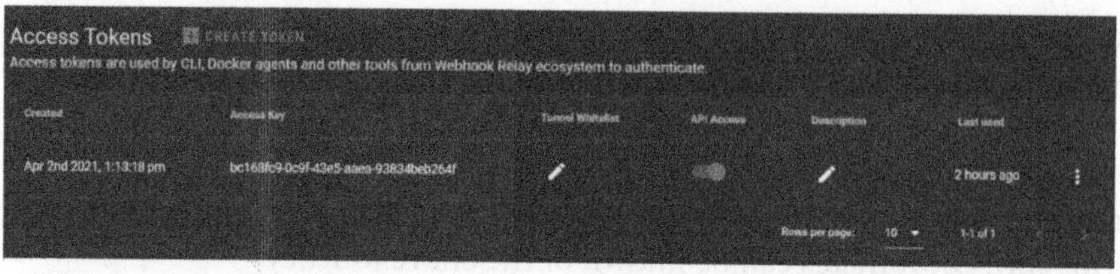

Figure 22.4 Access token.

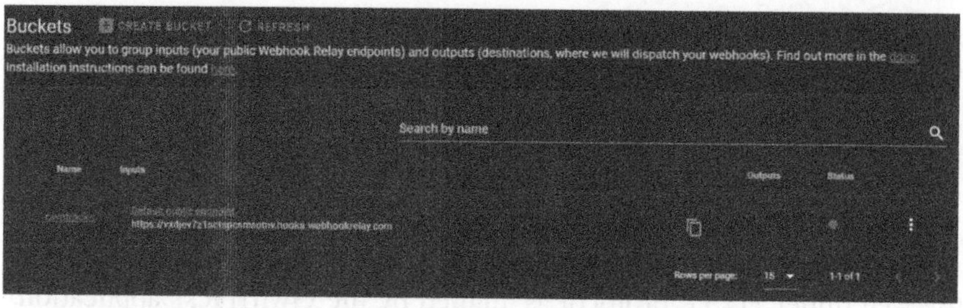

Figure 22.5 Data is received from the OwnTracks application.

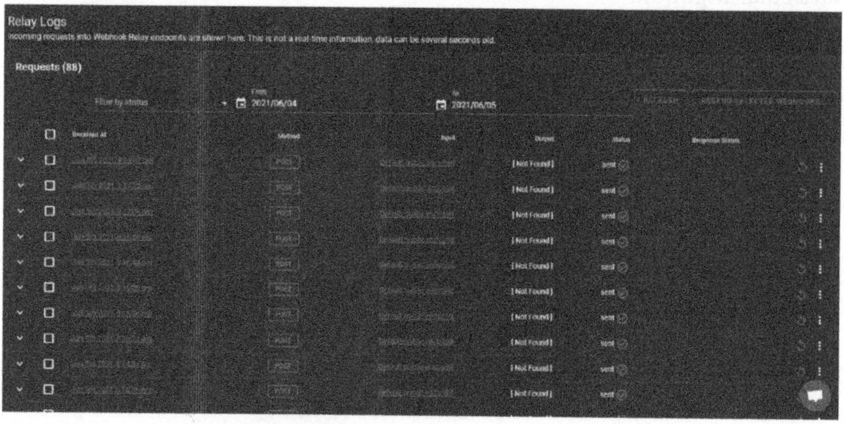

Figure 22.6 Logbook of Webhooks being relayed to the client end.

The bucket end contains the received information (Figures 22.5 and 22.6) in the form of logs and works in cohesion with the log end to relay the information to the client UI from the docking server.

3. Node-Red

We are using node-red for client and server integration by the means of visual programming (Figure 22.7) and IoT communication protocol.

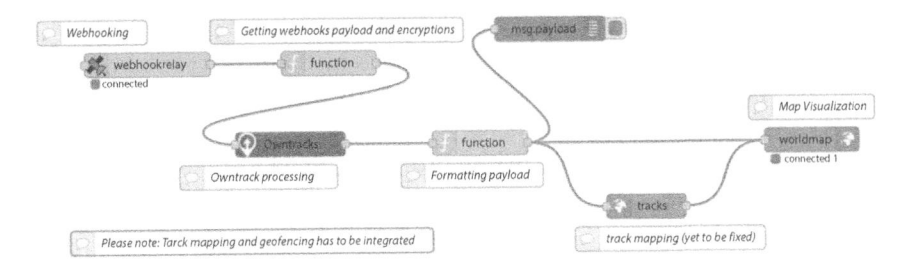

Figure 22.7 Node-red flow.

4. World Map

This is a node in node-red which is used to tap locations of different fleets carrying the vaccine using the world-map API (Figure 22.8).

5. Cellular Device

Device connected to an internet with OwnTracks application installed in it, in order to send the location coordinates.

All the five components in combination work to come up with the solution to establish communication between the charging van and the EV owner.

The order for the same is:

- The cellular device (used for tracking) installed on the mobile EV charging units triggers their location and relays it to the webhooks from the OwnTracks end to the server containing their current location and other relevant information like the charge they can deliver, if they are operational or not, etc.

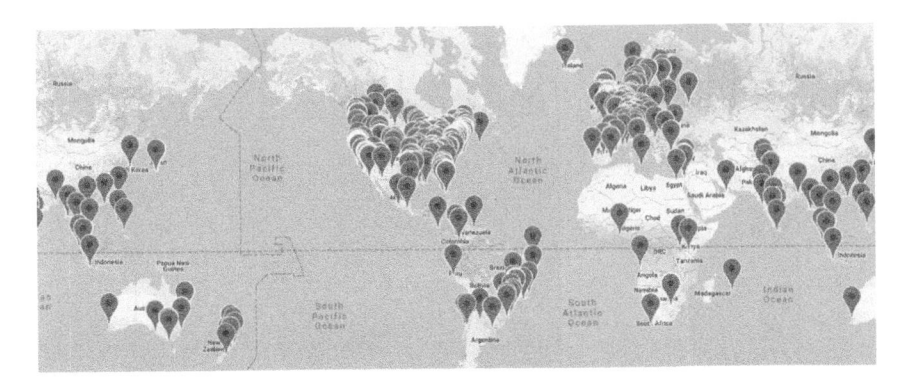

Figure 22.8 Sample of locations tapped by world map API.

- Once the data is stored in the pseudo server it is triggered to the client end upon the call generated by them.
- Then the particular location can be mapped down on the monitoring websites thereby giving us the exact location of the particular EV charging units along with the relevant parameters.

22.2.2 Block Diagrams

1) Proposed Technology (Figure 22.9)
2) Working of the application is shown below (Figure 22.10)

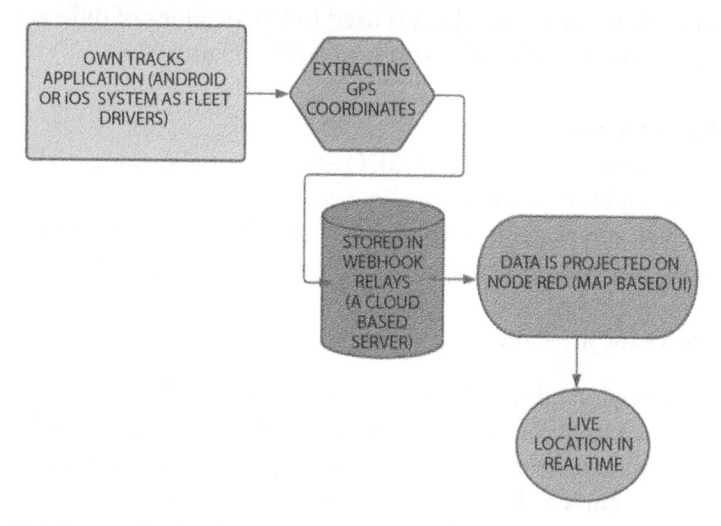

Figure 22.9 Proposed technology.

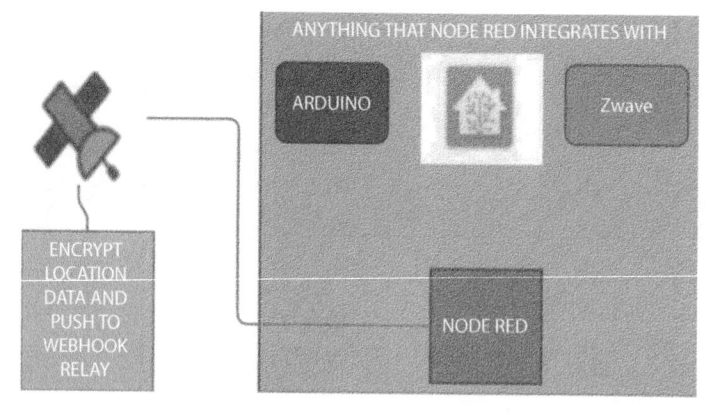

Figure 22.10 The flow of the project.

22.3 Result

Locations tapped: These are the locations tapped (Figure 22.11) and the path tracked (Figure 22.12). These are our own locations which we have obtained for testing purposes.

Figure 22.11 Locations tapped.

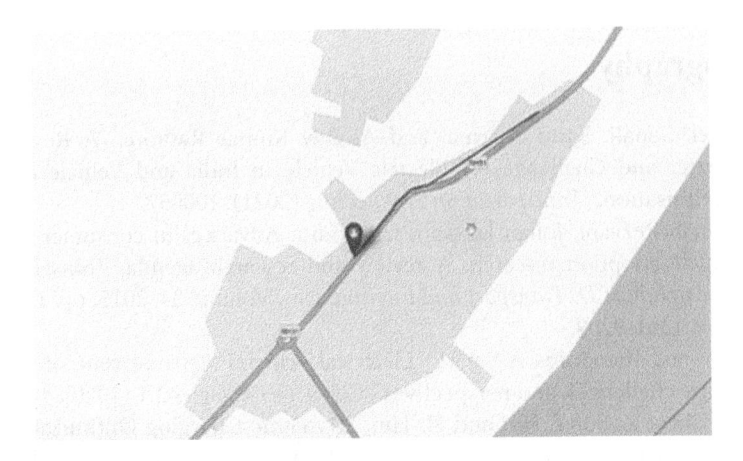

Figure 22.12 Path traced for charging stations.

22.4 Conclusions

The rate of consumption of non-renewable energy resources has resulted in the degradation of the ecosystem. The energy consumption rate is further increasing with the increasing demand and will grow exponentially in the coming times with upcoming technologies. As per the researchers, with this rate of usage of Natural Oil and Coal, the non-renewable resources will not last long enough. So, it is the need of the hour to shift to renewable energy resources and save what we have with us. With time, the demand for vehicles is also increasing, and with no doubt, it can be predicted that this will further increase in the coming times. More vehicles mean more consumption of energy, this is maybe renewable or non-renewable. Electric vehicles are a major trend shifter in this. Renewable energy-driven EVs provide a major cut through the overconsumption of non-renewable energy resources. But always there is an issue of locating charging stations for them, thus slowing down the adaption rate towards a better shift. With this project work, we have brought an IoT-based solution to locate the locations of the Charging Stations in a hassle-free manner. We have used the geo-locations of the world map to implement this. Using this, the user can know about the number of charging stations available in that particular route and can plan his or her journey before even starting it. Putting access to this technology in the hands of the user will promote more Electric Vehicles, as the difficulty of locating charging stations will have been removed. This will ensure the removal of the stumbling block to better development of mankind with a step towards Mother Earth.

Bibliography

1. Goel, Sonali, Renu Sharma, and Akshay Kumar Rathore. "A Review on Barrier and Challenges of Electric Vehicle in India and Vehicle to Grid Optimisation." *Transportation Engineering* (2021): 100057.
2. Zeinab Rezvani, Johan Jansson, Jan Bodin, Advances in consumer electric vehicle adoption research: A review and research agenda, *Transportation Research Part D: Transport and Environment*, Volume 34, 2015, pp. 122-136, ISSN 1361-9209,
3. Skouras, Theodoros A., *et al.* "Electrical vehicles: Current state of the art, future challenges, and perspectives." *Clean Technologies* 2.1 (2020): 1-16.
4. H. Chen, Z. Su, Y. Hui and H. Hui, "Dynamic Charging Optimization for Mobile Charging Stations in Internet of Things," in *IEEE Access*, vol. 6, pp. 53509-53520, 2018, doi: 10.1109/ACCESS.2018.2868937.

5. Bunce, Louise, Margaret Harris, and Mark Burgess. "Charge up then charge out? Drivers' perceptions and experiences of electric vehicles in the UK." *Transportation Research Part A: Policy and Practice* 59 (2014): 278-287.
6. R. Leou, "Optimal Charging/Discharging Control for Electric Vehicles Considering Power System Constraints and Operation Costs," in *IEEE Transactions on Power Systems*, vol. 31, no. 3, pp. 1854-1860, May 2016, doi: 10.1109/TPWRS.2015.2448722.
7. Arunkumar, P., and K. Vijith. "IOT Enabled smart charging stations for Electric Vehicle." *International Journal of Pure and Applied Mathematics* 119.7 (2018): 247-252.
8. A. Kampker, P. Burggräf and C. Nee, "Costs, quality and scalability: Impact on the value chain of electric engine production," *2012 2nd International Electric Drives Production Conference (EDPC)*, 2012, pp. 1-6, doi: 10.1109/EDPC.2012.6425089.
9. A. Y. S. Lam, Y.-W. Leung and X. Chu, "Electric Vehicle Charging Station Placement: Formulation, Complexity, and Solutions," in *IEEE Transactions on Smart Grid*, vol. 5, no. 6, pp. 2846-2856, Nov. 2014, doi: 10.1109/TSG.2014.2344684.
10. Huang, Yantao, and Kara M. Kockelman. "Electric vehicle charging station locations: Elastic demand, station congestion, and network equilibrium." *Transportation Research Part D: Transport and Environment* 78 (2020): 102179.
11. Anderson, John E., Marius Lehne, and Michael Hardinghaus. "What electric vehicle users want: Real-world preferences for public charging infrastructure." *International Journal of Sustainable Transportation* 12.5 (2018): 341-352.
12. X. Dong, Y. Mu, H. Jia, J. Wu and X. Yu, "Planning of Fast EV Charging Stations on a Round Freeway," in *IEEE Transactions on Sustainable Energy*, vol. 7, no. 4, pp. 1452-1461, Oct. 2016, doi: 10.1109/TSTE.2016.2547891.

About the Editors

Milind Shrinivas Dangate, PhD, is an assistant professor in the Department of Chemistry at Vellore University of Technology, Channai, India. He has authored several publications and has a grant and a fellowship to his credit, in addition to several postdoctoral appointments.

W. S. Sampath, PhD, is a professor in the Department of Mechanical Engineering, Colorado State University, Director for Next Generation Photovoltaics (NGPV) Laboratory at Colorado State University, and Site Director at NSF I/UCRC for Next Generation Photovoltaics. With over 30 years of industry experience, he has contributed significantly to the science of renewable energy.

O. V. Gnana Swathika, PhD, is an associate professor in the School of Electrical Engineering at VIT Chennai, India. She earned her PhD in electrical engineering at VIT University and completed her postdoc at the University of Moratuwa, Sri Lanka.

Sanjeevikumar Padmanaban, PhD, is a faculty member with the Department of Electrical Engineering, IT and Cybernetics, University of South-Eastern Norway, Porsgrunn, Norway. He received his PhD in electrical engineering from the University of Bologna, Italy. He has almost ten years of teaching, research and industrial experience and is an associate editor on a number of international scientific refereed journals. He has published more than 300 research papers and has won numerous awards for his research and teaching. He is currently involved in publishing multiple books with Wiley-Scrivener.

Index

Absorber, 115, 129, 133, 134, 138, 139
Agriculture, 130, 159, 160, 163, 180, 181, 186, 187, 188, 190, 263, 264, 267, 271, 288, 325
Alternative energy, 45, 54
Arduino, 13, 17, 20, 63, 68, 70, 71, 72, 73, 74, 75, 77, 78, 79, 82, 83, 84, 88, 97, 98, 99, 100, 141, 143, 144, 154, 244, 246, 247, 248, 253, 255, 256, 263, 270, 272, 299, 309, 311, 314, 315, 316, 321, 322, 324, 325, 334, 336, 346, 356
Automation, 11, 12, 13, 15, 17, 19, 21, 23, 24, 25, 26, 57, 61, 69, 70, 89, 90, 91, 92, 93, 94, 95, 96, 97, 99, 101, 102, 103, 113, 141, 142, 261, 287, 291, 292, 325, 327
Autonomous, 14, 61

Baby monitoring, 241, 259, 260, 261
Blynk, 12, 19, 20, 21, 23, 100, 243, 244, 314, 318, 320, 321, 322, 323

Capillary number, 119, 123
Cellular device, 14, 350, 355
Child security, 241
Circuits, 75, 79, 129, 131, 133, 138, 299, 300
CNN, 36, 63, 64, 65, 66, 67, 69, 263, 265, 267, 268, 269, 271, 273, 274, 275, 279, 281, 283, 287, 289, 291, 292, 294, 295, 297, 298, 299, 300, 302, 306, 307
Color capturing, 309

Combined heat and power, 59, 105, 106, 107, 110
Computer vision, 61, 269, 272, 286, 306, 307, 312, 313, 325

Deep learning, 69, 244, 263, 268, 269, 274, 286, 289, 291, 292, 293, 294, 295, 301, 306, 307, 313
Degree of freedom, 141, 150
DHT Sensor, 93, 98
Discounted, 229, 234
Distributed generation, 11, 12, 13
Drip irrigation, 182, 185, 186, 187, 189, 190, 191, 192, 193, 194, 195, 196, 197, 198, 199, 200, 201, 202, 203, 204
Dynamic billing system, 217, 219, 225

Economic growth, 157, 219
Efficient irrigation technologies, 185
Electrical Vehicle (EV), 349
Electricity, 5, 8, 9, 13, 23, 28, 45, 46, 49, 50, 51, 53, 54, 55, 56, 59, 71, 72, 73, 74, 75, 78, 80, 81, 82, 84, 87, 88, 89, 90, 93, 106, 107, 108, 111, 113, 114, 115, 116, 117, 159, 160, 161, 162, 163, 164, 166, 168, 169, 170, 171, 173, 175, 176, 177, 179, 180, 182, 215, 220, 223, 225, 227, 232, 327, 328, 329, 347
EMS, 89, 91, 216, 327, 328, 329, 330, 334, 335

363

Energy economics, 2, 9, 182, 229, 231, 233, 235, 237, 239, 240
Energy management, 25, 61, 71, 73, 74, 91, 113, 116, 117, 157, 185, 217, 218, 220, 222, 224, 227, 228, 327, 328, 329, 330, 331, 333, 334, 335, 337, 339, 341, 343, 345, 346, 347, 349
Energy policies, 185
Energy pricing, 157, 158, 159, 162, 163, 164, 165, 168, 169, 180, 181
Energy storage system, 27, 28

Face detection, 61, 70, 294, 295, 298, 307
Firebase, 61, 63, 64, 65, 66, 67, 68, 69
Flexible, 2, 13, 19, 23, 61, 129, 131, 133, 135, 137, 139, 216, 330
Fourth industrial revolution, 2
Freshwater resources, 3, 9
Fuel cell, 42, 45, 46, 47, 48, 49, 50, 51, 52, 53, 54, 55, 56, 57, 58, 59, 110, 111, 116, 117, 347

Gears, 208, 209, 211, 212, 214, 215, 316
GHG, 220, 327, 330, 350
Google assistant, 93, 94, 96, 98, 100, 102
Governmental policies, 2
GPS, 241, 242, 245, 247, 248, 249, 251, 253, 255, 256, 350
Green economy, 1, 2, 3, 5, 7, 9

High power density, 27
Home automation, 11, 12, 13, 23, 25, 61, 69, 70, 93, 94, 95, 96, 97, 99, 101, 102, 103, 327

IFTTT, 93, 96, 97, 102
Image processing, 62, 69, 155, 263, 266, 287, 294
Impact on energy policies, 185
Inclusive environment, 2

Incremental analysis, 229, 234
Integrated energy, 105, 106, 107, 109, 111, 113, 115, 117
Internet of things, 8, 11, 12, 25, 71, 72, 73, 75, 76, 78, 79, 80, 81, 89, 91, 93, 94, 95, 139, 217, 218, 228, 260, 261, 271, 287, 292, 295, 307, 309, 319, 324, 327, 358
Irrigation technology, 157, 182, 185, 186, 187, 188, 192, 193, 203, 204

Kilowatt (KW), 46, 56, 72, 82, 84, 108, 109, 113, 116, 161, 166, 170, 171, 209, 215, 216, 217, 220, 221, 223, 224, 225
Kilowatt hour (KWh), 56, 72, 82, 84, 113, 116, 161, 166, 170, 171, 217, 220, 223

Leaf disease, 263, 289
Lithium-ion batteries, (LIB) 27, 28, 29, 33, 34, 35, 39, 41
Load curve, 217, 219, 220, 221, 223, 224, 225, 227
Low-carbon, 1, 2, 4, 5, 8, 27, 28, 47

Main base structure, 208, 210, 212, 214
Mask detection, 291, 292, 293, 294, 295, 297, 298, 299, 300, 301, 303, 305, 306, 307
Metamaterial, 129, 130, 131, 132, 133, 134, 135, 137, 138, 139
Missing data prediction, 71
Money flow, 229, 234, 238, 239

Natural base, 119
Net present worth, 229, 233
Neural network training, 263
Node MCU, 15, 18, 93, 97
Node red, 217, 221, 222, 223, 224, 225, 228, 349, 351, 353, 354, 355, 356

Own tracks, 351, 356

Peak load, 12, 82, 217, 219, 220, 221, 222, 223, 224, 225, 226, 227, 228
Peak Load Management, (PLM), 217, 220
Plant disease detection, 263, 272, 275, 286
Power, BI 12, 23, 24, 54
Power consumption, 11, 13, 14, 17, 19, 22, 23, 71, 73, 74, 80, 82, 93, 95, 97, 98, 99, 101, 102, 103, 113, 208, 215, 218, 219, 220, 221, 223, 226, 328, 346
Preprocessing, 273, 286, 309, 313
Pro-conservation of energy sources, 185
PSO, 327, 332, 333, 334, 337, 340, 346

Quantitative analytics, 71, 86

RaspberryPi, 292
Renewable energy, 3, 5, 9, 10, 12, 14, 27, 28, 39, 50, 52, 55, 105, 106, 113, 115, 117, 139, 217, 329, 350, 358
Replacement analysis, 229, 237, 238, 239
Return on costs, 229
Robotic arm, 141, 142, 143, 145, 147, 149, 150, 151, 152, 153, 154, 155, 156
Rotating platform and rollers, 208, 212, 213

Self-checkout stores, 291, 292, 293, 295, 297, 299, 301, 303, 305, 307
SEM, 327, 332, 333, 334, 335, 336, 339, 342, 343, 344, 345, 346
SEMS, 327, 331, 334, 340, 341, 346
Sensitivity analysis, 229, 235, 236, 237, 240

Sensors, 8, 12, 13, 14, 23, 25, 62, 70, 71, 74, 96, 98, 99, 141, 142, 143, 144, 155, 242, 243, 245, 246, 247, 248, 249, 250, 253, 254, 256, 260, 265, 266, 267, 268, 288, 291, 292, 298, 302, 303, 306, 309, 310, 311, 312, 313, 314, 316, 319, 321, 325
Smart grid ,11, 12, 13, 14, 23, 71, 89, 90, 91, 141, 207, 218, 227, 228, 229, 327, 328, 329, 330, 347, 359
Smart metering, 71, 72, 73, 75, 77, 79, 81, 83, 85, 87, 88, 89, 90, 91
Sodium carbonate, 119, 124, 125, 126
Sodium-ion batteries (SIB), 27, 29, 33, 34, 35, 36, 37, 38, 39
Speed reduction, 208, 211
SSM, 327, 332, 334, 335, 336, 337, 339, 341, 342, 344, 345, 346
SVR, 327, 332, 333, 334, 337, 340, 346
Switch mode power supply, 208, 210, 216

ThingSpeak, 71, 76, 77, 78, 79, 80, 84, 85, 88, 93, 96, 97, 98, 102, 247, 248, 249, 252, 254, 256, 257, 270, 271, 281
Time of Day (TOD), 217, 219, 221
Time worth of cash, 229
Torque, 141, 142, 151, 152, 209, 210, 211

Unified system, 241, 242
Unsoften saline, 119, 122, 124

Wearable electronics, 130, 241, 259
Webhook relay, 349, 351, 353, 356
Wi-Fi, 12, 14, 15, 17, 18, 19, 20, 62, 71, 74, 77, 78, 81, 83, 84, 93, 99, 101, 243, 248, 252, 253, 254, 256, 271, 272, 313, 316, 317, 318, 321, 322, 334, 336, 337, 341
Wireless system, 139, 241

Also of Interest

Check out these other related titles from Scrivener Publishing

INTEGRATED GREEN ENERGY SOLUTIONS VOLUME 2, Edited by Milind Shrinivas Dangate, W. S. Sampath, O. V. Gnana Swathika, and Sanjeevikumar Padmanaban, ISBN: 9781394193660. This second volume in a two-volume set continues to present the state of the art for the concepts, practical applications, and future of renewable energy and how to move closer to true sustainability.

RENEWABLE ENERGY SYSTEMS: Modeling, Optimization, and Applications, edited by Sanjay Sharma, Nikita Gupta, Sandeep Kumar, and Subho Upadhyay, ISBN: 9781119803515. Providing updated and state-of-the-art coverage of a rapidly changing science, this groundbreaking new volume presents the latest technologies, processes, and equipment in renewable energy systems for practical applications.

Encyclopedia of Renewable Energy, by James G. Speight, ISBN 9781119363675. Written by a highly respected engineer and prolific author in the energy sector, this is the single most comprehensive, thorough, and up to date reference work on renewable energy.

INTELLIGENT RENEWABLE ENERGY SYSTEMS: *Integrating Artificial Intelligence Techniques and Optimization Algorithms*, Edited by Neeraj Priyadarshi, Akash Kumar Bhoi, Sanjeevikumar Padmanaban, S. Balamurugan, and Jens Bo Holm-Nielsen, ISBN: 9781119786276. This collection of papers on artificial intelligence and other methods for improving renewable energy systems, written by industry experts, is a reflection of the state of the art, a must-have for engineers, maintenance personnel, students, and anyone else wanting to stay abreast with current energy systems concepts and technology.

POWER ELECTRONICS FOR GREEN ENERGY CONVERSION, Edited by Mahajan Sagar Bhaskar, Nikita Gupta, Sanjeevikumar Padmanaban, Jens Bo Holm-Nielsen, and Umashankar Subramaniam, ISBN: 9781119786481. Written and edited by a team of renowned experts, this exciting new volume explores the concepts and practical applications of power electronics for green energy conversion, going into great detail with ample examples, for the engineer, scientist, or student.

INTEGRATION OF RENEWABLE ENERGY SOURCES WITH SMART GRIDS, Edited by A. Mahaboob Subahani, M. Kathiresh and G. R. Kanagachidambaresan, ISBN: 9781119750420. Provides comprehensive coverage of renewable energy and its integration with smart grid technologies.

Green Energy: Solar Energy, Photovoltaics, and Smart Cities, Edited by Suman Lata Tripathi and Sanjeevikumar Padmanaban, ISBN 9781119760764. Covering the concepts and fundamentals of green energy, this volume, written and edited by a global team of experts, also goes into the practical applications that can be utilized across multiple industries, for both the engineer and the student.

Microgrid Technologies, edited by C. Sharmeela, P. Sivaraman, P. Sanjeevikumar, and Jens Bo Holm-Nielsen, ISBN 9781119710790. Covering the concepts and fundamentals of microgrid technologies, this volume, written and edited by a global team of experts, also goes into the practical applications that can be utilized across multiple industries, for both the engineer and the student.

Energy Storage, edited by Umakanta Sahoo, ISBN 9781119555513. Written and edited by a team of well-known and respected experts in the field, this new volume on energy storage presents the state-of-the-art developments and challenges in the field of renewable energy systems for sustainability and scalability for engineers, researchers, academicians, industry professionals, consultants, and designers.

Biofuel Cells, Edited by Inamuddin, Mohd Imran Ahamed, Rajender Boddula, and Mashallah Rezakazemi, ISBN 9781119724698. This book covers the most recent developments and offers a detailed overview of fundamentals, principles, mechanisms, properties, optimizing parameters, analytical characterization tools, various types of biofuel cells, edited by one of the most well-respected and prolific engineers in the world and his team.

Biodiesel Technology and Applications, Edited by Inamuddin, Mohd Imran Ahamed, Rajender Boddula, and Mashallah Rezakazemi, ISBN 9781119724643. This outstanding new volume provides a comprehensive overview on biodiesel technologies, covering a broad range of topics and practical applications, edited by one of the most well-respected and prolific engineers in the world and his team.

Energy Storage 2nd Edition, by Ralph Zito and Haleh Ardibili, ISBN 9781119083597. A revision of the groundbreaking study of methods for storing energy on a massive scale to be used in wind, solar, and other renewable energy systems.

Hybrid Renewable Energy Systems, edited by Umakanta Sahoo, ISBN 9781119555575. Edited and written by some of the world's top experts in renewable energy, this is the most comprehensive and in-depth volume on hybrid renewable energy systems available, a must-have for any engineer, scientist, or student.

Progress in Solar Energy Technology and Applications, edited by Umakanta Sahoo, ISBN 9781119555605. This first volume in the new groundbreaking series, Advances in Renewable Energy, covers the latest concepts, trends, techniques, processes, and materials in solar energy, focusing on the state-of-the-art for the field and written by a group of world-renowned experts.

A Polygeneration Process Concept for Hybrid Solar and Biomass Power Plants: Simulation, Modeling, and Optimization, by Umakanta Sahoo, ISBN 9781119536093. This is the most comprehensive and in-depth study of the theory and practical applications of a new and groundbreaking method for the energy industry to "go green" with renewable and alternative energy sources.

Nuclear Power: Policies, Practices, and the Future, by Darryl Siemer, ISBN 9781119657781. Written from an engineer's perspective, this is a treatise on the state of nuclear power today, its benefits, and its future, focusing on both policy and technological issues.

Zero-Waste Engineering 2nd Edition: A New Era of Sustainable Technology Development, by M. M. Kahn and M. R. Islam, ISBN 9781119184898. This book outlines how to develop zero-waste engineering following natural pathways that are truly sustainable using methods that have been developed for sustainability, such as solar air conditioning, natural desalination, green building, chemical-free biofuel, fuel cells, scientifically renewable energy, and new mathematical and economic models.

Sustainable Energy Pricing, by Gary Zatzman, ISBN 9780470901632. In this controversial new volume, the author explores a new science of energy pricing and how it can be done in a way that is sustainable for the world's economy and environment.

Sustainable Resource Development, by Gary Zatzman, ISBN 9781118290392. Taking a new, fresh look at how the energy industry and we, as a planet, are developing our energy resources, this book looks at what is right and wrong about energy resource development. This book aids engineers and scientists in achieving a true sustainability in this field, both from an economic and environmental perspective.

The *Greening of Petroleum Operations,* by M. R. Islam *et al.,* ISBN 9780470625903. The state of the art in petroleum operations, from a "green" perspective.

Printed and bound by CPI Group (UK) Ltd, Croydon, CR0 4YY
21/06/2023
03229312-0003